Morphology of Electrochemically and Chemically Deposited Metals

Konstantin I. Popov • Stojan S. Djokić
Nebojša D. Nikolić • Vladimir D. Jović

Morphology of Electrochemically and Chemically Deposited Metals

Springer

Konstantin I. Popov
University of Belgrade
Belgrade, Serbia

Stojan S. Djokić
Elchem Consulting Ltd.
Edmonton, Alberta, Canada

Nebojša D. Nikolić
ICTM- Institute of Electrochemistry
University of Belgrade
Belgrade, Serbia

Vladimir D. Jović
Institute for Multidisciplinary Research
University of Belgrade
Belgrade, Serbia

ISBN 978-3-319-26071-6 ISBN 978-3-319-26073-0 (eBook)
DOI 10.1007/978-3-319-26073-0

Library of Congress Control Number: 2016932404

Printed on acid-free paper

This Springer imprint is published by Springer Nature
The registered company is Springer International Publishing AG Switzerland

Preface

The aim of this book is to bring to scientists, researchers, engineers, and students the newest achievements in electrochemically and chemically deposited metals and alloys. This book is to some degree a continuation of our former book *Fundamental Aspects of Electrometallurgy*, although there are significant differences. More particularly, this book is devoted to the surface morphology of deposited metals and/or alloys and offers an in-depth analysis of the influence of the parameters of electrodeposition or chemical deposition of metals and alloys, which could lead to technological advances in industrial settings worldwide. As such, the proposed book may equally attract attention from those working in electrometallurgical or electroplating plants and from members of research departments in industry or academia.

The surface morphology of electrochemically deposited metals and alloys is a very important property and thus may significantly influence their potential applications. Conditions of electrochemical deposition such as electrolyte composition, pH, temperature, stirring, time, deposition overpotential, and current density determine the surface morphology.

The morphology of electrodeposited metals is usually analyzed on the basis of the exchange current density and overpotential for hydrogen evolution. When electrodeposition of metals is characterized by *very large* exchange current densities, at lower overpotentials individual grains or boulders are formed, while at higher overpotentials formation of dendrites takes place. In the case of *large* exchange current densities, spongy deposits and dendrites are produced at lower or higher overpotentials, respectively. Finally, during electrodeposition characterized by *medium and low* exchange current densities, compact deposits are obtained at low overpotentials, while dendrites or spongy deposits are formed at high overpotentials.

A proper analysis of the polarization curves in correlation with the surface morphology could contribute to very important conclusions useful for technological operations in order to produce a metal deposit with the desired properties. Electrodeposition of metals at low overpotentials proceeds under ohmic control and at higher overpotentials under diffusion control when the process is characterized by the *extremely large* exchange current densities. In the ohmic control regime, granular or regular crystals are deposited, while under diffusion control, different shapes of dendrites are formed. When the electrodeposition of metals is characterized by *large* exchange current densities, diffusion control is observed in the whole range of overpotentials. Therefore, spongy deposits are formed at lower and dendrites at higher overpotentials. For metals where electrodeposition is characterized by *medium or low* exchange current densities, the process proceeds under activation control at low overpotentials. Activation-controlled electrodeposition produces large grains of metal with well-defined crystal shapes. This is observed for overpotentials in the regions of the Tafel linearity. For overpotentials larger than those consistent with Tafel linearity and smaller than those determining the limiting diffusion current density plateau (the mixed activation–diffusion control), the surface morphology is influenced by mass transfer. Compact or uniform deposits of metals are obtained under these conditions. Formation of dendrites is observed at overpotentials inside the plateau of the limiting diffusion current density. For metals where electrodeposition is characterized by *medium and low* exchange current densities, the shapes of dendrites are very different than those produced for metals where the electrodeposition is characterized by the *extremely large or large* exchange current densities. This observation raises many further questions from the fundamental point of view of material science in conjunction with electrochemistry and opens a new window for future studies.

Vigorous hydrogen evolution changes hydrodynamic conditions in the near-electrode layer. In this way the presence of significant hydrogen evolution may strongly affect the morphology of metal deposits. The typical structure obtained in the presence of vigorous hydrogen evolution is honeycomb-like, with holes or pores formed from detached hydrogen bubbles and surrounded by cauliflower-like agglomerates of metal grains. Complexing agents and additives (organic or inorganic) present in the electroplating solutions may also strongly affect the morphology of deposits. The difference in the morphology of electrodeposited metals relative to the one obtained by the electrodepositions from simple salt solutions is caused by the change of the kinetic parameters of electrodeposition due to the effect of additives or complexing agents. The deposition of compact metals and alloys, bright galvanic coatings, and dispersed and powdered precipitates is discussed in detail. Also, the effects of periodically changing currents and/or potentials during the electrodeposition on the morphology and related properties of deposits are thoroughly presented in this book.

All the remarks mentioned above are applicable as well to the electrodeposition of alloys. Using Brenner's classification of alloy electrodeposition, e.g., equilibrium, regular, anomalous, etc., all existing combinations of deposition parameters and their influence on the alloy morphology are analyzed. Interestingly, certain features, which are not recognized in the electrodeposition of pure metals, are observed in the alloy deposition processes. An example includes the *spatiotemporal structures*, which is discussed in this book.

Chemically deposited metals and alloys from aqueous solutions have been significantly less studied than their electrodeposition counterparts. Both chemical deposition, or as frequently termed in the literature *electroless* deposition, and electrodeposition are charge transfer processes and as such can be considered as *electrochemical*. Practically, all metals and/or alloys that can be electrodeposited from aqueous solutions can be deposited by the electroless means using appropriate reducing agents. Electroless deposition of various metals has been widely used in industry since the 1950s. Metals of interest include Ni, Cu, Ag, Au, Pd, and Co. Many alloys of these metals find applications in different industries. Both galvanic and autocatalytic types of electroless deposition in relation with the surface morphology are analyzed in this book.

The electrodeposition portion of this book is written by Popov, Nikolić, and Jović. Chapters 1, 2, and 3 are written by Popov and Nikolić, Chap. 4 is written by Popov, Chap. 5 is written by Nikolić, Chap. 6 is written by Nikolić and Popov, and Chaps. 7 and 8 are written by Jović. The chapter related to the chemical deposition is written by S. Djokić. Chapters 1, 2, 3, 4, 5, 6, 7, and 8 of this book are based on classical studies performed at the Department of Physical Chemistry and Electrochemistry at the Faculty of Technology and Metallurgy, University of Belgrade; Department of Electrochemistry at the Institute of Chemistry, Technology and Metallurgy (ICTM), University of Belgrade; and at the Institute for Multidisciplinary Research, University of Belgrade, Serbia. K. I. Popov, N.D. Nikolić, and V.D. Jović would like to acknowledge contributions and inspiration by Professor A.R. Despić who initiated research in the area of electrodeposition at the University of Belgrade. Popov and Nikolić (Chaps. 1, 2, 3, 4, 5, and 6) are thankful for help from their colleagues Professors M.D. Maksimović, M.G. Pavlović, N.V. Krstajić, B.N. Grgur, P.M. Živković, S.K. Zečević, and B.J. Lazarević. Also they acknowledge contributions from Drs R.M Stevanović, S.M Pešić, Z. Rakočević, Lj.J. Pavlović, G. Branković, V.M. Maksimović, and S.B. Krstić, as well as from numerous colleagues and students who participated in their research.

V.D. Jović (Chaps. 7 and 8) is thankful to U.Č. Lačnjevac and B.M. Jović from the Institute for Multidisciplinary Research University of Belgrade, Serbia, for contributions in the published chapters in *Modern Aspects of Electrochemistry* series. He further expresses his gratitude to Prof. Ivan Krastev, Institute of Physical Chemistry, Bulgarian Academy of Science, for providing necessary literature and explanation of specific structural phenomena in electrodeposition of alloys, given in Sect. 2.3 of Chap. 7. Popov, Nikolić, and Jović are also indebted to the Ministry of Education, Science and Technological Development of the Republic of Serbia, for the financial support of this work.

Djokić is very thankful to many colleagues across the globe (Europe, North America, China, and India) for inspiring him to further investigate chemical deposition processes. Special thanks go to all of his students, clients, and his family for the establishment of a few electroless processes in industry. Further, Djokić expresses his thanks to Dr. Kenneth Howell of Springer for many very helpful advice during the preparation of this book. As well, Djokić acknowledges the help of Ms. Marion Pritchard and Ms. Nada Djokić (both from University of Alberta, Canada) for their unselfish help in the preparation of this manuscript.

We hope that the book will be of a particular interest to the individuals or groups dealing with electrochemistry of metals or more specifically with the electrodeposition phenomena in relation with the surface morphology. The research departments in the automotive, aerospace, electronics, energy device, and perhaps in the biomedical fields may find this book as a very useful source in developments of their future programs. Professors and students in the university settings worldwide, when learning, investigating, or lecturing various electrodeposition processes at both undergraduate or graduate levels, will find this book as a very valuable source for their courses and/or projects. We believe that for the university environments, the book can be attractive to the engineering students. Perhaps, chemistry or physics students will find the book as a very useful source in their studies.

Belgrade, Serbia — Konstantin I. Popov
Edmonton, Alberta, Canada — Stojan S. Djokić
Belgrade, Serbia — Nebojša D. Nikolić
Belgrade, Serbia — Vladimir D. Jović
August, 2015

Contents

Short Biographies

Konstantin I. Popov Professor Konstantin I. Popov received his BSc, MSc, and PhD degrees in the field of electrochemical science, from the Faculty of Technology and Metallurgy, University of Belgrade, Serbia. He joined the Faculty of Technology and Metallurgy as a teaching assistant and advanced to the position of full professor of electrometallurgy at the Department of Physical Chemistry and Electrochemistry, at the same institution. At the same time, Professor Popov closely collaborated with the Institute of Chemistry, Technology and Metallurgy (ICTM), Department of Electrochemistry, University of Belgrade, Serbia. During his rich academic carrier, Professor Popov has published over 230 research papers, twelve book chapters, and coauthored the book *Fundamental Aspects of Electrometallurgy*. In addition, Popov has published several textbooks devoted to the electrochemistry/electrometallurgy fields in the Serbian language which are used by the university students and professionals there. He had numerous presentations, including invited lectures, at national and international conferences. Professor Popov successfully supervised more than 80 students in their BSc, MSc, and PhD research studies. The main scientific interests of Professor Popov include various fundamental and applied aspects of electrochemical deposition and dissolution of metals, constant and periodically changing regimes of electrolysis, production and characterization of metal powders by electrolysis, the current density distribution effects, bright coatings, and open porous electrodes (the honeycomb-like structures). Professor Popov is a member of the Serbian Chemical Society and Serbian Society of Corrosion and Materials Protection. He is a recipient of several national awards, including Medal of Serbian Chemical Society for enduring and outstanding contributions to science. Currently, Professor Popov is an emeritus scientist of the Republic of Serbia.

Stojan S. Djokić Dr. Djokić obtained his BSc degree in chemical engineering/analytical chemistry from the University of Belgrade, Yugoslavia. He holds an MSc in electrochemical energy conversion and a PhD in electrochemistry/materials science from the University of Belgrade. At the University of Ottawa, as a research fellow, he has successfully collaborated with Alcan International Ltd. in the field of molten salts electrochemistry. He was a senior scientist/electrochemist at Sherritt Inc., Westaim Corporation, Westaim Ambeon, Superior MicroPowders, etc., working on various projects related to electronics, automotive/aerospace, corrosion, and metallurgical and biomedical applications. Dr. Djokić has published more than 90 papers including seven book chapters and coauthored a book *Fundamental Aspects of Electrometallurgy,* and he has edited four volumes of *Modern Aspects of Electrochemistry*. He holds eight US patents. Dr. Djokić is an active member of the Electrochemical Society, and he has successfully organized and chaired several international symposiums. He is as well a registered professional engineer with the APEGA. Stojan's professional experience covers both industrial and academic environments within Europe, North America, and China in the fields of electrochemistry, electrometallurgy, polymeric composite materials, biomaterials, materials science, and analytical chemistry. Dr. Djokić has developed several processes and materials which are currently used in industry worldwide. His current interests include electroless and electrodeposition, kinetics, disinfectants, electronic materials, and biomaterials used as infection-resistant surfaces or in immunodiagnostics. Currently, Dr. Djokić is the director of Elchem Consulting Ltd. and an adjunct professor of chemical and materials engineering at the University of Alberta.

Nebojša D. Nikolić Dr. Nebojša D. Nikolić graduated from the Faculty of Technology and Metallurgy, University of Belgrade, Serbia, where he earned his BSc, MSc, and PhD degrees in the field of electrochemical science. Presently, he is a senior research scientist (full research professor) at the Institute of Chemistry, Technology and Metallurgy (ICTM), Department of Electrochemistry, University of Belgrade, Serbia. Dr. Nikolić had an intensive scientific collaboration with Laboratorio de Fisica de Sistemas Pequnos y Nanotechnologia, CSIC, Madrid, Spain, and Departamento de Fisica, Universidade Federal de Santa Catarina (UFSC), Florianopolis–Santa Catarina, Brazil. The main research interests of Dr. Nikolić encompass various aspects of electrochemical metal deposition and dissolution, including bright coating formation, electrochemical production of metal powders, application of periodically changing regimes of electrolysis, and other related phenomena. A special topic of his investigation involves the synthesis and characterization of open porous structures of copper (the honeycomb-like deposits), using constant and periodically changing regimes of electrolysis. Dr. Nikolić has been author or coauthor of ten chapters in relevant books (including eight chapters in *Modern Aspects of Electrochemistry*) as well as 102 original scientific papers. In addition, he had 76 presentations, including invited lectures, at national and international conferences. He is a member of the ISE (International Society of Electrochemistry), Serbian Chemical Society, Serbian Society of Corrosion and Materials Protection, and Association of the Chemical Engineers of Serbia (AChE).

Vladimir D. Jović Dr. Vladimir D. Jović obtained his BSc, MSc, and PhD in the field of electrochemical science and technology from the University of Belgrade, Serbia. He has an extensive academic experience at the University of Belgrade working as a researcher at the Institute of Technical Sciences of the Serbian Academy of Science and Arts and Institute for Multidisciplinary Research, University of Belgrade, Serbia. As well, Dr. Jović was a guest scientist at the National Institute of Standards and Technology, Materials Science and Engineering Laboratory, Gaithersburg, MD, USA, and a research associate at the Department of Materials Engineering, Drexel University, Philadelphia, USA. His main interests in the field of electrochemistry include the underpotential deposition of Pb, Tl, and Cd onto Ag and Cu single crystals from aqueous solutions; anion adsorption onto Ag and Cu single crystals from aqueous solutions; UPD of Al onto Cu single crystals from room-temperature AlCl3–MeEtimCl molten salt; characterization of electrodeposited binary alloys by electrochemical ALSV technique; electrodeposition of metals, powders, alloys, and composite structures; electrochemistry of MAX phases; and electrodeposited catalysts for industrial electrolysis. Dr. Jović published more than 130 scientific papers (including chapters), 3 patents, and over 200 poster presentations and was an invited speaker on numerous international occasions. Currently, Dr. Jović is a retired senior research fellow from the Institute for the Multidisciplinary Research, University of Belgrade, Serbia.

Chapter 1
The Cathodic Polarization Curves in Electrodeposition of Metals

1.1 Introduction

Morphology is probably the most important property of electrodeposited metals. It depends mainly on the kinetic parameters of the deposition process and on the deposition overpotential or current density. The morphology of an electrodeposited metal also depends on the deposition time until the deposit has attained its final form.

Morphology of electrodeposited metal strongly depends on the nature of metals, and electrodeposition processes are usually classified in the dependence of the exchange current density of deposition process [1]. For example, the individual grains or boulders formed at lower overpotentials and dendrites at higher ones are characteristics of the electrodeposition processes characterized by very large exchange current densities. Formation of spongy deposits at lower overpotentials and dendrites at higher ones is a characteristic of the electrodeposition processes characterized by large exchange current densities. Finally, compact deposits are obtained at lower overpotentials, while both dendritic and spongy-dendritic deposits are formed at higher overpotentials during electrodeposition of metals characterized by medium and low exchange current densities.

Obviously, morphology of metal electrodeposits depends on the type of control of the electrodeposition process. For example, the activation-controlled electrodeposition of copper produces large grains with relatively well-defined crystal shapes. This happens at overpotentials belonging to the region of the Tafel linearity [2, 3]. At overpotentials that are between the end of the Tafel linearity and the beginning of the limiting diffusion current density plateau (the mixed activation–diffusion control), morphological forms are influenced by the mass transfer conditions, and large grains are not formed [4, 5]. Dendrites are formed at overpotentials inside the plateau of the limiting diffusion current density and at the higher ones at which there is no hydrogen evolution or it is not enough to affect hydrodynamic conditions in the near-electrode layer [6, 7]. The above consideration is valid for all

K.I. Popov et al., *Morphology of Electrochemically and Chemically Deposited Metals*, DOI 10.1007/978-3-319-26073-0_1

electrodeposition processes characterized by medium and low values of the exchange current density in the absence of the noticeable hydrogen evolution reaction [7]. In the presence of strong hydrogen evolution, the honeycomb-like deposits are formed [7].

On the other hand, formation of spongy deposits at the lower, and dendrites at the higher, overpotentials was explained by long-standing assumption that the electrodeposition processes characterized by large values of the exchange current density are diffusion controlled at all overpotentials [8]. However, formation of individual grains or granules at low overpotentials was not possible to explain by this assumption. Recently, it was discussed [9] the linear dependence of current density on the overpotential for the fast electrochemical processes, and it was shown that this linear dependence can be ascribed to the ohmic control of the electrodeposition process [10–13].

There is a close correlation between the morphology of metal electrodeposits and the electrodeposition process control. Besides, the shape of the polarization curve depends also on the electrodeposition process control. Hence, it can be expected that the morphology of deposits can be correlated with the shape of the polarization curves. This chapter introduces the basic correlations between polarization and morphological characteristics of metal deposits.

1.2 Polarization Curves for the Case of Massive Active Cathodes

1.2.1 Polarization Curves Without Included Ohmic Potential Drop

1.2.1.1 Concentration Dependence of the Exchange Current Density Is Not Taken into Account

The general form of current density–overpotential relationship in electrodeposition of metals for the reaction

$$\mathrm{Ox} + n\mathrm{e}^- \rightleftarrows \mathrm{Red} \tag{1.1}$$

taking cathodic current density and overpotential as positive, is given by

$$i = i_0\left[\left(\frac{a_\mathrm{i}^\mathrm{O}}{a_\mathrm{e}^\mathrm{O}}\right)f_\mathrm{c} - \left(\frac{a_\mathrm{i}^\mathrm{R}}{a_\mathrm{e}^\mathrm{R}}\right)f_\mathrm{a}\right] \tag{1.2}$$

where i_0 is the exchange current density and a_i is the activity of the oxidized (Ox) or reduced (Red) state at a current density i and a_e is the activity in the equilibrium state [14].

On the other hand,

$$f_{\mathrm{c}} = \exp\left(\frac{\alpha_{\mathrm{c}} F \eta}{RT}\right) = \exp\left(\frac{2.3}{b_{\mathrm{c}}}\eta\right) \tag{1.3}$$

and

$$f_{\mathrm{a}} = \exp\left(\frac{-\alpha_{\mathrm{a}} F \eta}{RT}\right) = \exp\left(\frac{-2.3}{b_{\mathrm{a}}}\eta\right) \tag{1.4}$$

where α_{c} and α_{a} are the cathodic and anodic transfer coefficient, b_{c} and b_{a} are corresponding Tafel slopes, and η is the overpotential, and

$$b_{\mathrm{c}} = \frac{2.3RT}{\alpha_{\mathrm{c}} F} \tag{1.5}$$

and

$$b_{\mathrm{a}} = \frac{2.3RT}{\alpha_{\mathrm{a}} F} \tag{1.6}$$

where R is gas constant, T is temperature, and F is Faraday constant.

The ratio of the activities for the cathodic reaction may be written as

$$\frac{a_{\mathrm{i}}^{\mathrm{O}}}{a_{\mathrm{e}}^{\mathrm{O}}} = 1 - \frac{i}{i_{\mathrm{L}}} \tag{1.7}$$

where i_{L} is the limiting diffusion current density, and for the reverse anodic reaction as [14]

$$\frac{a_{\mathrm{i}}^{\mathrm{R}}}{a_{\mathrm{e}}^{\mathrm{R}}} = \exp \frac{2\gamma V}{RT r_{\mathrm{el}}} \tag{1.8}$$

taking into account the Kelvin term which becomes appreciable at low values of electrode radii [15]. In Eq. (1.8), γ is the interfacial energy between metal and solution, V is the molar volume of the electrodeposited metal, and r_{el} is the radius of the spherical electrode. Equation (1.8) is valid for two electron reactions [14], while the other possibilities are discussed in Ref. [16].

For a spherical electrode, Eq. (1.2) can be written as

$$i_{\mathrm{spher}} = i_0 \left[\left(1 - \frac{i}{i_{\mathrm{L,spher}}}\right) f_{\mathrm{c}} - f_{\mathrm{a}} \exp\left(\frac{2\gamma V}{RT r_{\mathrm{el}}}\right)\right] \tag{1.9}$$

or

$$i_{\text{spher}} = \frac{i_0\left[f_{\text{c}} - f_{\text{a}}\exp\left(\frac{2\gamma V}{RT r_{\text{el}}}\right)\right]}{1 + \frac{i_0 f_{\text{c}}}{i_{\text{L,spher}}}} \tag{1.10}$$

where

$$i_{\text{L,spher}} = \frac{nFDC_0}{r_{\text{el}}} \tag{1.11}$$

and n is the number of electrons involved in the electrode reaction, C_0 is the bulk concentration, and D is the coefficient of diffusion of a depositing ions. A somewhat modified Eq. (1.10) is necessary for an understanding of electrodeposition on the tip of dendrites inside the diffusion layer of a macroelectrode (see Chap. 2) and in the case of electrodeposition at a periodically changing rate (see Chap. 4) [1, 17].

For sufficiently large r_{el} to make surface energy term negligible, Eq. (1.10) can be rewritten in the form:

$$i_{\text{spher}} = \frac{i_0(f_{\text{c}} - f_{\text{a}})}{1 + \frac{i_0 f_{\text{c}}}{i_{\text{L,spher}}}} \tag{1.12.}$$

For flat electrodes and sufficiently large spherical electrodes, Eq. (1.10) becomes

$$i = \frac{i_0(f_{\text{c}} - f_{\text{a}})}{1 + \frac{i_0 f_{\text{c}}}{i_{\text{L}}}} \tag{1.13}$$

where

$$i_{\text{L}} = \frac{nFDC_0}{\delta} \tag{1.14}$$

and δ is the diffusion layer thickness.

1.2.1.2 Concentration Dependence of the Exchange Current Density Is Taken into Account

In this case, the current density is given by

$$i = i_{0,\text{s}}(f_{\text{c}} - f_{\text{a}}) \tag{1.15}$$

where $i_{0,\text{s}}$ is the exchange current density on the electrode surface at current density i, being determined by Eq. (1.16) [18]:

$$i_{0,\mathrm{s}} = \left(\frac{C_\mathrm{s}}{C_0}\right)^{\xi} i_0 \tag{1.16}$$

where C_s is the surface concentration, i_0 is the exchange current density for a surface concentration C_0 equal to the one in the bulk, and $i_{0,\mathrm{s}}$ is the exchange current density for a surface concentration C_s, and

$$\xi = \frac{\mathrm{d}\log i_0}{\mathrm{d}\log C_0} \tag{1.17}$$

Using relation

$$\frac{C_\mathrm{s}}{C_0} = 1 - \frac{i}{i_\mathrm{L}} \tag{1.18}$$

Eq. (1.15) can be rewritten in the form:

$$i = i_0\left(1 - \frac{i}{i_\mathrm{L}}\right)^{\xi}(f_\mathrm{c} - f_\mathrm{a}) \tag{1.19}$$

or, for $\zeta = 1$, after rearranging

$$i = \frac{i_0(f_\mathrm{c} - f_\mathrm{a})}{1 + \frac{i_0(f_\mathrm{c} - f_\mathrm{a})}{i_\mathrm{L}}} \tag{1.20}$$

It is necessary to note that Eq. (1.20) is an approximation because the value of ζ is lower than unity. This approximation is widely used in qualitative discussions, because it permits the simple mathematical treatment of electrochemical processes with relatively small errors and with clear physical meaning. If ζ different than unity is included in the derivation of the general polarization curve equation, simple analytical solutions are not available and numerical solutions are required.

1.2.1.3 Comparison of Eqs. (1.13) and (1.20) for Different Exchange Current Density to the Limiting Diffusion Current Density Ratios

Equations (1.13) and (1.20) are compared taking $i_0/i_\mathrm{L} = 100$, 10, 1, 0.1, and 0.01. $f_\mathrm{c} = 10^{\frac{\eta}{120}}$ and $f_\mathrm{a} = 10^{-\frac{\eta}{120}}$ and the diagrams presented in Fig. 1.1 are obtained by digital simulation. It can be seen that for $i_0/i_\mathrm{L} \le 1$, diagrams computed using Eqs. (1.13) and (1.20) are the same.

At larger values of i_0/i_L ratio, the polarization curves calculated by Eq. (1.13) are the same as follows from Eq. (1.13) when $i_0/i_\mathrm{L} \to \infty$, for $f_\mathrm{c} > f_\mathrm{a}$. On the other

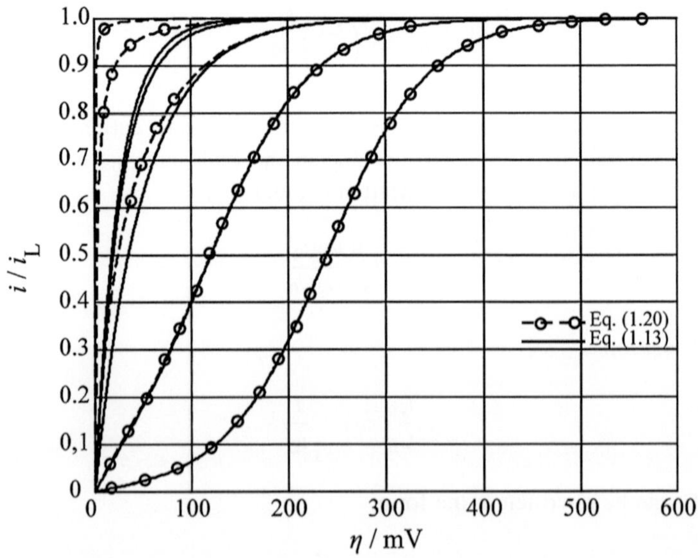

Fig. 1.1 The comparison of polarization curves calculated using Eqs. (1.13) and (1.20). From left to right side of diagram, i_0/i_L ratio corresponds to 100, 10, 1, 0.1, and 0.01, respectively; $f_c = 10^{\frac{\eta}{120}}$, and $f_a = 10^{-\frac{\eta}{120}}$ (Reprinted from Ref. [19] with permission from Elsevier and Ref. [9] with kind permission from Springer)

hand, the polarization curves calculated using Eq. (1.20) under the same conditions become

$$i \cong i_L \tag{1.21}$$

at all overpotentials for $i_0/i_L \rightarrow \infty$. It can be seen from Fig. 1.1 that the increase of the value of i_0/i_L ratio leads to the decrease of the electrochemical overpotential. The activation part of overpotential is lost at i_0/i_L values larger than 10, while both activation and diffusion overpotential vanish at i_0/i_L values larger than 100 (Fig. 1.1). In the second case, the ohmic-controlled electrochemical reaction can occur (see Sects. 1.2.1.5 and 1.2.2.2).

1.2.1.4 The Approximations for $i_0/i_L << 1$

Although Eqs. (1.13) and (1.20) are generally valid, it is more convenient to use some approximative relations derived from them [1]. For a flat surface, if

$$f_c > f_a \text{and } \frac{i_0 f_c}{i_L} << 1 \tag{1.22}$$

Eqs. (1.13) and (1.20) can be rewritten in the form:

$$i = i_0(f_c - f_a) \tag{1.23}$$

which becomes

$$i = i_0 \frac{nF}{RT} \eta \tag{1.24}$$

at very low overpotentials by expanding the exponential terms in Eq. (1.23) and retaining the two terms of expansion of each exponential terms, if

$$\alpha_c = \beta n \tag{1.25}$$

and

$$\alpha_a = (1 - \beta)n \tag{1.26}$$

where β is the symmetry factor.

When

$$f_c >> f_a \text{ and } \frac{i_0 f_c}{i_L} << 1 \tag{1.27}$$

the relation

$$i = i_0 f_c \tag{1.28}$$

or

$$\eta = \frac{b_c}{2.3} \ln \frac{i}{i_0} \tag{1.29}$$

is valid.

If $f_c >> f_a$, Eqs. (1.13) and (1.20) become

$$i = \frac{i_0 f_c}{1 + \frac{i_0 f_c}{i_L}} \tag{1.30}$$

or

$$\eta = \frac{b_c}{2.3} \ln \frac{i}{i_0} + \frac{b_c}{2.3} \ln \frac{1}{1 - \frac{i}{i_L}} \tag{1.31}$$

For

$$f_c > f_a \text{ and } \frac{i_0 f_c}{i_L} >> 1 \tag{1.32}$$

Eq. (1.13) can be rewritten in the form:

$$i = i_L\left(1 - \frac{f_a}{f_c}\right) = i_L\left[1 - \exp\left(-\frac{nF\eta}{RT}\right)\right] \tag{1.33}$$

which is valid if Eqs. (1.25) and (1.26) are valid. Finally, if

$$f_c >> f_a \text{ and } \frac{i_0 f_c}{i_L} >> 1 \tag{1.34}$$

Eqs. (1.13) and (1.20) become

$$i = i_L \tag{1.35}$$

The range of validity of Eqs. (1.23), (1.24), (1.28), and (1.29) as well as (1.30), (1.31), and (1.35) can be easily determined from $\eta - \log(i)$ and $\eta - i$ plots in Fig. 1.2. Equations (1.28) and (1.29) are valid from the beginning to the end of Tafel linearity (Tafel line). At lower overpotentials, Eqs. (1.23) and (1.24) are valid,

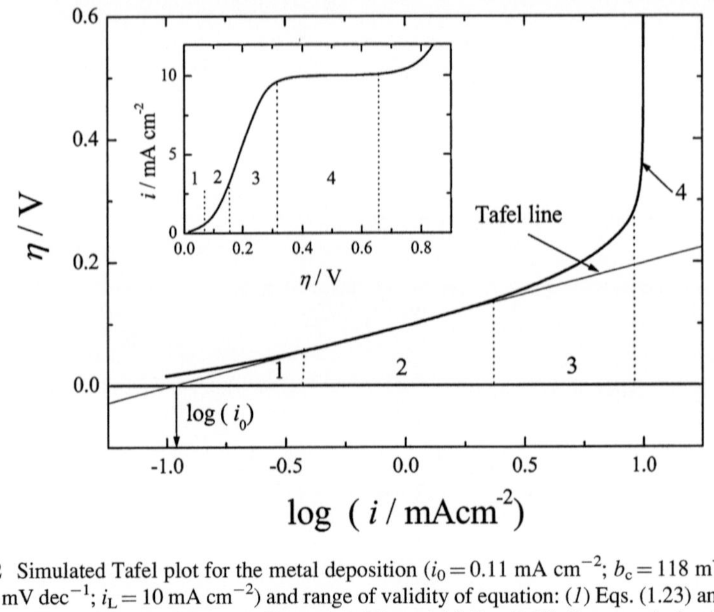

Fig. 1.2 Simulated Tafel plot for the metal deposition ($i_0 = 0.11$ mA cm^{-2}; $b_c = 118$ mV dec^{-1}; $b_a = 40$ mV dec^{-1}; $i_L = 10$ mA cm^{-2}) and range of validity of equation: (*1*) Eqs. (1.23) and (1.24), (*2*) Eqs. (1.28) and (1.29), (*3*) Eqs. (1.30) and (1.31), and (*4*) Eq. (1.35). Insert: polarization curve (Reprinted from Ref. [1] with kind permission from Springer)

while at higher ones, Eqs. (1.30), (1.31), and (1.35) are valid. In Fig. 1.2, the simulated polarization curve for cathodic metal electrodeposition together with the Tafel plot and the range of validity of mentioned equations is shown. Usually, in electrochemistry the marked regions are called (1 and 2) activation-controlled region, (3) mixed activation–diffusion-controlled region, and (4) pure diffusion-controlled region.

1.2.1.5 The Shape of the Calculated Polarization Curves as Function of i_0/i_L Ratio Without Included Ohmic Potential Drop

Using the current density–overpotential relationships and the procedure for the determination of the ohmic potential drop, the polarization curves for electrodeposition processes can be successfully simulated [9, 20].

The current density–overpotential curve equation (Eq. (1.20)), derived by taking the concentration dependence of i_0 into account and the linear dependence of i_0 on the C_s/C_0 ratio, can be rewritten in the form:

$$\frac{i}{i_L} = \frac{\frac{i_0}{i_L}(f_c - f_a)}{1 + \frac{i_0}{i_L}(f_c - f_a)} \tag{1.36}$$

being valid for all i_0/i_L ratios and overpotentials. Because of this, this form will be used in discussion of the effect of the i_0/i_L ratio on the shape of the polarization curves. In Eq. (1.36), i, i_0, i_L, C_s, and C_0 have already mentioned meanings, while f_c and f_a can also be written in the form:

$$f_c = 10^{\frac{\eta}{b_c}} \quad \text{and} \quad f_a = 10^{-\frac{\eta}{b_a}} \tag{1.37}$$

where b_c and b_a are the cathodic and anodic Tafel slopes. Equation (1.36) is modified for use in electrodeposition of metals using the values of the cathodic current density and overpotential as positive. In Eq. (1.36), the ohmic potential drop is not included in the overpotential values. Equation (1.36) is operative if IR error is eliminated by using the electronic devices that distinguish the superfast change in potential that occurs in this IR portion, when the current density is switched on and off from the slower change of the electrode potential itself, where the charging of the interfacial capacitor takes time [20]. In this way, the polarization curves which do not include the ohmic potential drop can be simulated using Eq. (1.36).

The polarization curves without including the ohmic potential drop for different i_0/i_L ratio values for both, one and two electron reactions, are shown in Fig. 1.3 [21]. They are obtained by using Eq. (1.36) for the different i_0/i_L ratios and f_c and f_a values obtained for different η, b_a, and b_c in the dependence of the mechanism of electrodeposition reactions. From Fig. 1.3, it is a clear that these dependencies are similar to each other for the large values of the i_0/i_L ratio and at any low value of overpotential. Because of this, the polarization curves without including the ohmic potential drop are not suitable and, for that reason, will be not treated further.

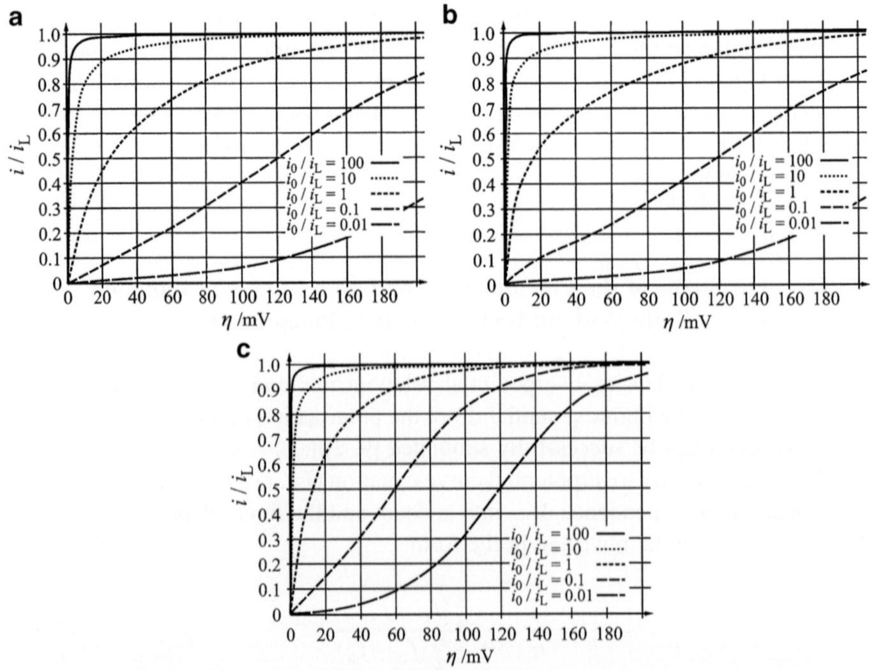

Fig. 1.3 Dependencies $i/i_L - \eta$ calculated using Eqs. (1.36) and (1.37) for the different values of i_0/i_L ratio: (**a**) $b_a = 120$ mV dec^{-1}, $b_c = 120$ mV dec^{-1}; (**b**) $b_a = 40$ mV dec^{-1}, $b_c = 120$ mV dec^{-1}; and (**c**) $b_a = 60$ mV dec^{-1}, $b_c = 60$ mV dec^{-1} (Reprinted from Ref. [21] with permission from the Serbian Chemical Society)

1.2.2 *Polarization Curves with Included Ohmic Potential Drop*

1.2.2.1 The Structure of Polarization Curves

If IR error is not eliminated, the measured value of overpotential, η_m, includes the ohmic potential drop and it is given by Eq. (1.38) [20]:

$$\eta_m = \eta + i\frac{L_c}{\kappa} \tag{1.38}$$

where i is the current density, L_c is the length of the electrolyte column between the tip of a liquid capillary and the electrode, and κ is the specific conductivity of the electrolyte. The values of i/i_L for selected overpotentials η are obtained using Eq. (1.36), and multiplication with i_L produces the values of i_0. η and i are then substituted in Eq. (1.38), and overpotentials which included ohmic potential, η_m, drop are obtained. Now, $i/i_L - \eta_m$ dependencies can be plotted. Hence, the polarization curves which include ohmic potential drop can be simulated by using Eqs. (1.36) and (1.38) [9].

1.2.2.2 The Shape of the Calculated Polarization Curves as Function of the i_0/i_L Ratios and the Kind of the Electrodeposition Process Control

The shapes of the polarization curves depend strongly on the i_0/i_L ratios. It is obvious that the continuous change of the exchange current density to the limiting diffusion current density ratio for processes of the metal electrodeposition is only possible by the digital simulation. In this way, the relations between i_0/i_L ratios, the shape of polarization curve, and the electrodeposition process control can be established.

The simulated polarization curves which included the ohmic potential drop are shown in Fig. 1.4. They are obtained using the same data as those shown in Fig. 1.3 and for $i_L = 50$ mA cm^{-2}, $L_c = 0.2$ cm, and $\kappa = 0.1$ S cm^{-1}. In all cases, for $i_0/i_L = 100$, there is linear dependence of the current density on overpotential up to $i = 45$ mA cm^{-2} or to $i/i_L \approx 0.9$. In these cases, the overpotential without included ohmic potential drop (see Fig. 1.3) is very low, and the measured overpotential is practically equal to the ohmic potential drop between the working and the reference electrodes. Hence, for $i_0/i_L > 100$, there is the ohmic control of the electrodeposition process. It is obvious that the shape of the linear part of polarization curve does not depend on the mechanism of the electrode reaction.

The polarization curves consist of two parts in the mixed ohmic–diffusion-controlled electrodeposition [12]. The first part corresponds to the ohmic control

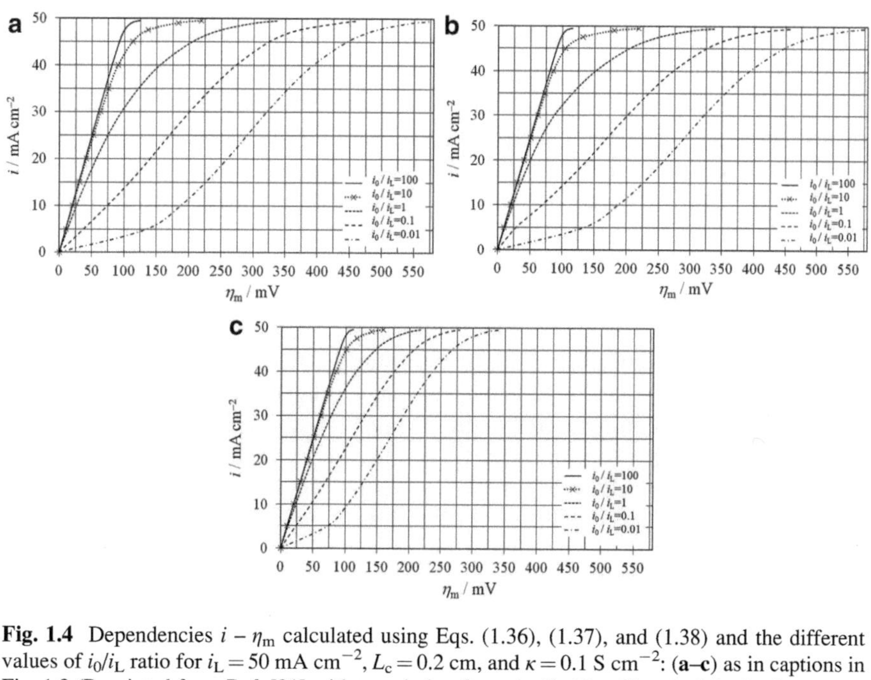

Fig. 1.4 Dependencies $i - \eta_m$ calculated using Eqs. (1.36), (1.37), and (1.38) and the different values of i_0/i_L ratio for $i_L = 50$ mA cm^{-2}, $L_c = 0.2$ cm, and $\kappa = 0.1$ S cm^{-2}: (**a–c**) as in captions in Fig. 1.3 (Reprinted from Ref. [21] with permission from the Serbian Chemical Society)

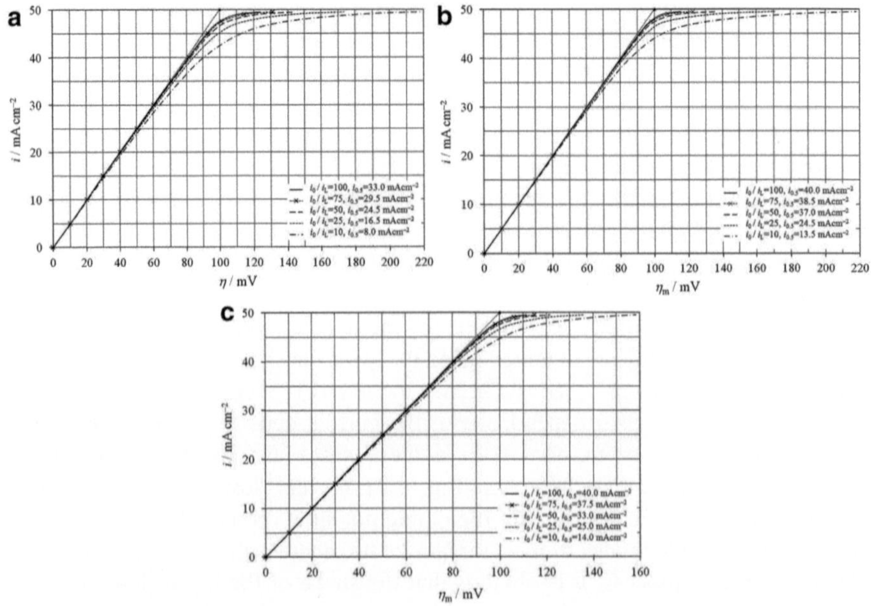

Fig. 1.5 Dependencies $i - \eta_m$ for $10 \leq i_0/i_L \leq 100$ calculated as those in Fig. 1.4. (**a**) $b_a = 120$ mV dec^{-1}, $b_c = 120$ mV dec^{-1}; (**b**) $b_a = 40$ mV dec^{-1}, $b_c = 120$ mV dec^{-1}; and (**c**) $b_a = 60$ mV dec^{-1}, $b_c = 60$ mV dec^{-1} (Reprinted from Ref. [21] with permission of the Serbian Chemical Society)

(it is the linear part), and the second one corresponds to the diffusion control (Fig. 1.5). The length of the ohmic part at the polarization curve decreases with decreasing the i_0/i_L ratio values. Assuming that the diffusion control of the electrodeposition process becomes visible at $\eta_m \geq 0.5$ mV, the values of $i_{0.5}/i_L$ at the end of the linear part of polarization curves are calculated and shown in the function of i_0/i_L ratio in Fig. 1.6 and given in Table 1.1. It can be seen from Fig. 1.6 and Table 1.1 that the linear part of the polarization curve vanishes at $i_0/i_L = 1$. Hence, there is the mixed ohmic–diffusion control in the interval of $1 < i_0/i_L \leq 100$.

At values of i_0/i_L ratio lower than 1, the complete diffusion control of the electrodeposition process arises at all overpotentials. The lower limit of the region of the complete diffusion control can be determined as follows: it is obvious that the convex shape of the polarization curve characterizes the diffusion control of deposition process and the concave one the activation control of deposition process. The i/i_L ratio as function of η is shown in Fig. 1.3 and the i as a function of η_m in Fig. 1.4. In both cases, the convex shape of curves changes in the concave one at approximately $i_0/i_L \sim 0.1$, meaning that the diffusion control changes in the activation one at the beginning of the polarization curve at low η and η_m. At larger overpotentials, the diffusion control occurs. Hence, the diffusion control at all overpotentials appears at $0.1 < i_0/i_L \leq 1$, while the activation control appears at $i_0/i_L \leq 0.1$ at low overpotentials.

It can be seen from Figs. 1.3 and 1.4 that the same conclusion can be derived for the polarization curves with and without included ohmic overpotential drop. The simulated polarization curves with included ohmic drop are calculated using data for

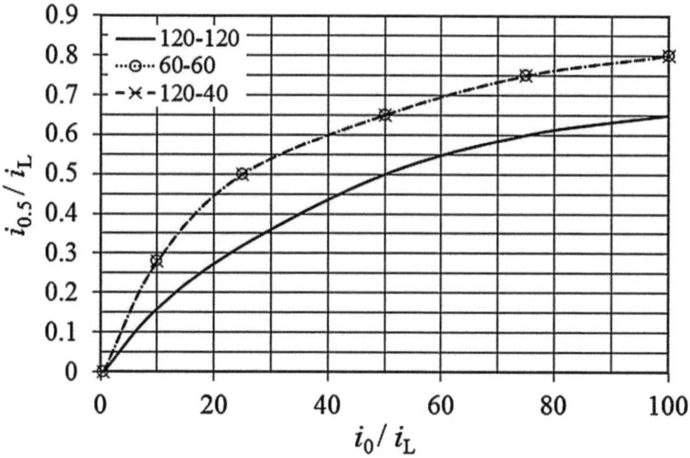

Fig. 1.6 Dependencies $i_{0.5}/i_L$ ratio on the i_0/i_L ratio. —— $b_a = 120$ mV dec^{-1}, $b_c = 120$ mV dec^{-1}; ···○··· $b_a = 60$ mV dec^{-1}, $b_c = 60$ mV dec^{-1}; ---x--- $b_a = 40$ mV dec^{-1}, $b_c = 120$ mV dec^{-1} (Reprinted from Ref. [21] with permission of the Serbian Chemical Society)

Table. 1.1 Calculated value of $i_{0.5}$ as function of i_0/i_L ratio (a, b) and (c) as in captions in Fig. 1.3.

	$i_{0.5}$ / mA cm^{-2}		
i_0/i_L	a	b	c
10	8.0	13.5	14.0
25	16.5	24.5	25.0
50	24.5	37.0	33.0
75	29.5	38.5	37.5
100	33.0	40.0	40.0

Reprinted from Ref. [21] with permission from the Serbian Chemical Society

1 M solution of typically fully dissociated electrolyte without supporting electrolyte [20]. It is obvious that for less concentrated solutions with supporting electrolyte (mainly some acid or base), the ohmic potential drop can be neglected. Because of this fact, η instead of η_m will be used, except for some special cases indicated.

1.3 Experimental Measured Polarization Curves

1.3.1 Polarization Curves for the Different Kinds of the Electrodeposition Process Control

1.3.1.1 The Ohmic and the Mixed Ohmic–Diffusion Control of Electrodeposition Process

Electrodeposition processes characterized by extremely high values of the exchange current density ($i_0 \rightarrow \infty$) belong to the fast electrochemical processes, and they usually occur under mixed ohmic–diffusion or even full ohmic control of the electrodeposition [10–13]. The typical ohmic–diffusion and full ohmic polarization curves recorded from the solutions of different concentrations of Pb^{2+} ions are shown in Fig. 1.7a. The basic characteristic of the full ohmic control is the linear dependence of the current density on the overpotential up to the inflection point (Fig. 1.7b for 0.45 M $Pb(NO_3)_2$ in 2.0 M $NaNO_3$) [12, 22, 23]. On the other hand, the polarization curve characterizing the mixed ohmic–diffusion control consists of

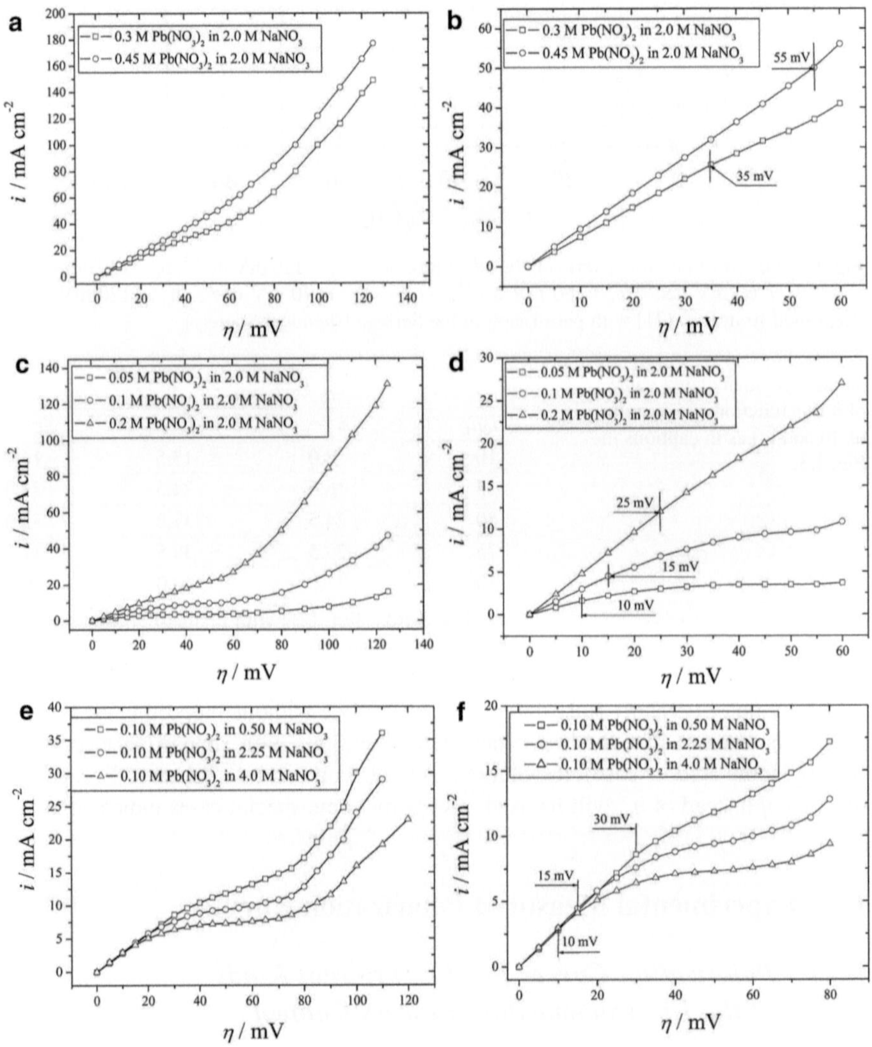

Fig. 1.7 The polarization curves for electrodeposition of Pb from (**a, b**) 0.30 and 0.45 M $Pb(NO_3)_2$ in 2.0 M $NaNO_3$, (**c, d**) 0.050, 0.10 and 0.20 M $Pb(NO_3)_2$ in 2.0 M $NaNO_3$, and (**e, f**) 0.10 M $Pb(NO_3)_2$ in 0.50, 2.25, and 4.0 M $NaNO_3$. The cathodic current densities are taken as positive ones (Reprinted from Ref. [12] with permission from Elsevier and Ref. [23] with kind permission from Springer)

ohmic and diffusion portions in the part of the polarization curve up to the inflection point (Fig. 1.7b for 0.30 M $Pb(NO_3)_2$ in 2.0 M $NaNO_3$) [12, 23].

The ratio of the ohmic to the overall control of the electrodeposition increases with increasing concentration of Pb^{2+} so that the electrodeposition process becomes a full ohmic-controlled one at high concentrations of Pb^{2+} (Fig. 1.7a–d; the ends of the ohmic control are denoted by the vertical lines in Fig. 1.7b, d) [12, 23]. For the solutions containing the different concentrations of Pb^{2+} ions in 2.0 M $NaNO_3$, the inflection point corresponds to an overpotential of 55 mV. Effect of increasing the supporting electrolyte concentration on the polarization characteristics of Pb was equivalent to the effect of the decreasing concentration of Pb^{2+} (at the constant concentration of the supporting electrolyte) (Fig. 1.7e, f).

The typical surface morphologies of Pb formed in each of types of electrodeposition control are shown in Fig. 1.8. The regular hexagonal crystals of Pb as those

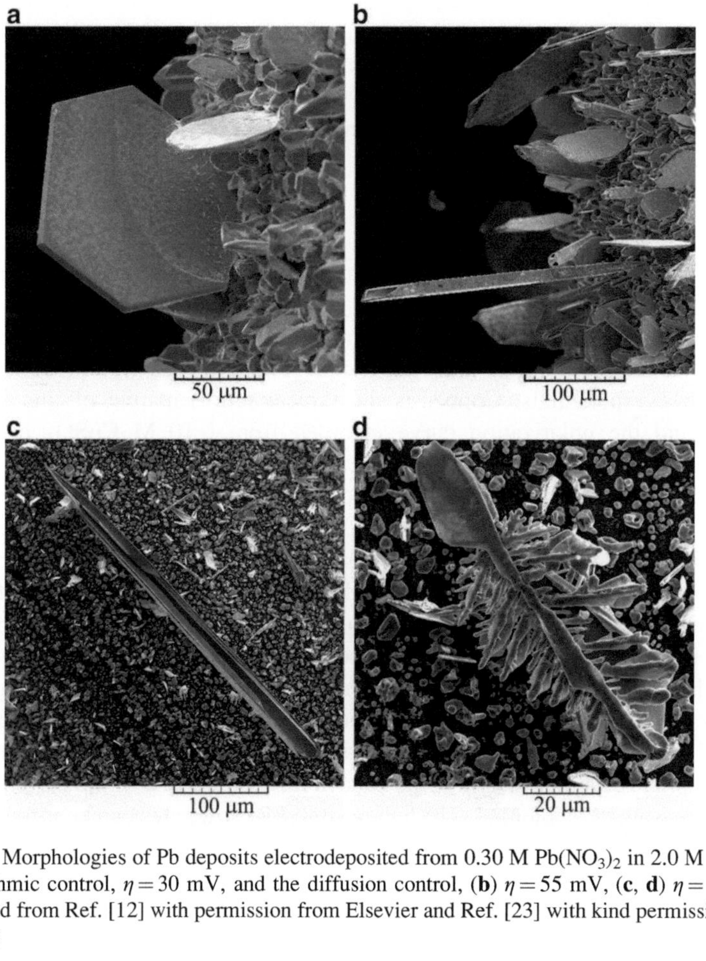

Fig. 1.8 Morphologies of Pb deposits electrodeposited from 0.30 M $Pb(NO_3)_2$ in 2.0 M $NaNO_3$: (**a**) the ohmic control, $\eta = 30$ mV, and the diffusion control, (**b**) $\eta = 55$ mV, (**c**, **d**) $\eta = 120$ mV (Reprinted from Ref. [12] with permission from Elsevier and Ref. [23] with kind permission from Springer)

shown in Fig. 1.8a are the characteristic of the ohmic control. The dendrites of the different shape are characteristic of the diffusion-controlled electrodeposition (Fig. 1.8b–d). Metals characterized by the high values of the exchange current density also show high overpotentials for hydrogen discharge [24], and the electrodeposition system remains under the diffusion control after the inflection point in the zone of the rapid increase of the current density with increasing the overpotential (Fig. 1.8c, d).

1.3.1.2 The Diffusion Control of Electrodeposition Process

The processes of zinc electrodeposition also belong to the fast electrochemical processes. However, unlike of Pb, after low nucleation rate the electrodeposition process immediately becomes the diffusion-controlled one. The polarization curve for Zn electrodeposition from 0.10 M $ZnSO_4$ in 2.0 M NaOH is shown in Fig. 1.9a. As seen from Fig. 1.9a, the well-defined plateau of the limiting diffusion current density is obtained starting from 65 mV. The spongy deposit is obtained in the range of overpotentials up to reaching of the plateau of the limiting diffusion current density (Fig. 1.9b), while the dendrites are electrodeposited at the plateau (Fig. 1.9c).

1.3.1.3 The Activation–Diffusion Control of Electrodeposition Process

The activation–diffusion control of electrodeposition process is a characteristic of metals characterized by the medium exchange current density values and lower hydrogen overpotentials. Copper is the typical representative of this group of metals, and the polarization curve recorded from 0.10 M $CuSO_4$ in 0.50 M H_2SO_4 is shown in Fig. 1.10a.

The four parts can be detected in this polarization curve: (a) activation, (b) mixed activation–diffusion, (c) diffusion, and (d) the zone of the fast increase of current density with overpotential. Electrodeposition in each of these regions produces characteristic surface morphologies. The large grains are obtained in the activation-controlled electrodeposition (Fig. 1.10b). The carrot-like and globular forms are formed in the mixed activation–diffusion control (Fig. 1.10c). The dendrites formed inside the plateau of the limiting diffusion current density are shown in Fig. 1.10d. Due to the lower overpotentials for hydrogen evolution reaction, the effect of evolved hydrogen as parallel reaction to copper electrodeposition at the high overpotentials becomes visible in the zone of the fast increase of the current density with increasing overpotential. The honeycomb-like structures can be formed in this zone (Fig. 1.10e).

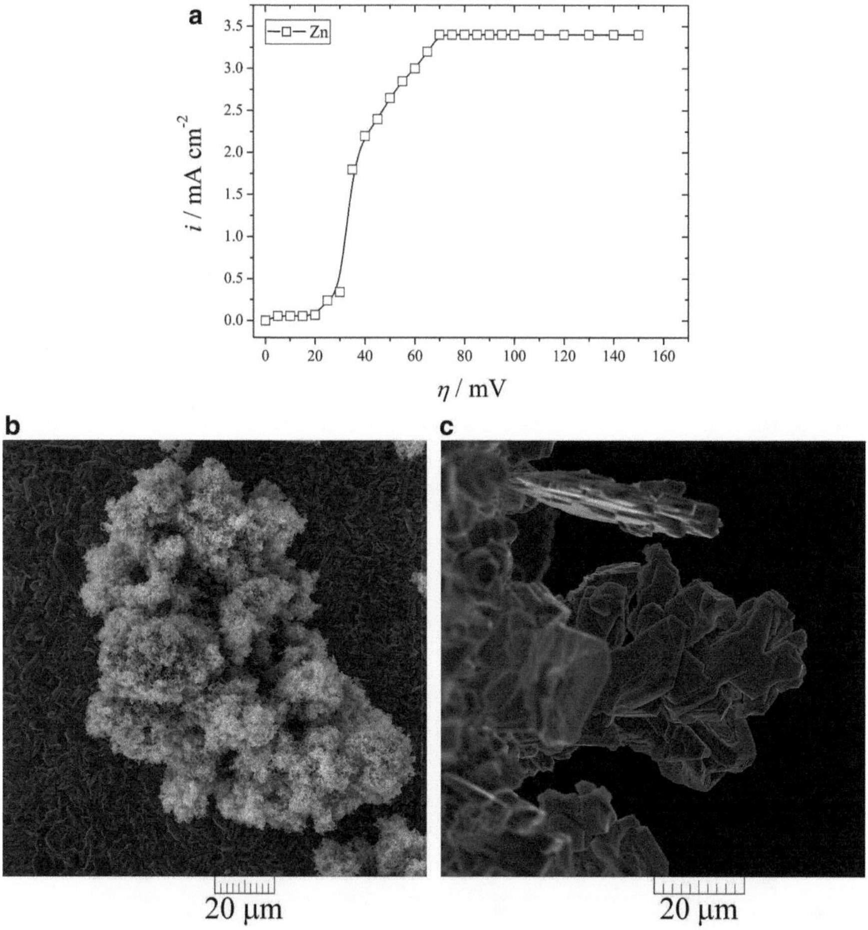

Fig. 1.9 (**a**) The polarization curve for Zn electrodeposition from 0.10 M $ZnSO_4$ in 2.0 M NaOH and the typical surface morphologies obtained at overpotentials of (**b**) $\eta = 45$ mV (the zone of the fast increase in the current density with increasing overpotential) and (**c**) $\eta = 100$ mV (the plateau of the limiting diffusion current density). The cathodic current densities in Fig. 1.9a are taken as positive ones (Reprinted from Ref. [21] with permission from the Serbian Chemical Society)

Effect of the Different Parameters of Electrolysis on the Shape of Polarization Curves

Figure. 1.11 shows Cu polarization curves obtained from solutions of different concentrations of Cu^{2+} ions (Fig. 1.11a), the supporting electrolyte, *i.e.*, H_2SO_4 (Fig. 1.11b), and at temperatures lower and higher than the room temperatures (Fig. 1.11c).

The increase in the concentration of Cu^{2+} ions (at the constant H_2SO_4 concentration) leads to a shift in overpotentials at which the limiting diffusion current

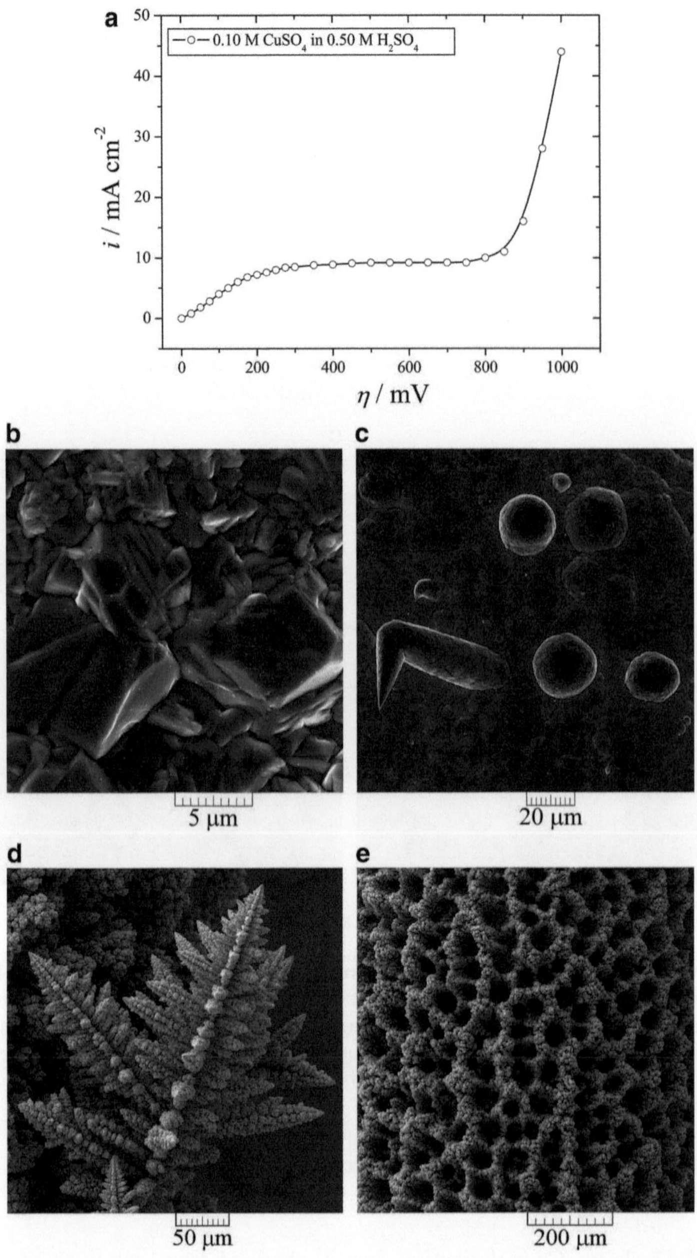

Fig. 1.10 (**a**) The polarization curve for Cu electrodeposition from 0.10 M $CuSO_4$ in 0.50 M H_2SO_4 and the typical surface morphologies obtained in the different types of control: (**b**) activation; $\eta = 90$ mV; (**c**) activation–diffusion, $\eta = 210$ mV; (**d**) diffusion, $\eta = 650$ mV; and (**e**) zone of the fast increase of current density with overpotential, $\eta = 1000$ mV. The cathodic current densities in Fig. 1.10a are taken as positive ones (Reprinted from Ref. [21] with permission from the Serbian Chemical Society)

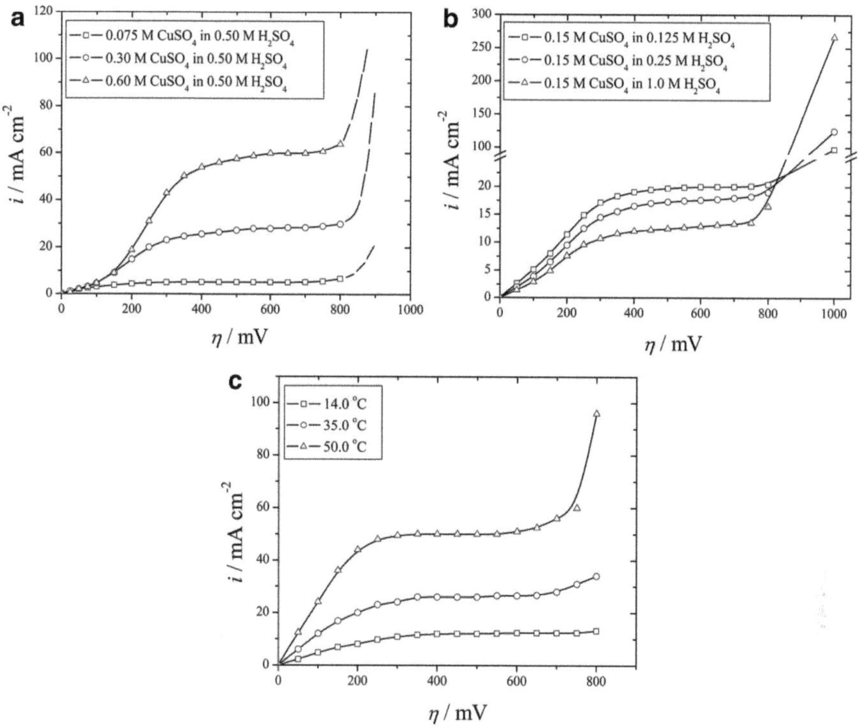

Fig. 1.11 The polarization curves for Cu electrodeposition from (**a**) 0.075, 0.30, and 0.60 M $CuSO_4$ in 0.50 M H_2SO_4; (**b**) 0.15 M $CuSO_4$ in 0.125, 0.25, and 1.0 M H_2SO_4; and (**c**) 0.15 M $CuSO_4$ in 0.50 M H_2SO_4. Temperatures: 14.0, 35.0, and 50.0 ± 0.5 °C. The cathodic current densities are taken as positive ones (Reprinted from Ref. [25] with permission from MDPI, Ref. [26] with permission from Elsevier, Ref. [27] with permission from the Serbian Chemical Society and Ref. [7] with kind permission from Springer)

density plateaus initiate toward the higher overpotentials, while the end of these plateaus remains practically constant (Fig. 1.11a) [7, 25]. At the constant concentration of Cu^{2+} ions, the beginning of the plateau of the limiting diffusion current density was slightly shifted to higher electrodeposition overpotentials with the decreasing concentration of H_2SO_4 (Fig. 1.11b) [7, 26]. The ends of the plateau of the limiting diffusion current density practically did not depend on the concentration of H_2SO_4. The limiting diffusion current density value decreased with the increasing concentration of H_2SO_4. The decrease in the values of the limiting diffusion current density with increasing H_2SO_4 concentration is a consequence of the decrease of the Cu^{2+} ion activity with increasing acidity of the solution [7, 26]. Increasing the temperature leads to an increase of the limiting diffusion current density as well as to a shift of both the beginning and the end of the plateau of the limiting diffusion current density towards lower electrodeposition overpotentials [7, 27].

1.3.1.4 The Polarization Curves with Parallel Processes of Metal Electrodeposition and Hydrogen Evolution

Due to low values of overpotentials for hydrogen evolution, electrodeposition of some metals like cobalt occurs with simultaneous hydrogen evolution, and it is practically impossible to determine the diffusion limiting current density from the polarization curves without applied IR drop compensation technique [28, 29]. As can be seen from Fig. 1.12a after IR drop correction (solid line, i_{tot}), significantly different current response is obtained than the one measured without IR drop correction (dot line, i_m), being characterized by a sudden increase of current density at the commencement of the electrodeposition process. As far as the deposition started, hydrogen evolution also started and the process of Co deposition was accompanied by hydrogen evolution in the whole range of investigated potentials. As a consequence, extremely high currents were recorded (since at the pH 9.2–9.5 hydrogen is evolving from water molecules), and accordingly correction for IR drop caused significant change in the shape of the polarization diagrams. The current density for Co electrodeposition (i_{Co}) is obtained after a subtraction of the current density for hydrogen evolution ($i(H_2)$) from the measured (corrected for IR drop) current density values (i_{tot}) (Fig. 1.12b). The method of determination of the current density for hydrogen evolution reaction is described elsewhere [28, 29].

1.3.2 *Polarization Curves Measured for Different* i_0/i_L *Ratios*

The polarization curves for nickel, copper, and cadmium electrodeposition (in all solutions, concentrations of depositing ions were 0.070 M) are shown in Fig. 1.13, while corresponding Tafel plots and the results of linear polarization experiments

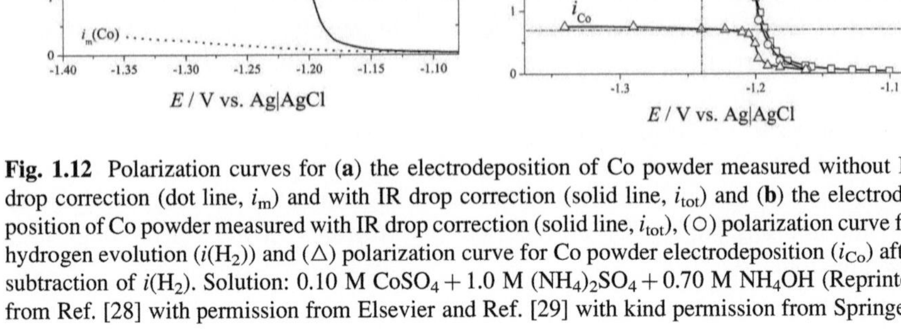

Fig. 1.12 Polarization curves for (**a**) the electrodeposition of Co powder measured without IR drop correction (dot line, i_m) and with IR drop correction (solid line, i_{tot}) and (**b**) the electrodeposition of Co powder measured with IR drop correction (solid line, i_{tot}), (○) polarization curve for hydrogen evolution ($i(H_2)$) and (△) polarization curve for Co powder electrodeposition (i_{Co}) after subtraction of $i(H_2)$. Solution: 0.10 M $CoSO_4$ + 1.0 M $(NH_4)_2SO_4$ + 0.70 M NH_4OH (Reprinted from Ref. [28] with permission from Elsevier and Ref. [29] with kind permission from Springer)

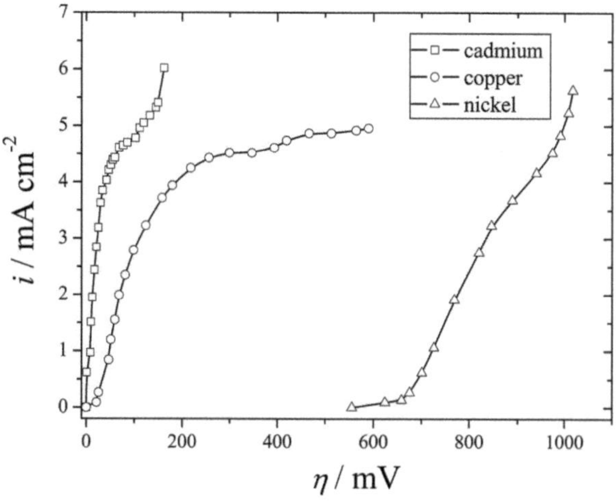

Fig. 1.13 Polarization curves for cadmium (□), copper (○), and (Δ) nickel depositions (Reprinted from Ref. [30] with permission from the Serbian Chemical Society and Ref. [31] with kind permission from Springer)

Table 1.2 The exchange current density and i_L/i_0 ratios for Cd, Cu, and Ni deposition processes

Metal	$i_0/A\ cm^{-2}$	i_L/i_0
Cadmium	1.5×10^{-3}	3.0
Copper	3.2×10^{-4}	14.4
Nickel	1.6×10^{-9}	2.9×10^{6}

Reprinted from Ref. [30] with permission from the Serbian Chemical Society and Refs. [1, 31] with kind permission from Springer

are given in Ref. [30]. The limiting diffusion currents in all cases are practically the same (about 4.80 mA cm^{-2}), but the exchange current densities (Table 1.2) are very different.

The shape of polarization curves is qualitatively in accordance with polarization curves presented in Fig. 1.4 due to the different i_0/i_L ratios.

References

1. Popov KI, Djokić SS, Grgur BN (2002) Fundamental aspects of electrometallurgy. Kluwer Academic/Plenum Publishers, New York
2. Popov KI, Pavlović MG, Pavlović LjJ, Čekerevac MI, Removič GŽ (1988) Electrode surface coarsening in pulsating overpotential copper electrodeposition. Surf Coat Technol 34:355–363
3. Damjanović A (1965) On the mechanism of metal electrocrystallization. Plating 52:1017–1026
4. Popov KI, Grgur BN, Pavlović MG, Radmilović V (1993) The morphology of copper electrodeposits: I. The mechanism of copper cauliflower-like electrodeposit formation. J Serb Chem Soc 58:1055–1062
5. Popov KI, Radmilović V, Grgur BN, Pavlović MG (1994) The morphology of copper electrodeposits II. The mechanism of carrot-like electrodeposit formations. J Serb Chem Soc 59:47–52

6. Popov KI, Maksimović MD, Trnjančev JD, Pavlović MG (1981) Dendritic electrocrystallization and the mechanism of powder formation in the potentiostatic electrodeposition of metals. J Appl Electrochem 11:239–246
7. Nikolić ND, Popov KI (2010) Hydrogen co-deposition effects on the structure of electrodeposited copper. In: Djokić SS (ed) Electrodeposition: theory and practice, modern aspects of electrochemistry, vol 48. Springer, New York, pp 1–70
8. Popov KI, Krstajić NV (1983) The mechanism of spongy electrodeposits formation on inert substrate at low overpotentials. J Appl Electrochem 13:775–782
9. Popov KI, Živković PM, Nikolić ND (2010) The effect of morphology of activated electrodes on their electrochemical activity. In: Djokić SS (ed) Electrodeposition: theory and practice, modern aspects of electrochemistry, vol 48. Springer, New York, pp 163–213
10. Popov KI, Živković PM, Grgur BN (2007) Physical and mathematical models of an inert macroelectrode modified with active hemispherical microelectrodes. Electrochim Acta 52:4696–4707
11. Popov KI, Živković PM, Krstić SB, Nikolić ND (2009) Polarization curves in the ohmic controlled electrodeposition of metals. Electrochim Acta 54:2924–2931
12. Nikolić ND, Popov KI, Živković PM, Branković G (2013) A new insight into the mechanism of lead electrodeposition: ohmic–diffusion control of the electrodeposition process. J Electroanal Chem 691:66–76
13. Nikolić ND, Popov KI, Ivanović ER, Branković G, Stevanović SI, Živković PM (2015) The potentiostatic current transients and the role of local diffusion fields in formation of the 2D lead dendrites from the concentrated electrolyte. J Electroanal Chem 739:137–148
14. Diggle JW, Despić AR, Bockris JO'M (1969) The mechanism of the dendritic electrocrystallization of zinc. J Electrochem Soc 116:1503–1514
15. Barton JL, Bockris JO'M (1962) The electrolytic growth of dendrites from ionic solutions. Proc R Soc A268:485–505
16. Despić AR, Popov KI (1972) Transport controlled deposition and dissolution of metals. In: Conway BE, Bockris JO'M (eds) Modern aspects of electrochemistry, vol 7. Plenum Press, New York, pp 199–313
17. Popov KI, Stojilković ER, Radmilović V, Pavlović MG (1997) Morphology of lead dendrites electrodeposited by square-wave pulsating overpotential. Powder Technol 93:55–61
18. Newman JS (1973) Electrochemical systems. Prentice-Hall Inc., Engelwood Cliffs, p 177
19. Popov KI, Nikolić ND, Živković PM, Branković G (2010) The effect of the electrode surface roughness at low level of coarseness on the polarization characteristics of electrochemical processes. Electrochim Acta 55:1919–1925
20. Bockris JO'M, Reddy AKN, Gamboa–Aldeco M (2000) Modern electrochemistry 2A, fundamentals of electrodics, 2nd edn. Kluwer Academic/Plenum Publishers, New York, p 1107
21. Popov KI, Živković PM, Jokić B, Nikolić ND (2016) The shape of the polarization curve and diagnostic criteria for the metal electrodeposition process control. J Serb Chem Soc 81: in press
22. Nikolić ND, Branković G, Lačnjevac U (2012) Formation of two-dimensional (2D) lead dendrites by application of different regimes of electrolysis. J Solid State Electrochem 16:2121–2126
23. Nikolić ND, Popov KI (2014) A new approach to the understanding of the mechanism of lead electrodeposition. In: Djokić SS (ed) Electrodeposition and surface finishing, modern aspects of electrochemistry, vol 57. Springer, New York, pp 85–132
24. Winand R (1994) Electrodeposition of metals and alloys – new results and perspectives. Electrochim Acta 39:1091–1105
25. Nikolić ND, Popov KI, Pavlović LjJ, Pavlović MG (2007) Determination of critical conditions for the formation of electrodeposited copper structures suitable for electrodes in electrochemical devices. Sensors 7:1–15
26. Nikolić ND, Branković G, Pavlović MG, Popov KI (2008) The effect of hydrogen codeposition on the morphology of copper electrodeposits II. Correlation between the properties of electrolytic solutions and the quantity of evolved hydrogen. J Electroanal Chem 621:13–21

27. Nikolić ND, Pavlović LjJ, Pavlović MG, Popov KI (2007) Effect of temperature on electrodeposition of disperse copper deposits. J Serb Chem Soc 72:1369–1381
28. Jović VD, Jović BM, Pavlović MG (2006) Electrodeposition of Ni, Co and Ni–Co alloy powders. Electrochim Acta 51:5468–5477
29. Jović VD, Nikolić ND, Lačnjevac UČ, Jović BM, Popov KI (2012) Morphology of different electrodeposited pure metal powders. In: Djokić SS (ed) Electrochemical production of metal powders, modern aspects of electrochemistry, vol 54. Springer, New York, pp 63–123
30. Popov KI, Grgur BN, Stojilković ER, Pavlović MG, Nikolić ND (1997) The effect of deposition process exchange current density on the thin metal films formation on inert substrate. J Serb Chem Soc 62:433–442
31. Popov KI, Nikolić ND (2012) General theory of disperse metal electrodeposits formation. In: Djokić SS (ed) Electrochemical production of metal powders, modern aspects of electrochemistry, vol 54. Springer, New York, pp 1–62

Chapter 2
Mechanisms of Formation of Some Forms of Electrodeposited Pure Metals

2.1 Electrodeposition on Native Substrate

2.1.1 *Macroelectrodes and Microelectrodes*

As already given in Chap. 1, the most frequently used form of the cathodic polarization curve equation for flat or large spherical electrode of massive metal is given by:

$$i = \frac{i_0(f_c - f_a)}{1 + \frac{i_0 f_c}{i_L}} \tag{1.13}$$

where i, i_0, and i_L are the current density, exchange current density, and limiting diffusion current density, respectively, and

$$f_c = 10^{\frac{\eta}{b_c}} \text{ and } f_a = 10^{-\frac{\eta}{b_a}} \tag{1.37}$$

where b_c and b_a are the cathodic and anodic Tafel slopes and η is the overpotential. Equation (1.13) is modified for use in electrodeposition of metals by taking cathodic current density and overpotential as positive. Derivation of the Eq. (1.13) is performed under assumption that the concentration dependence of i_0 can be neglected [1–4].

The electrochemical processes on microelectrodes in bulk solution can be under activation control at overpotentials which correspond to the limiting diffusion current density plateau of the macroelectrode. The cathodic limiting diffusion current density for steady-state spherical diffusion, $i_{L,spher}$, is given by:

$$i_{L,\,spher} = \frac{nFDC_0}{r_{el}} \tag{1.11}$$

K.I. Popov et al., *Morphology of Electrochemically and Chemically Deposited Metals*, DOI 10.1007/978-3-319-26073-0_2

and for steady-state linear diffusion, i_L, by:

$$i_L = \frac{nFDC_0}{\delta} \tag{1.14}$$

It follows from Eqs. (1.11) and (1.14) that

$$\frac{i_{L,spher}}{i_L} = \frac{\delta}{r_{el}} \tag{2.1}$$

An electrode around which the hydrodynamic diffusion layer can be established, being considerably lower than dimensions of it, could be considered as a macroelectrode. An electrode, mainly spherical, whose diffusion layer is equal to the radius of it, satisfying

$$\delta >> r_{el} \tag{2.2}$$

can be considered as a microelectrode [5].

According to the Eq. (1.13) for $f_c >> f_a$ and

$$\frac{i_0 f_c}{i_L} >> 1 \tag{2.3}$$

the cathodic process on the macroelectrode enters full diffusion control, *i.e.*,

$$i = i_L \tag{1.35}$$

Simultaneously, the cathodic current density on the spherical microelectrode, i_{spher}, is given by[1]:

$$i_{spher} = \frac{i_0(f_c - f_a)}{1 + \frac{i_0 f_c}{i_{L,spher}}} \tag{1.12}$$

or, because of the Eq. (2.1):

$$i_{spher} = \frac{i_0(f_c - f_a)}{1 + \frac{i_0}{i_L} \cdot \frac{r_{el}}{\delta} \cdot f_c} \tag{2.4}$$

and if a condition of the Eq. (2.3) is also valid, but

[1] The reversible potential of a surface with radius of curvature r_{cur} would depart from that of a planar surface by the quantity $\Delta E_r = 2\gamma V/(nFr_{cur})$, where γ is the interfacial energy between metal and solution and V molar volume of metal [5]. It is valid at extremely low r_{cur}, being of the order of few millivolts, and it can be neglected except in some special cases, like the stability of the shape of the tips of dendrites [5].

$$\frac{r_{el}}{\delta} \rightarrow 0 \tag{2.5}$$

the Eq. (2.4) can be rewritten in the form

$$i = i_0 f_c \tag{2.6}$$

This means that the process on the microelectrode in the bulk solution can be under complete activation control at the same overpotential at which the same process on the macroelectrode is simultaneously under full diffusion control.

The different behavior of macroelectrodes and microelectrodes under the same conditions of electrodeposition causes the disperse deposit formation.

2.1.2 Active Microelectrodes Placed Inside Diffusion Layer of the Active Macroelectrode

2.1.2.1 Mathematical Model

Naturally, the microelectrodes can be placed on the macroelectrodes inside their diffusion layers. Let us consider the model of surface irregularities shown in Fig. 2.1. The electrode surface irregularities are buried deep in the diffusion layer, which is characterized by a steady linear diffusion to the flat portion of the surface [1, 6, 7], and the limiting diffusion current density is given by Eq. (1.14).

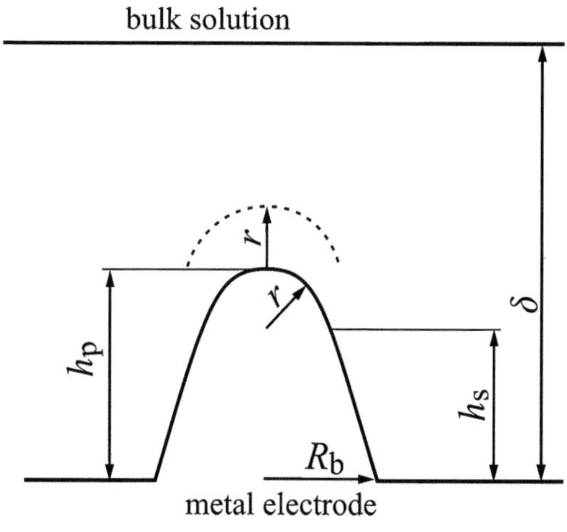

Fig. 2.1 Model of a paraboloidal surface protrusion; h_p is the height of the protrusion relative to the flat portion of the surface, h_s is the corresponding local side elongation, r is the radius of the protrusion tip, R_b is the radius of the protrusion base, and δ is the thickness of the diffusion layer and $\delta >> h_p$ (Reprinted from Refs. [1, 8] with kind permission from Springer and Ref. [6] with permission from Elsevier)

At the side of an irregularity, the limiting diffusion current density, $i_{L,s}$, is given as:

$$i_{\mathrm{L,s}} = \frac{nFDC_0}{\delta - h_\mathrm{s}} = i_\mathrm{L}\frac{\delta}{\delta - h_\mathrm{s}} \tag{2.7}$$

Obviously, this is valid if the protrusion height does not affect the outer limit of the diffusion layer, and if a possible lateral diffusion flux supplying the reacting ions can be neglected. At the tip of an irregularity, the lateral flux cannot be neglected and the situation can be approximated by assuming a spherical diffusion current density, $i_{L,tip}$, given by the Eq. (2.8) [7]:

$$i_{\mathrm{L,tip}} = \frac{nFDC^*}{r} \tag{2.8}$$

where C^* is the concentration of the diffusing species at a distance r from the tip, assuming that around the tip a spherical diffusion layer having a thickness equal to the radius of the protrusion tip is formed [5]. Obviously, if $r > \delta$, the spherical diffusion layer around the tip of protrusion cannot be formed and the Eq. (2.9) is valid:

$$i_{\mathrm{L,tip}} = \frac{nFDC_0}{\delta - h_\mathrm{p}} \tag{2.9}$$

If deposition to the macroelectrode is under full diffusion control, the distribution of the concentration C inside the linear diffusion layer is given by the Eq. (2.10) [3]:

$$C = C_0\frac{h_{\mathrm{difl}}}{\delta} \tag{2.10}$$

where $0 \leq h_{\mathrm{difl}} \leq \delta$. Hence,

$$C^* = C_0\frac{h_\mathrm{p} + r}{\delta} \tag{2.11}$$

and

$$i_{\mathrm{L,tip}} = i_\mathrm{L}\left(1 + \frac{h_\mathrm{p}}{r}\right) \tag{2.12}$$

because of Eqs. (1.14), (2.8), and (2.11).

The tip radius of the paraboloidal protrusion is given by the Eq. (2.13) [3, 5, 9]:

$$r = \frac{{R_\mathrm{b}}^2}{2h_\mathrm{p}} \tag{2.13}$$

and substitution of r from the Eq. (2.13) in the Eq. (2.12) gives

$$i_{\mathrm{L,tip}} = i_{\mathrm{L}}\left(1 + \frac{2h_{\mathrm{p}}^{2}}{R_{\mathrm{b}}^{2}}\right) \tag{2.14}$$

or

$$i_{\mathrm{L,tip}} = i_{\mathrm{L}}\left(1 + 2k_{\mathrm{p}}^{2}\right) \tag{2.15}$$

where

$$k_{\mathrm{p}} = \frac{h_{\mathrm{p}}}{R_{\mathrm{b}}} \tag{2.16}$$

If $h_{\mathrm{p}} = R_{\mathrm{b}}, k_{\mathrm{p}} = 1$, hence for a hemispherical protrusion is

$$i_{\mathrm{L,tip}} = 3i_{\mathrm{L}} \tag{2.17}$$

If $h_{\mathrm{p}} << R_{\mathrm{b}}, k_{\mathrm{p}} \rightarrow 0$, and

$$i_{\mathrm{L,tip}} \rightarrow i_{\mathrm{L}} \tag{2.18}$$

and if $R_{\mathrm{b}} << h_{\mathrm{p}}, k_{\mathrm{p}} \rightarrow \infty$ and

$$i_{\mathrm{L,tip}} \rightarrow \infty \tag{2.19}$$

Substitution of $i_{\mathrm{L,tip}}$ from Eq. (2.15) instead of i_{L} in Eq. (1.13) and further rearranging give Eq. (2.20):

$$i_{\mathrm{tip}} = \frac{i_0(f_{\mathrm{c}} - f_{\mathrm{a}})}{1 + \frac{i_0}{i_{\mathrm{L}}} \cdot \frac{1}{1+2k_{\mathrm{p}}^2} f_{\mathrm{c}}} \tag{2.20}$$

The current density on the tip of a protrusion, i_{tip}, is determined by k_{p}, hence by the shape of the protrusion. If $k_{\mathrm{p}} \rightarrow 0$, $i_{\mathrm{tip}} \rightarrow i$ (see the Eq. (1.13)) and if $k_{\mathrm{p}} \rightarrow \infty$, $i_{\mathrm{tip}} \rightarrow i_0\,(f_{\mathrm{c}} - f_{\mathrm{a}}) >> i$. The electrochemical process on the tip of a sharp needle-like protrusion can be under pure activation control outside the diffusion layer of the macroelectrode. Inside it, the process on the tip of a protrusion is under mixed control, regardless it is under complete diffusion control on the flat part of the electrode for $k_{\mathrm{p}} \rightarrow 0$ (see section "Model of the spherical diffusion around the tip of a surface protrusion–deposition to the point"). If $k_{\mathrm{p}} = 1$, hence for hemispherical protrusion, i_{tip} will be somewhat larger than i, but the kind of control will not be changed. It is important to note that the current density to the tip of hemispherical protrusion does not depend on the size of it if $k_{\mathrm{p}} = 1$. This makes a substantial difference between spherical microelectrodes in bulk solution and microelectrodes inside diffusion layer of the macroelectrode [3, 9, 10]. In the first case, the limiting diffusion current density depends strongly on the radius of the microelectrode.

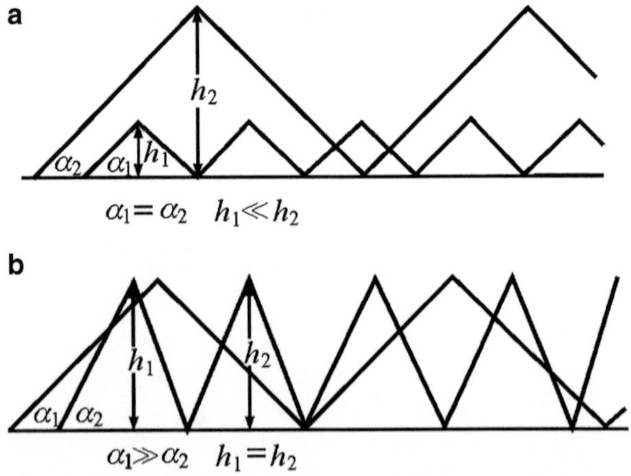

Fig. 2.2 Models of surfaces with (**a**) the same surface roughness and different coarseness and (**b**) vice versa (Reprinted from Ref. [11] with kind permission from Springer)

Electrode Surface Coarseness

Any solid metal surface that represents a substrate for metal deposition possesses a certain roughness. In addition, it may appear coarse or smooth, and this is not necessarily related to the roughness. Figure 2.2 shows cases of surfaces with (a) the same roughness and profoundly different coarseness and (b) vice versa.

It is the level of coarseness which determines the appearance of metal deposits, while even with considerable roughness, if below the visual level, the surface may appear smooth.

It is convenient to define the surface coarseness as the difference in thickness of the metal at the highest and lowest points above an arbitrary reference plane facing the solution.

Non-dendritic Amplification of Surface Coarseness

The current densities, i_s and i_{tip} to different points at the electrode surface, can be obtained by substitution of i_L in the Eq. (1.13) by appropriate values from the Eqs. (2.7) and (2.12) for the side and the tip of the protrusion, around which the spherical diffusion layer is formed, respectively. Hence:

$$i_s = \frac{i_0(f_c - f_a)}{1 + \left(\frac{i_0 f_c}{i_L}\right)\left(\frac{\delta - h_s}{\delta}\right)} \tag{2.21}$$

and if the spherical diffusion layer around the tip of a protrusion can be formed ($r < \delta$), it will be valid as in Eq. (2.22):

$$i_{tip} = \frac{i_0(f_c - f_a)}{1 + \left(\frac{i_0 f_c}{i_L}\right)\left(\frac{r}{r + h_p}\right)} \tag{2.22}$$

However, if the spherical diffusion layer around the tip of a protrusion cannot be formed ($r > \delta$), it will be valid:

$$i_{\mathrm{tip}} = \frac{i_0(f_{\mathrm{c}} - f_{\mathrm{a}})}{1 + \left(\frac{i_0 f_{\mathrm{c}}}{i_{\mathrm{L}}}\right)\left(\frac{\delta - h_{\mathrm{p}}}{\delta}\right)} \tag{2.23}$$

The effective rate of growth of the side elevation is equal to the rate of motion of the side elevation relative to the rate of motion of the flat surface [11]. Hence:

$$\frac{\mathrm{d}h_{\mathrm{s}}}{\mathrm{d}t} = \frac{V}{nF}(i_{\mathrm{s}} - i) \tag{2.24}$$

and substitution of i_{s} from the Eq. (2.21) and i from the Eq. (1.13) in the Eq. (2.24) and further rearrangement give the Eq. (2.25) [12]:

$$\frac{\mathrm{d}h_{\mathrm{s}}}{\mathrm{d}t} = \frac{i^2 V h_{\mathrm{s}}}{i_{\mathrm{L}} nF\delta} \tag{2.25}$$

if $\delta >> h_{\mathrm{s}}$ and $f_{\mathrm{c}} >> f_{\mathrm{a}}$, or in the integral form:

$$h_{\mathrm{s}} = h_{0,\mathrm{s}} \exp\left(\frac{i}{i_{\mathrm{L}}}\frac{Q}{Q_{0,\mathrm{D}}}\right) \tag{2.26}$$

where $h_{0,\mathrm{s}}$ is the initial height of the local side elevation (see also Eqs. (2.32), (2.33), and (2.34)) and Q is given by the Eq. (2.27):

$$Q = i \cdot t \tag{2.27}$$

and

$$Q_{0,\mathrm{D}} = \frac{nF\delta}{V} \tag{2.28}$$

Hence, an increase in the surface coarseness can be expected with increasing quantity of deposited metal for the same deposition current density, as well as with increasing current density for the same quantity of electrodeposited metal.

In the same way, the propagation rate of the protrusion tip can be obtained by substituting i_{tip} from the Eq. (2.22) and i from the Eq. (1.13) into the Eq. (2.24), where i_{s} and h_{s} are substituted by i_{tip} and h_{p}, on further rearrangement the following expression is obtained:

$$\frac{\mathrm{d}h_{\mathrm{p}}}{\mathrm{d}t} = \frac{V i\, i_{tip} h_{\mathrm{p}}}{nF i_{\mathrm{L}}(r + h_{\mathrm{p}})} \tag{2.29}$$

It should be noted that the Eq. (2.29) is only valid if the radius of the protrusion tip is sufficiently large to make the surface energy term negligible [5].

It is obvious from Eqs. (2.25) and (2.29) that

$$\frac{dh_p}{dt} > \frac{dh_s}{dt} \tag{2.30}$$

because $i_{tip} > i$ and $h_p/(r + h_p) > h_s/\delta$, which means that the tip propagation protrusion will be larger under spherical diffusion control.

2.1.2.2 Physical Model

To test the validity of the above equations, Popov et al. [7, 11, 13, 14] carried out experiments on diffusion-controlled metal electrodeposition on a well-defined, triangularly shaped surface profile, through a diffusion layer of well-defined thickness $\delta >> h_p$. A phonograph disk negative was used as the substrate upon which a layer of an agar-containing copper sulfate–sulfuric acid solution was placed and left to solidify, as illustrated in Fig. 2.3a.

As current was passed and the layer was depleted of copper ions, an increase in the height of the triangular ridges was observed. Metallographic samples were made in wax, and the cross sections of the deposit were photographed under a microscope (Fig. 2.3b) [14]. The three parts of the surface can be noticed in Fig. 2.3b: the flat part of the electrode surface and the sides and the tips of irregularities, providing an excellent physical illustration of the mathematical model.

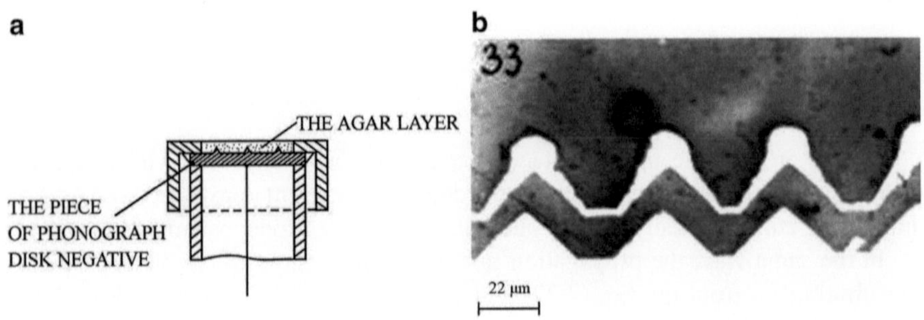

Fig. 2.3 (**a**) The model of an electrode for diffusion-controlled electrodeposition of metals and (**b**) cross section of Cu deposit obtained from 0.50 M $CuSO_4$ in 0.50 M H_2SO_4 in a system from (**a**). The thickness of the agar diffusion layer was 1.0 mm. Deposition overpotential: 300 mV; deposition time: 120 min. The substrate is a piece of a phonograph disk negative (Reprinted from Refs. [7, 11, 13] with kind permission from Springer)

2.1.2.3 Real Systems

The Effect of the Deposition Current Density on the Electrode Surface Coarseness

The effect of deposition current density on the increase in surface coarseness for a fixed quantity of electrodeposited metal is illustrated by Fig. 2.4. As expected, the surface coarseness increases strongly with the increasing current density in the region of mixed activation–diffusion control where obtained deposits are polycrystalline (Fig. 2.4b, c) [15, 16]. It should be noted that the theories describing the increase of electrode surface coarseness are valid for $h_p \ll \delta$, *i.e.*, for the short deposition times. For qualitative investigations, the effect of current density on the increase of surface coarseness and large quantities of metal were used, as it was assumed that the qualitative picture of the phenomenon would not be changed. In quantitative investigations, the deposition times in which the approximation $\delta \gg h_p$ is valid [17] must be determined and taken into account. In the region of the activation-controlled electrodeposition, at the same initial surface, the situation is quite different as can be seen from Fig. 2.4a.

The Deposits with a Minimal Coarseness

Activation-controlled deposition of copper produces large grains with relatively well-defined crystal shapes. This can be explained by the fact that the values of the

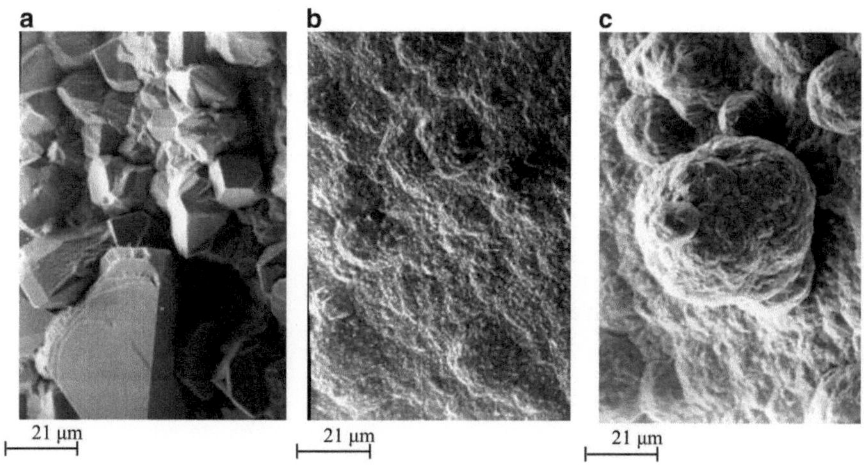

Fig. 2.4 Copper deposits obtained from 0.10 M $CuSO_4$ in 0.50 M H_2SO_4. Quantity of electricity, Q: 20 mA h cm^{-2}: (**a**) activation-controlled electrodeposition: $\eta = 90$ mV, initial current density: 3.3 mA cm^{-2}, (**b**) mixed activation–diffusion-controlled electrodeposition: $\eta = 140$ mV, initial current density: 4.2 mA cm^{-2}, and (**c**) dominant diffusion-controlled electrodeposition: $\eta = 210$ mV, initial current density: 6.5 mA cm^{-2} (Reprinted from Refs. [7, 8, 13] with kind permission from Springer and Ref. [16] with permission from Elsevier)

exchange current densities on different crystal planes are quite different, whereas the reversible potential is approximately the same for all planes [13, 18]. This can lead to preferential growth of some crystal planes, because the rate of deposition depends only on the orientation, which leads to the formation of a large-grained rough deposit. However, even at low degrees of diffusion control, the formation of large, well-defined grains is not to be expected, because of irregular growth caused by mass-transport limitations. Hence, the current density which corresponds to the very beginning of the mixed control (a little larger than this at the end of the Tafel linearity) will be the optimum one for compact metal deposition [13, 16].

All the above facts are illustrated in Fig. 2.4 [13, 16].

Cauliflower-Like Forms

It can be seen from Fig. 2.4c that the surface protrusions are globular and cauliflower-like ones. If the initial electrode surface protrusions are of ellipsoidal shape, they can be characterized by the base radius R_0 and the height h_p as shown in Fig. 2.5a [13, 19].

The tip radius is then given by:

$$r = \frac{R_0^2}{h_p} \tag{2.31}$$

The initial electrode surface protrusion is characterized by $h_p \to 0$ and $r \to \infty$ if $R_0 \neq 0$. In this situation, a spherical diffusion layer cannot be formed around the tip of the protrusion if $r > \delta$, and if the linear diffusion control occurs, leading to an increase in the height of the protrusion relative to the flat surface.

The rate of growth of the tip of a protrusion for $r > \delta$ is equal to the rate of motion of the tip relative to the rate of motion of the flat surface. Hence, in the diffusion-controlled electrodeposition:

$$\frac{\mathrm{d}h_p}{\mathrm{d}t} = \frac{V}{nF}\left(i_{L,\mathrm{tip}} - i_L\right) \tag{2.32}$$

Substitution of $i_{L,\mathrm{tip}}$ from the Eq. (2.9) and i_L from the Eq. (1.14) in the Eq. (2.32) and further rearranging gives

$$\frac{\mathrm{d}h_p}{\mathrm{d}t} = \frac{i^2 V h_p}{i_L nF\delta} \tag{2.33}$$

or

$$h_p = h_{p,0}\exp\left(\frac{i\ Q}{i_L Q_{0,D}}\right) \tag{2.34}$$

where Q and $Q_{0,D}$ are given by Eqs. (2.27) and (2.28).

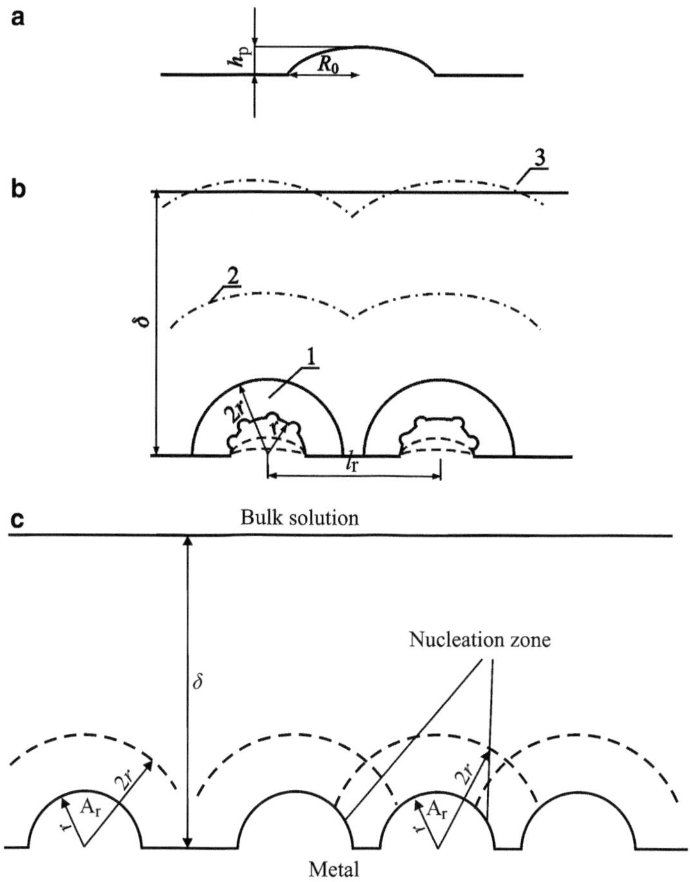

Fig. 2.5 Schematic representations of (**a**) the initial electrode surface protrusion, (**b**) the establishment of spherical diffusion layers around independently growing protrusions (1. $r < (\delta - h_p)$ and $r < 1/4l_r$ spherical diffusion zones are formed; 2. $r < (\delta - h_p)$ and $r > 1/4l_r$ spherical diffusion zones overlap); 3. $r > (\delta - h_p)$, spherical diffusion zones are not formed, and (**c**) spherical diffusion layers and their overlap around growing protrusions (Reprinted from Refs. [7, 8, 13] with kind permission from Springer and Ref. [19] with permission from the Serbian Chemical Society)

If the surface protrusions are sufficiently far one from other, then the spherical diffusion control can be operative around the whole surface of protrusion when h_p increases and r decreases, as illustrated by Fig. 2.5b. In this situation, second-generation protrusions can grow inside the diffusion layer of first-generation protrusions in the same way as first-generation protrusions grow inside the diffusion layer of the macroelectrode, and so on.

A cauliflower-like deposit is formed under such conditions, as is shown in Fig. 2.6a, b. It can be seen from Fig. 2.6a that the distance between the cauliflower-like grains is sufficiently large to permit the formation of spherical diffusion zones around each of them. Simultaneously, second-generation

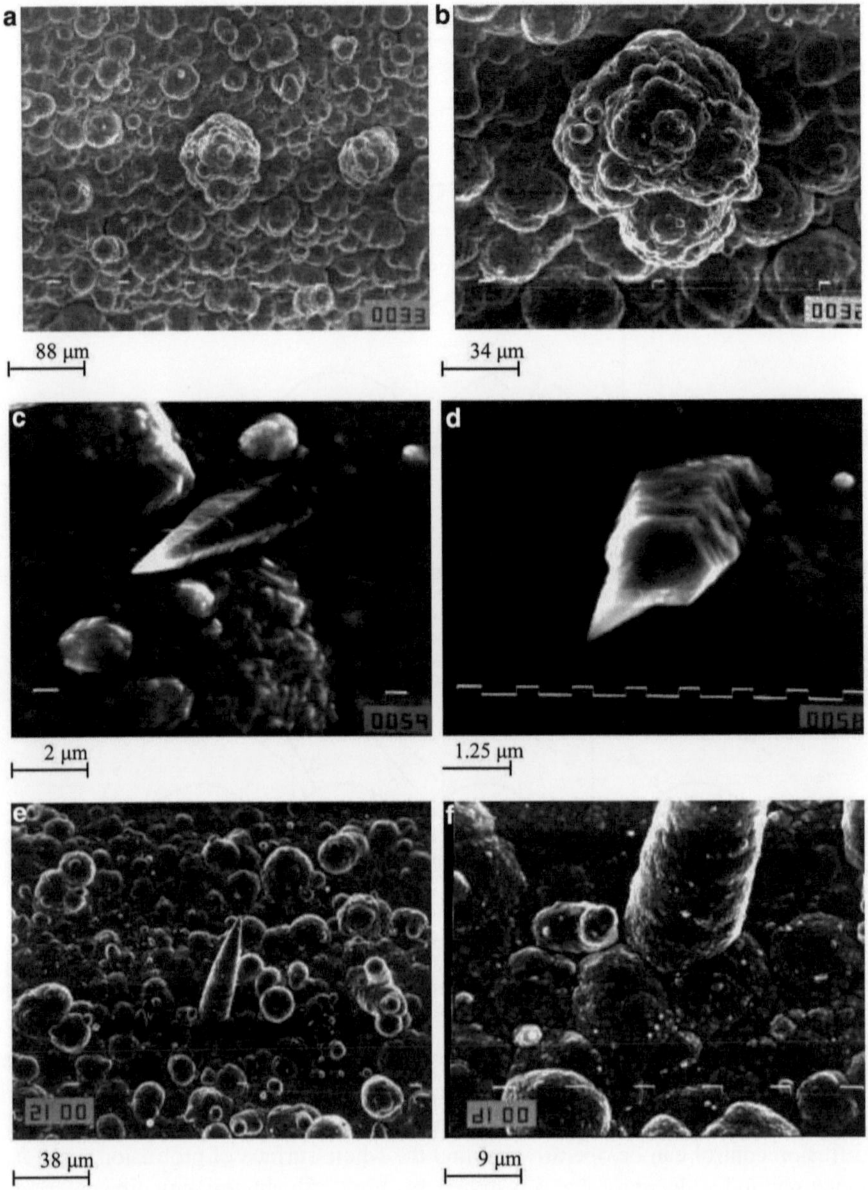

Fig. 2.6 Copper deposits obtained from 0.30 M $CuSO_4$ in 0.50 M H_2SO_4 by electrodeposition under mixed activation–diffusion control. Deposition overpotential: 220 mV. Quantity of electricity: (**a**) 40 mA h cm^{-2}, (**b**) the same as in (**a**), (**c**) 10 mA h cm^{-2}, (**d**) 10 mA h cm^{-2}, (**e**) 20 mA h cm^{-2}, and (**f**) the root of the carrot from (**e**) (Reprinted from Refs. [7, 8, 13] with kind permission from Springer and Ref. [20] with permission from the Serbian Chemical Society)

protrusions grow in all directions, as shown in Fig. 2.6b. This confirms the assumption that the deposition takes place in a spherically symmetric fashion.

To a first approximation, the rate of propagation can be taken to be practically the same in all directions, meaning that the cauliflower-type deposit formed by spherically symmetric growth inside the diffusion layer of the macroelectrode will be hemispherical, as is illustrated in Fig. 2.6a, b.

This type of protrusion is much larger than that formed by linearly symmetric growth inside the diffusion layer of the macroelectrode (Fig. 2.6a, b).

This is because a spherical diffusion layer cannot be formed around closely packed protrusions, their diffusion fields overlap, and they grow in the diffusion layer of the macroelectrode.

Carrot-Like Forms

If spherical diffusion layer can be established around the tip of a protrusion, the limiting diffusion current to the tip is given by Eq. (2.12) or by:

$$i_{\mathrm{L,tip}} = i_{\mathrm{L}}\frac{h_{\mathrm{p}}}{r} \tag{2.35}$$

for

$$\frac{h_{\mathrm{p}}}{r} >> 1 \tag{2.36}$$

It can also be seen from Fig. 2.6c–f that the growth of such protrusions produces carrot-like forms, another typical form obtained in copper deposition under mixed activation–diffusion control. This happens under the condition $k_{\mathrm{p}} << 1$, when activation control takes place only around the tip of the protrusion, as is illustrated in Fig. 2.6c, d. In this case, Eq. (2.20) can be rewritten in the form:

$$i_{\mathrm{tip}} = i_{0,\mathrm{tip}}(f_{\mathrm{c}} - f_{\mathrm{a}}) \tag{2.37}$$

meaning that deposition on the protrusion tip can also be under activation control at overpotentials lower than the critical one for the initiation of dendritic growth (see section "Model of the spherical diffusion around the tip of a surface protrusion–deposition to the point").

This happens if the nuclei have a shape like that in Fig. 2.6c, d. The assumption that the protrusion tip grows under activation control is confirmed by the regular crystallographic shape of the tip [20] just as in the case of grains growing on the macroelectrode under activation control (see Fig. 2.4a).

The maximum growth rate at a given overpotential corresponds to activation-controlled deposition. As a result, the propagation rate at the tip will be many times larger than that in other directions, resulting in protrusions like that in Fig. 2.6e.

The final form of the carrot-like protrusion is shown in Fig. 2.6e. It can be concluded from the parabolic shape that such protrusions grow as moving paraboloids in accordance with the Barton–Bockris theory [5], the tip radius remaining constant because of the surface energy effect. It can be concluded from Fig. 2.6f that thickening of such a protrusion is under mixed activation–diffusion control because the deposit is seen to be of the same quality as that on the surrounding macroelectrode surface. It can be seen from the Fig. 2.6d that activation control takes place only at the very tip of the protrusion.

The Position of New Nuclei

It is seen from Fig. 2.6c, d that new nuclei are mainly formed at the bottom of the already growing protrusions or between them. If the diffusion zones of two or more hemispherical protrusions partially overlap, as illustrated by Fig. 2.5c, the current densities i at each point outside the overlapping field will be those required by Eq. (2.38):

$$\eta = b_c \log\left(\frac{i}{i_0}\right) + b_c \log\left(\frac{1}{1 - \frac{i}{i_L}}\right) \tag{2.38}$$

for $f_c >> f_a$ and where i_L is the limiting diffusion current density (linear or spherical).

In the overlapping field, the current density i' at the points where the maximum possible current density is i in the absence of overlapping field will be lower than i. In this way, the substrate becomes partially inert, and the formation of new nuclei is enhanced at these points if the entire electrode surface is at same potential because

$$\eta = b_c \log\left(\frac{i'}{i_0}\right) + b_c \log\left(\frac{1}{1 - \frac{i'}{i_L}}\right) + \Delta\eta \tag{2.39}$$

where $\Delta\eta$ is the nucleation overpotential [7, 19].

Some of the new nuclei are the precursors of carrot-like protrusions, depending on their crystal orientation. In this case, they are in the form of small hexagonal pyramids, as shown in Fig. 2.6c, d. Based on their morphology and because copper has a face-centered cubic crystal structure, it is reasonable to assume that they are truncated by a high-Miller index plane. According to Pangarov and Vitkova [21, 22], the orientation of nuclei is related to the overvoltage used. It is reasonable to expect that the appearance of precursors of carrot-like protrusions has its own overvoltage range.

Different Possible Forms

The tips of growing carrot-like protrusions can change the direction of protrusions. This is mainly observed in a later stage of protrusion growth. Thus, the smaller the protrusion height, the lower is the probability of a change in the growth direction. The reason for this behavior is not yet clear. However, it is probably caused by local perturbations on the growth front on the protrusion tip. This effect and the further nucleation on already growing protrusions cause the growth of different forms.

The growth of surface protrusions in galvanostatic deposition has been considered only for deposition in the limiting diffusion current density range [23, 24].

In summary, deposition under mixed activation–diffusion control causes the formation of a number of growth forms and the increase of surface coarseness, this increase being more pronounced at higher current densities. It should be noted that electrodeposition at a periodically changing rate can change considerably the morphology of the deposits [13, 25].

2.1.3 Dendritic Deposits

2.1.3.1 Basic Facts

Two phenomena seem to distinguish dendritic from carrot-like growth [7, 11, 13]:

1. A certain well-defined critical overpotential value appears to exist below which dendrites do not grow.
2. Dendrites exhibit a highly ordered structure and grow and branch in well-defined directions. According to Wranglen [26], a dendrite is a skeleton of a monocrystal and consists of a stalk and branches, thereby resembling a tree.

   It is known that dendritic growth occurs selectively at three types of growth sites [11]:

   (a) Dendritic growth occurs at screw dislocations. Swordlike dendrites with pyramidal tips are formed by this process [3, 11].
   (b) Many investigations of the crystallographic properties of dendrites have reported the existence of twin structures [27–30]. In the twinning process, a so-called indestructible reentrant groove is formed. Repeated one-dimensional nucleation in the groove is sufficient to provide for growth extending in the direction defined by the bisector of the angle between the twin planes [11].
   (c) It is a particular feature of a hexagonal close-packed lattice that growth along a high-index axis does not lead to the formation of low-index planes. Grooves containing planes are perpetuated, and so is the chance for extended growth by the one-dimensional nucleation mechanism [31].

In all the above cases, the adatoms are incorporated into the lattice by repeated one-dimensional nucleation. On the other hand, deposition to the tip of screw dislocations can be theoretically considered as deposition to a point; in the other two cases, the deposition is to a line. Hence, in the case under (a), the conditions of spherical diffusion are fulfilled, while the conditions of the cylindrical diffusion are fulfilled in the case under (b).

From the electrochemical point of view, a dendrite can be defined as an electrode surface protrusion that grows under activation or mixed control, while deposition to the flat part of the electrode surface is under complete diffusion control [3, 7, 9, 11, 13].

2.1.3.2 Dendrite Precursors: Schematic Presentation

The monoatomic high-step edges, the microsteps, are required for continuous metal electrocrystallization. Possible sources of microsteps on a surface are shown in Figs. 2.7, 2.8a, and 2.9a, *i.e.*, the low-index planes, two-dimensional nuclei, emergent screw dislocations, and indestructible reentrant grooves [11, 32].

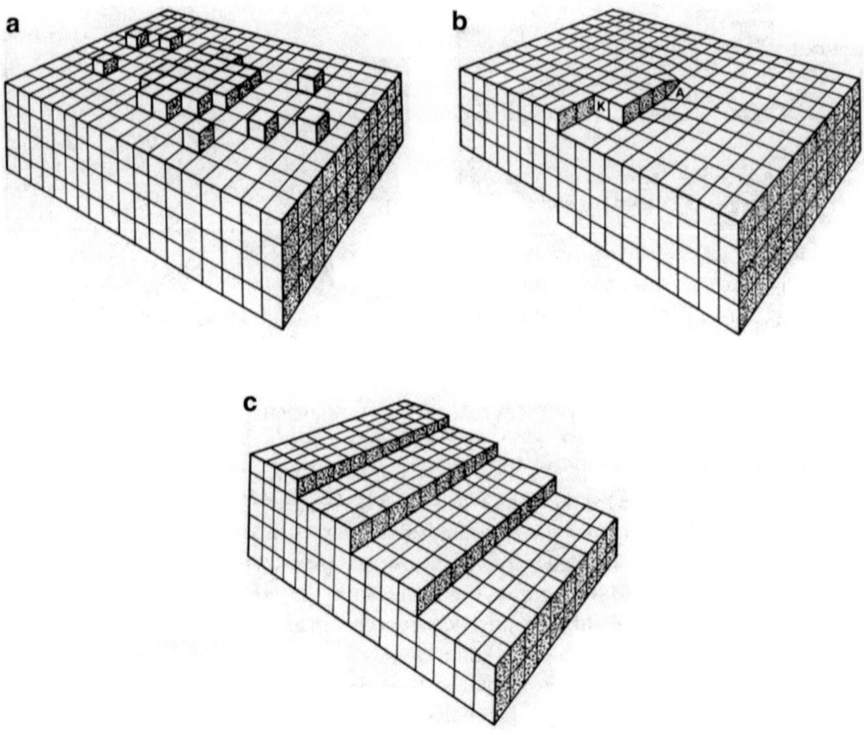

Fig. 2.7 Models of different sources of microsteps on a surface: (**a**) a two-dimensional nucleus, (**b**) an emergent screw dislocation, (**c**) misorientation of the surface with respect to the ideal low-index plane (Reprinted from Refs. [13, 32] with kind permission from Springer)

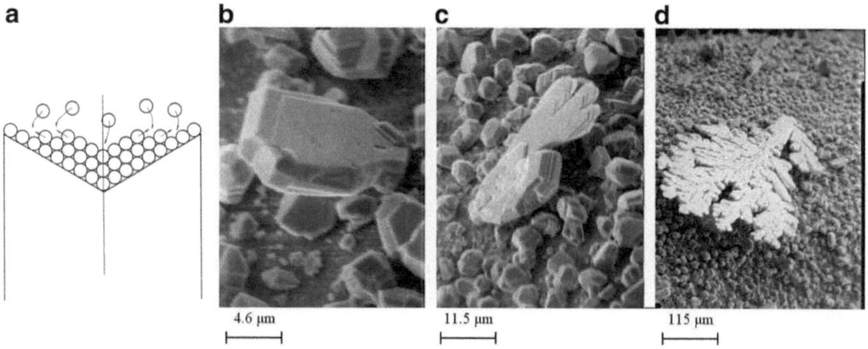

Fig. 2.8 (**a**) Schematic representation of the formation of an "indestructible reentrant groove" and the cadmium deposits obtained by deposition from 0.10 M $CdSO_4$ in 0.50 M H_2SO_4 onto a cadmium electrode. Deposition overpotential: 50 mV. Deposition times: (**b**) 2 min, (**c**) 2 min, and (**d**) 10 min (Reprinted from Refs. [7, 11, 13] with kind permission from Springer and Ref. [33] with permission from Elsevier)

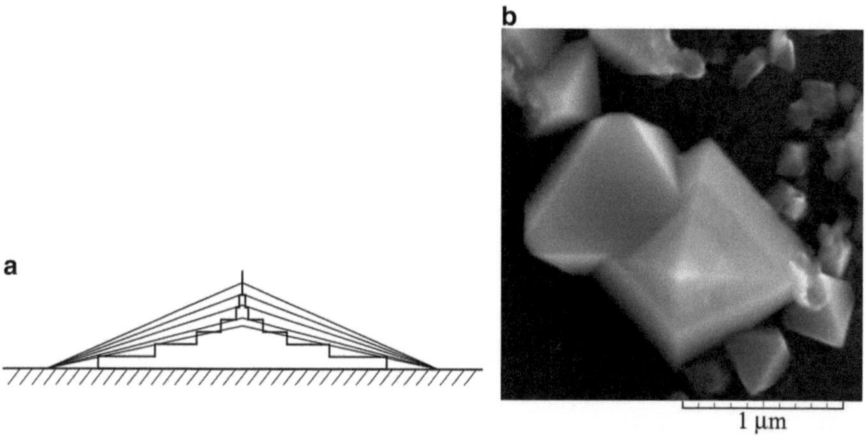

Fig. 2.9 (**a**) Schematic representation of the growth at screw dislocation and (**b**) the initial stage of Pb electrodeposition from 0.40 M $Pb(NO_3)_2$ in 2.0 M $NaNO_3$ at an overpotential of 100 mV with a time of 0.5625 s (According to [3] and reprinted from Ref. [34] with permission from Elsevier)

It is obvious from Fig. 2.7a, c that after the formation of a low-index plane, two-dimensional nucleation is necessary for the growth to be continued.

In the case of a reentrant groove (Fig. 2.8a), the growth of new layers can be started by one-dimensional nucleation. The growth rates in the cases of one-dimensional and two-dimensional nucleation as rate-determining steps can be compared to each other by considering the growth of two-dimensional flat cadmium dendrites from Fig. 2.8b–d.

The tip of the twined cadmium dendrite precursor from Fig. 2.8b represents the physical equivalent of the scheme of the growth site from Fig. 2.8a. As shown in Fig. 2.8a, a layer of atoms advance in the direction determined by twining laws, an edge is constantly renewed, in which the new layers can be started by one-dimensional nucleation. Further growth and branching of precursor like that from Fig. 2.8b produces the dendrites shown in Fig. 2.8c, d. The deposition on the lateral flat dendrite surfaces takes place by repeated two-dimensional nucleation, as in deposition on dislocation free surface [33]. This makes the deposition rate in the direction of tip motion many times larger, which results in dendrite shape like that from Fig. 2.8d.

In the case of a screw dislocation (Fig. 2.9a), the step provokes the growth by retaking itself with one end fixed at the point where the screw dislocation emerges. The initial stage of growth at the screw dislocation is shown in Fig. 2.9b. The needle-like dendrites as those shown in Fig. 1.8b, c are formed by this type of growth [34].

Aside from dendrites of the different shapes [34, 35], during electrodeposition of Pb in the diffusion control, crystals of irregular shapes were also formed (Fig. 2.10). These irregular crystals do not follow the classical Wranglen definition of a dendrite [26], but they behave as dendrites with regard to their electrochemical characteristics. For this reason, these crystals formed under diffusion control are denoted as precursors of dendrites [34].

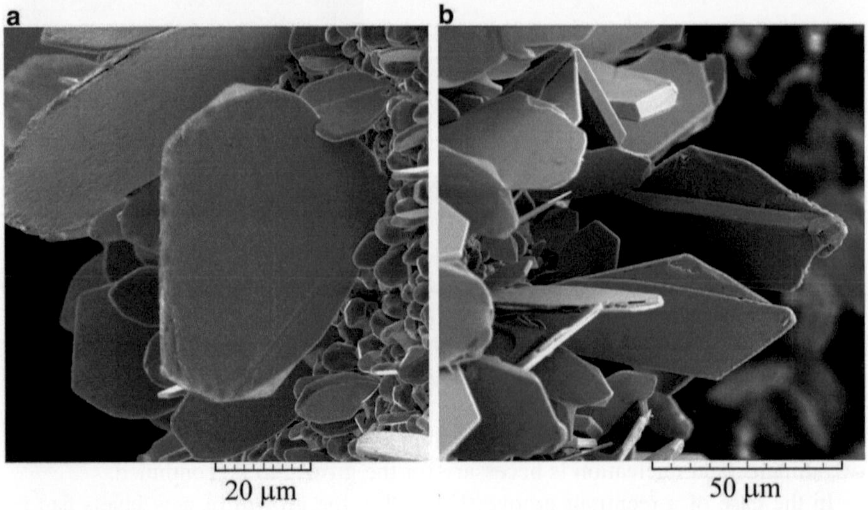

Fig. 2.10 Lead crystals formed by electrodeposition from 0.40 M $Pb(NO_3)_2$ in 2.0 M $NaNO_3$ at overpotentials of (**a**) 50 mV and (**b**) 75 mV with quantity of the electricity of 0.95 mA h cm^{-2} (Reprinted from Ref. [36] with permission from the Serbian Chemical Society)

2.1.3.3 Mathematical Model Considering the Conditions of Spherical and Cylindrical Diffusion

Model of the Spherical Diffusion Around the Tip of a Surface Protrusion–Deposition to the Point

Considering the model of surface irregularities shown in Fig. 2.1, the surface irregularities are buried deep in the diffusion layer, which is characterized by a steady linear diffusion to the flat portion of completely active surface and by the spherical diffusion field formed around the tip of a surface protrusion. The conditions of spherical diffusion are fulfilled around the tips of both needle-like protrusion (Fig. 2.9b) and irregular crystals (Fig. 2.10b).

If the protrusion does not affect the outer limit of the diffusion layer, *i.e.*, if $\delta \gg h_p$, the limiting diffusion current density to the tip of the protrusion from Fig. 2.1, $i_{L,tip}$, is given by:

$$i_{L,tip} = i_L\left(1 + \frac{h_p}{r}\right) \qquad (2.12)$$

Substitution of $i_{L,tip}$ from Eq. (2.12) into Eq. (1.13) produces for $h_p/r >> 1$:

$$i_{tip} = i_{0,tip}(f_c - f_a) \qquad (2.37)$$

where $i_{0,tip}$ is the exchange current density at the tip of a protrusion.

Obviously, deposition to the tip of such protrusion inside the diffusion layer is activation-controlled relative to the surrounding electrolyte, but it is under mixed activation–diffusion control relative to the bulk solution.

If deposition to the flat part of electrode is diffusion-controlled process and assuming a linear concentration distribution inside diffusion layer (see Eq. (2.10)), the concentration C_{tip} at the tip of a protrusion can be given:

$$C_{tip} = C_0\frac{h_p}{\delta} \qquad (2.40)$$

According to Newman [37], the exchange current density at the tip of a protrusion, $i_{0,tip}$, is given by:

$$i_{0,tip} = i_0\left(\frac{C_{tip}}{C_0}\right)^{\xi} \qquad (2.41)$$

where

$$\xi = \frac{d \log i_0}{d \log C_0} \qquad (1.17)$$

or

$$i_{0,\mathrm{tip}} = i_0 \left(\frac{h_\mathrm{p}}{\delta}\right)^{\xi} \tag{2.42}$$

because of Eq. (2.40).

Taking into account the Eq. (2.37), the current density to the tip of a protrusion is then given by:

$$i_\mathrm{tip} = i_0 \left(\frac{h_\mathrm{p}}{\delta}\right)^{\xi} (f_\mathrm{c} - f_\mathrm{a}) \tag{2.43}$$

being under mixed control due to the $(h_\mathrm{p}/\delta)^{\xi}$ term, which takes into account the concentration dependence of $i_{0,\mathrm{tip}}$, expressing in this way a mixed controlled electrodeposition process.

Outside the diffusion layer $h_\mathrm{p} \geq \delta$, and the Eq. (2.43) becomes:

$$i_\mathrm{tip} = i_0 (f_\mathrm{c} - f_\mathrm{a}) \tag{2.44}$$

indicating pure activation control, as the $(h_\mathrm{p}/\delta)^{\xi}$ term is absent.

For the dendrite growth, the current density to the tip of a protrusion formed on the flat part of the electrode surface growing inside the diffusion layer should be larger than the corresponding limiting diffusion current density [38]. Hence, if

$$i_\mathrm{L} < i_\mathrm{tip} \tag{2.45}$$

the protrusion grows as a dendrite.

In accordance with the Eq. (2.45), instantaneous dendrite growth is possible after reach of steady-state conditions at overpotentials larger than some critical value, η_c, which can be derived from the Eq. (2.43) as shown in Refs [7, 13, 38]:

$$\eta_\mathrm{c} = \frac{b_\mathrm{c}}{2.3} \ln \frac{i_\mathrm{L}}{i_0} \left(\frac{\delta}{h_\mathrm{p}}\right)^{\xi} \tag{2.46}$$

and minimum overpotential at which dendritic growth is still possible, for $h_\mathrm{p} = \delta$, η_i is given by:

$$\eta_\mathrm{i} = \frac{b_\mathrm{c}}{2.3} \ln \frac{i_\mathrm{L}}{i_0} \tag{2.47}$$

for $f_\mathrm{c} >> f_\mathrm{a}$. For very fast processes, when $i_0/i_\mathrm{L} >> 1$, and if $f_\mathrm{c} \approx f_\mathrm{a}$ but $f_\mathrm{c} > f_\mathrm{a}$, the Eq. (2.46) becomes:

$$\eta_\mathrm{c} = \frac{RT}{nF} \frac{i_\mathrm{L}}{i_0} \left(\frac{\delta}{h_\mathrm{p}}\right)^{\xi} \tag{2.48}$$

and the Eq. (2.47):

$$\eta_{\mathrm{i}} = \frac{RT}{nF}\frac{i_{\mathrm{L}}}{i_0} \tag{2.49}$$

meaning that in the case of ohmic-controlled reactions, dendritic growth can be expected at very low overpotentials, or better to say, if $i_0 \rightarrow \infty$, instantaneous dendritic growth is possible at all overpotentials if only mass transfer limitations are taken into consideration.

It should be noted that all the above derivations are valid, if the protrusion tip radius is sufficiently large to make the effect of the surface energy term negligible [5].

It follows from Eq. (2.49) that for systems with $i_0 \rightarrow \infty$, dendritic growth is possible at all overpotentials. Experimentally, some critical overpotential of dendritic growth initiation exists in all cases, being of the order of a few millivolts [5, 39, 40]. Assuming that under complete diffusion and surface energy control ($i_0 \rightarrow \infty$) the current density to the macroelectrode is given by [5]:

$$i = \eta\frac{(nF)^2 D C_0}{\delta RT} \tag{2.50}$$

and assuming that Eq. (2.40) is valid, the current density on the tip of a dendrite growing inside the diffusion layer of a macroelectrode, i_{d}, is given by:

$$i_{\mathrm{d}} = \eta^2\frac{(nF)^3 D C_0}{8\gamma VRT}\frac{h_{\mathrm{p}}}{\delta} \tag{2.51}$$

then, it is possible to derive the relationships:

$$\eta_{\mathrm{c}} = \frac{8\gamma V}{nF h_{\mathrm{p}}} \tag{2.52}$$

using the same procedure as in derivation of Eqs. (2.46) and (2.47), where γ is the interfacial energy between metal and solution and V is the molar volume of the metal, and minimum overpotential at which dendritic growth is still possible, η_{i} is given by:

$$\eta_{\mathrm{i}} = \frac{8\gamma V}{nF\delta} \tag{2.53}$$

Hence, a critical overpotential for initiation dendritic growth is also expected in such cases, being of the order of few millivolts [7, 13, 38].

Model of the Cylindrical Diffusion Around the Top Edge of a Surface Protrusion–Deposition to the Line

Unlike the conditions of spherical diffusion fulfilled around the tips of surface protrusions, the conditions of cylindrical diffusion are fulfilled around the top edges of the protrusions or crystals shown in Figs. 2.8b and 2.10a.

The stationary diffusion current, I, between two electrodes in a cylindrical diffusion field (see Fig. 2.11a) is given by the Eq. (2.54) [41]:

$$I = \frac{2\pi n\, l_{\mathrm{cyl}}\, FDC_0}{\ln \frac{R_{\mathrm{cyl}}}{r_{\mathrm{cyl}}}} \tag{2.54}$$

where l_{cyl} is a length of a cylindrical electrode and n, F, D, and C_0 have their already mentioned meanings. For the geometry consisted of two cylindrical electrodes

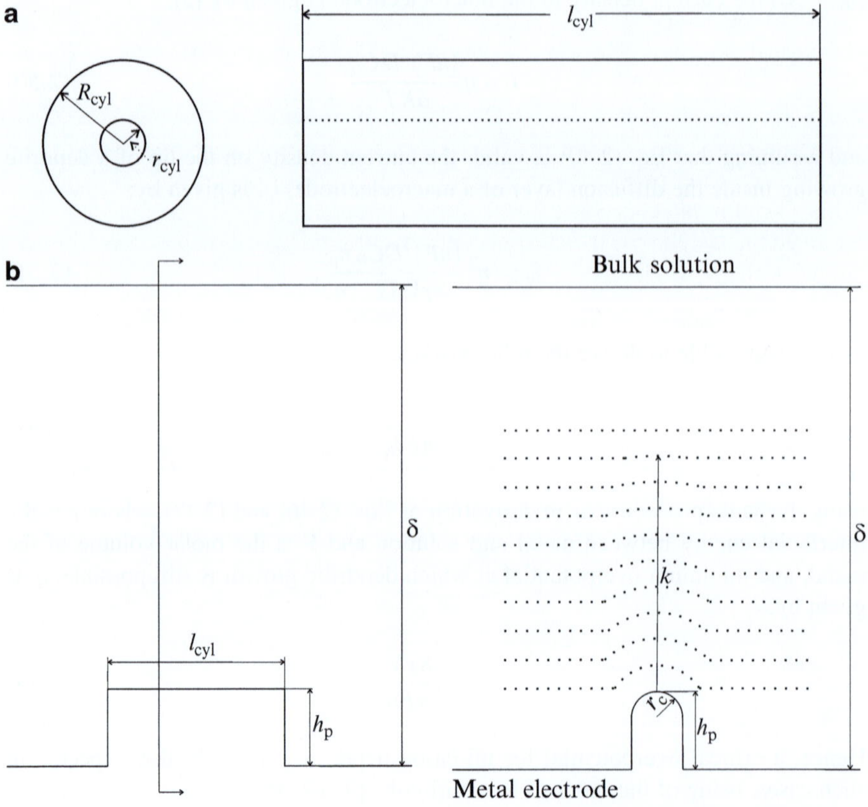

Fig. 2.11 The cylindrical diffusion: (**a**) geometry of the electrode and (**b**) schematic presentation of a protrusion growing inside the diffusion layer of the macroelectrode under the conditions for this type of diffusion (Reprinted from Ref. [34] with permission from Elsevier)

shown in Fig. 2.11a, the current density to the electrode characterized by a radius r_{cyl} will be:

$$i = \frac{nFDC_0}{r_{\mathrm{cyl}} \ln \frac{R_{\mathrm{cyl}}}{r_{\mathrm{cyl}}}} \tag{2.55}$$

where the concentration at the surface of an electrode characterized by a radius r_{cyl} is close to zero, while a concentration around a surface of electrode characterized by radius R_{cyl} is C_0.

The limiting diffusion current density to a cylindrical electrode, $i_{\mathrm{L,c}}$, in the bulk solution is given by Eq. (2.56) [33, 41]:

$$i_{\mathrm{L,c}} = \frac{nFDC_0}{r_{\mathrm{cyl}} \ln \frac{r_{\mathrm{cyl}} + d_{\mathrm{cyl}}}{r_{\mathrm{cyl}}}} \tag{2.56}$$

where $d_{\mathrm{cyl}} = (\pi Dt)^{0.5}$ and t is the time.

In order to simplify, the top edges of the protrusions can be approximated by a cylindrical protrusion growing inside the diffusion layer of the macroelectrode shown in Fig. 2.11b, where r_{c} is the radius of the top edge of a surface protrusion, $l_{\mathrm{cyl}} >> r_{\mathrm{c}}$, $\delta >> h_{\mathrm{p}}$ and $\delta >> r_{\mathrm{c}}$. The closest case of this approximation is crystal shown in Fig. 2.10a. Model of formation of iso-concentric lines around the top edges of growing protrusions was similar to the one proposed by Scharifker and Hills [42] for iso-concentric lines formed around the spherical nuclei. It was assumed that a cylindrical diffusion control of electrodeposition process is initiated by the last line deformed by a top edge of the growing protrusion and that the distribution of the concentration inside the diffusion layer of the macroelectrode does not vary with time. Hence, $k = m \cdot r_{\mathrm{c}}$, where m is a number of deformed lines around a top edge of the protrusion. Approximately, if Eq. (2.40) is still valid, the concentration at the outer limit of the cylindrical diffusion layer inside the diffusion layer of the macroelectrode will be:

$$C^{**} = C_0 \frac{h_{\mathrm{p}} + mr_{\mathrm{c}}}{\delta} \tag{2.57}$$

or

$$C^{**} = C_0 \frac{mr_{\mathrm{c}}}{\delta} \tag{2.58}$$

if the following condition is satisfied:

$$mr_{\mathrm{c}} >> h_{\mathrm{p}} \tag{2.59}$$

then, Eq. (2.60) will be valid for the limiting diffusion current density under the conditions of cylindrical diffusion:

$$i_{\mathrm{L,c}}^{**} = \frac{nFDC^{**}}{r_{\mathrm{c}} \ln \frac{r_{\mathrm{c}}+m r_{\mathrm{c}}}{r_{\mathrm{c}}}} \tag{2.60}$$

Using the Eq. (2.58), the Eq. (2.60) can be rewritten in the form:

$$i_{\mathrm{L,c}}^{**} = \frac{nFDC_0 \frac{m\, r_{\mathrm{c}}}{\delta}}{r_{\mathrm{c}} \ln \frac{r_{\mathrm{c}}+m\, r_{\mathrm{c}}}{r_{\mathrm{c}}}} \tag{2.61}$$

or

$$i_{\mathrm{L,c}}^{**} = \frac{m\, i_{\mathrm{L}}}{\ln m} \tag{2.62}$$

for $m >> 1$.

Substitution of $i_{\mathrm{L,c}}^{**}$ from Eq. (2.62) into Eq. (1.13), after rearranging, gives the Eq. (2.37) as in the case of spherical diffusion control, and the further derivation remains the same. Hence, the dendritic growth can be initiated at the same deposition overpotential under conditions of both spherical and cylindrical diffusion.

Comparison of the Rates of Growth Under Conditions of Spherical and Cylindrical Diffusion

The rate of growth of the protrusions under the conditions of spherical and cylindrical diffusion can be compared as follows. As shown earlier, the limiting diffusion current density at the tips of protrusions growing under the conditions of spherical diffusion (the needle-like dendrite or dendrite precursor), $i_{\mathrm{L,tip}}$, inside the diffusion layer of the macroelectrode is given by:

$$i_{\mathrm{L,tip}} = i_{\mathrm{L}}\left(1+\frac{h_{\mathrm{p}}}{r}\right) \approx i_{\mathrm{L}}\frac{h_{\mathrm{p}}}{r} \tag{2.63}$$

where $r << h_{\mathrm{p}}$, and the limiting diffusion current density under the conditions of cylindrical diffusion is given by Eq. (2.62).

The current density under a mixed activation–diffusion-controlled electrodeposition is given by Eq. (1.13), and substitution of the corresponding limiting diffusion current density from Eqs. (2.62) and (2.63) into the Eq. (1.13) produces after rearrangement:

$$i_{\mathrm{tip}} = \frac{i_0(f_{\mathrm{c}} - f_{\mathrm{a}})}{1 + \frac{i_0 f_{\mathrm{c}}}{i_{\mathrm{L}}}\frac{r}{h_{\mathrm{p}}}} \tag{2.64}$$

and

$$i_{\mathrm{tip,c}} = \frac{i_0(f_\mathrm{c} - f_\mathrm{a})}{1 + \frac{i_0 f_\mathrm{c}}{i_\mathrm{L}} \frac{\ln m}{m}} \tag{2.65}$$

where i_{tip} and $i_{\mathrm{tip,c}}$ are the current densities to the tips of dendrites growing under conditions of the spherical and cylindrical diffusion control, respectively.

The radius at the tip of protrusion growing in the conditions of spherical diffusion, r is extremely small [5], and, hence, the following is valid:

$$\frac{r}{h_\mathrm{p}} < \frac{\ln m}{m} \tag{2.66}$$

Thus,

$$i_{\mathrm{tip}} > i_{\mathrm{tip,c}} \tag{2.67}$$

or

$$\frac{i_0(f_\mathrm{c} - f_\mathrm{a})}{1 + \frac{i_0 f_\mathrm{c}}{i_\mathrm{L}} \frac{r}{h_\mathrm{p}}} > \frac{i_0(f_\mathrm{c} - f_\mathrm{a})}{1 + \frac{i_0 f_\mathrm{c}}{i_\mathrm{L}} \frac{\ln m}{m}} \tag{2.68}$$

Hence, during electrodeposition at a selected overpotential belonging to diffusion control, faster growth occurs under the conditions of spherical control of the electrodeposition than under those of cylindrical control.

2.1.3.4 Physical Model of Dendritic Growth

The cross sections of the copper deposits obtained in a model system described earlier (Fig. 2.3a) are shown in Fig. 2.12.

Deposits obtained at 300 mV are compact, while those obtained at 600 mV are dendritic ones. Since both overpotentials correspond to the plateau of the limiting diffusion current density [14], it is clear that dendrites are formed at overpotentials larger than certain critical value, as required by the Eq. (2.47). It is seen that the current density to the tips of dendrites depends on the h_p/δ ratio (see the Eq. (2.43)), so that larger dendrites are produced at more elevated points of the electrode surface. This is because the effective height of the dendrite precursor in the modeled diffusion layer is equal to the sum of the height of the precursor and the height of the point at which nucleation took places relative to the flat part of the electrode surface. In the same way, for nuclei formed on the tip of a protrusion (Fig. 2.12b), η_c (see Eq. (2.46)) is lower than that formed on the flat surface, and a dendrite is formed at the tip of the protrusion while dendrites are not formed on the flat part of the electrode at the same overpotential.

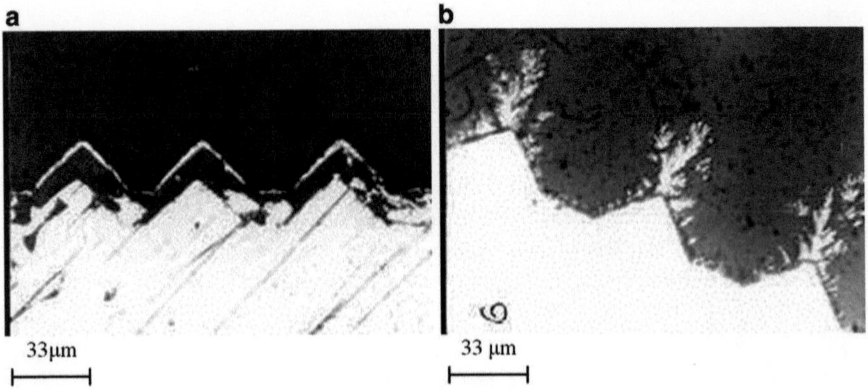

Fig. 2.12 Cross sections of copper deposits obtained in the same manner as those in Fig. 2.3b from 0.10 M $CuSO_4$ in 0.50 M Na_2SO_4. The thickness of the agar–agar diffusion layer was 1.0 mm. Deposition time: 30 min; deposition overpotential: (**a**) 300 mV and (**b**) 600 mV (Reprinted from Refs. [7, 13] with kind permission from Springer)

2.1.3.5 Dendritic Growth Inside Diffusion Layer of the Active Macroelectrode and Ohmic Diffusion and Activation–Diffusion-Controlled Deposition and Determination of η_i and η_c

In these cases, the dendritic growth starts at overpotentials larger than the one which corresponds to the beginning of the limiting diffusion current density plateau [7, 13].

There is an induction period before the initiation of dendritic growth [5]. During this induction period, dendrite precursors are formed and become sufficiently high to satisfy the Eq. (2.46) at a given overpotential, as illustrated in Fig. 2.13. The cross-like grains seen in Fig. 2.13a, b further develop into dendrite precursors (Fig. 2.13a, c).

The propagation of this structure by branching (Fig. 2.13d) produces dendrites as shown in Fig. 2.13e.

The initiation of dendritic growth is followed by a change in the slope of the current density–time curves [7, 11, 13], indicating a change in the growth mechanism of the deposit.

The slopes of these dependencies are similar to each other and independent of the deposition overpotential during the non-dendritic amplification of the surface coarsening according to the Eq. (2.34).

The change of the slope of the current-time dependencies due to the dendritic growth initiation will be treated here in somewhat simplified way.

The limiting diffusion current density to the elevated points of a surface protrusion, $i_{L,e}$, is given by:

$$i_{L,e} = \frac{nFDC_0}{\delta - h_e} \tag{2.69}$$

where h_e is the height of the elevated point relative to the flat surface.

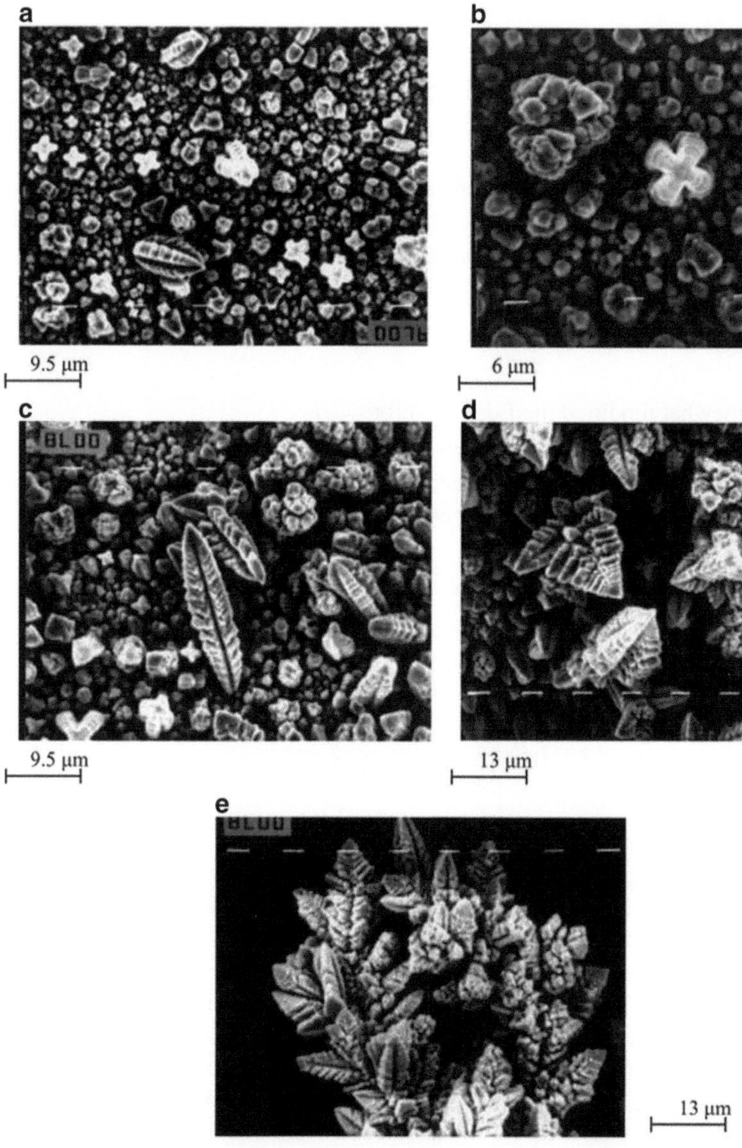

Fig. 2.13 SEM micrographs of copper deposits obtained by electrodeposition from 0.30 M $CuSO_4$ in 0.50 M H_2SO_4 onto a copper wire electrode. Deposition overpotential: 550 mV. Quantity of electricity: (**a**) 2 mA h cm^{-2}, (**b**) 2 mA h cm^{-2}, (**c**) 5 mA h cm^{-2}, (**d**) 10 mA h cm^{-2}, and (**e**) 10 mA h cm^{-2} (Reprinted from Ref. [19] with permission from the Serbian Chemical Society and Refs. [7, 8, 13] with kind permission from Springer)

Then, the limiting diffusion current density will be given by:

$$i_{\mathrm{L}} = \theta \frac{nFDC_0}{\delta} + \frac{1-\theta}{\mathrm{N}} \sum_{i=1}^{i=\mathrm{N}} \frac{nFDC_0}{\delta - h_{\mathrm{i,e}}} \tag{2.70}$$

where θ is the part of the electrode surface uncovered by dendrites, and N is the number of elevated points on the electrode surface, and $h_{\mathrm{i,e}}$ changes with time according to Eq. (2.71):

$$h_{\mathrm{i,e}} = h_{\mathrm{i,0,e}} \exp\left(\frac{VDC_0}{\delta^2} t\right) \tag{2.71}$$

which is somewhat modified the Eq. (2.34) for $i = i_{\mathrm{L}}$.

It is obvious that $\frac{\mathrm{d}i_{\mathrm{L}}}{\mathrm{d}t}$ does not depend on overpotential. After initiation of dendritic growth, the slopes become dependent on the overpotential. The overpotential η and current density i_{tip} on the tip of a dendrite are related by:

$$i_{\mathrm{tip}} = i_0 \frac{h_{\mathrm{p}}}{\delta} f_{\mathrm{c}} \tag{2.72}$$

for $f_{\mathrm{c}} \gg f_{\mathrm{a}}$ and the Eq. (2.70) can be rewritten in the form:

$$i = i_{\mathrm{L}}\theta + (1-\theta) i_0 \frac{f_{\mathrm{c}}}{\mathrm{N}} \sum_{i=1}^{i=\mathrm{N}} \frac{h_{\mathrm{p,i}}}{\delta} \tag{2.73}$$

and $\frac{\mathrm{d}i}{\mathrm{d}t}$ in this case depends on overpotential.

Hence, the maximum overpotential at which the slope of the apparent current density–time dependence remains constant and equal to that in non-dendritic amplification of the surface roughness corresponds to η_{i}. The minimum overpotential at which this slope cannot be recorded corresponds to η_{c}.

In this way, η_{i} and η_{c} can be estimated. It is known that the *i*-*t* dependence is different from case to case owing to different mechanisms of dendritic growth initiation and dendritic growth [13]. As a result of this, the analytical approach to the determination of η_{i} and η_{c} must be specific for each system under consideration; the procedure for one particular case is as follows.

Typical log (current)–time dependencies obtained for Cu electrodeposition from 0.20 M $CuSO_4$ in 0.50 M H_2SO_4 at overpotentials belonging to the limiting diffusion current plateau are shown in Fig. 2.14a. According to the above discussion, it is clear that the intersection points of the two linear dependencies determine the induction time of dendritic growth initiation [38].

The induction times for dendritic growth initiation extracted from the graphs in Fig. 2.14a can be presented as a function of overpotential, and the critical overpotential for instantaneous dendritic growth can be obtained by extrapolation to zero induction time [43].

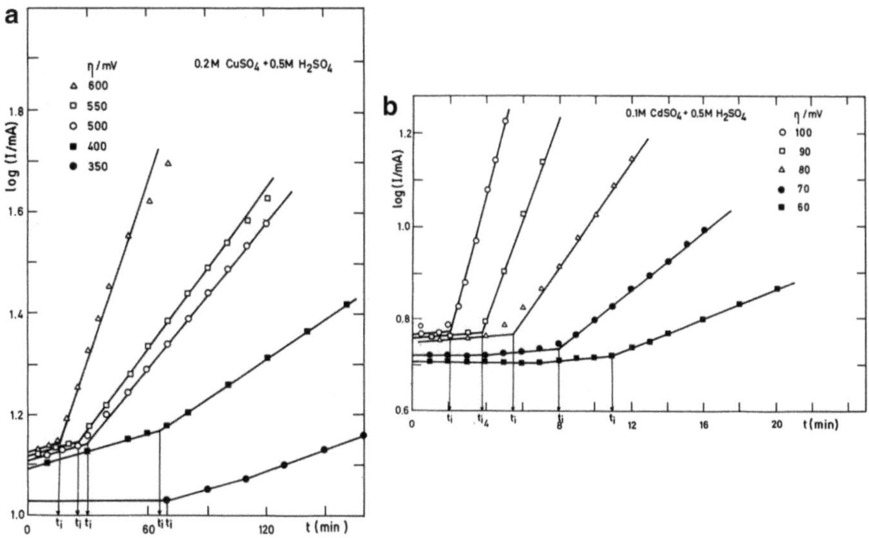

Fig. 2.14 log I as a function of time for (**a**) copper and (**b**) cadmium deposition (Reprinted from Refs. [13, 38, 43] with kind permission from Springer)

The critical overpotential of dendritic growth initiation can be determined by plotting the logarithm of the slopes of the straight lines from Fig. 2.14a as a function of overpotential, and the intersection point of the two straight lines determines η_i. The same procedure was followed for the deposition of cadmium from 0.10 M $CdSO_4$ in 0.50 M H_2SO_4 (Fig. 2.14b) [43].

The η_i and η_c of 260 and 660 mV for copper deposition (lower i_0 value) and 27 and 110 mV for cadmium deposition (larger i_0 value) are successfully determined using the above given procedure, being in perfect agreement with experimental findings as can be seen from Fig. 2.15 [13, 38, 43].

The cross sections of the copper and cadmium deposits obtained at $\eta < \eta_i$, $\eta_i < \eta < \eta_c$, and $\eta > \eta_c$ are shown in Fig. 2.15a, b, respectively. It can be seen that there is no dendrite formation when $\eta < \eta_i$, both compact and dendritic deposits are formed when $\eta_i < \eta < \eta_c$ and only dendritic metal is deposited when $\eta > \eta_c$. This is in perfect agreement with findings of Calusaru [44] for the morphology of deposits of the same metals deposited at overpotentials corresponding to full diffusion control.

The shapes of the polarization curves presented in Fig. 2.15a, b are in accordance with values of the exchange current density to the limiting diffusion current density ratios.

It is known [45] that, apart from decreasing the concentration of the depositing ion, the formation of a dendritic deposit can also be enhanced by increasing the concentration of the supporting electrolyte, increasing the viscosity of the solution, decreasing the temperature, and decreasing the velocity of motion of

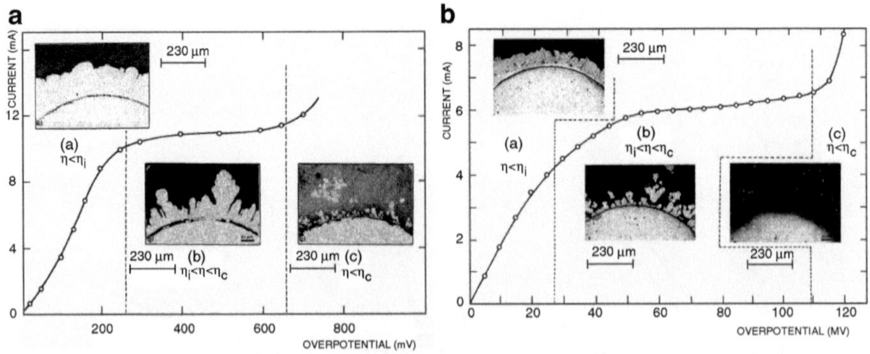

Fig. 2.15 Polarization curves for the potentiostatic deposition of (**a**) copper from 0.20 M $CuSO_4$ in 0.50 M H_2SO_4 and the cross sections of Cu deposits obtained on Cu wire electrodes previously plated with Ni: (*a*) overpotential: 200 mV, deposition time: 6 h; (*b*) overpotential: 300 mV, deposition time: 5 h, and (*c*) overpotential: 700 mV, deposition time: 2 min, and (**b**) cadmium from 0.10 M $CdSO_4$ in 0.50 M H_2SO_4 and the cross sections of Cd deposits obtained on Cu wire electrode: (*a*) overpotential: 20 mV, deposition time: 8 h; (*b*) overpotential: 40 mV, deposition time: 2 h, and (*c*) overpotential: 120 mV, deposition time: 9 min (Reprinted from Refs. [13, 38, 43] with kind permission from Springer)

the solution. Practically, all the above facts can be explained by the Eqs. (2.47) and (2.49), assuming that a decrease in η_i means enhanced dendrite formation because of the lower electrical work required to produce the dendrites. The possibility of obtaining dendrites of Pb [39] and Sn [40] from aqueous solutions at lower overpotentials than required for the formation of dendrites of Ag from aqueous solutions can also be explained by Eq. (2.53) owing to the much lower melting points of these metals, *i.e.*, their lower surface energy at room temperature. Dendrites of silver can be obtained from molten salts at overpotentials of a few millivolts [5], as in the case of Pb and Sn deposition from aqueous solutions [39, 40], because the difference between the melting point of silver and the working temperature for deposition from molten salts is not very different from the difference between the melting point of lead or tin and room temperature. On the other hand, dendrites grow from screw dislocation and nuclei of higher indices or twinned ones only [7, 11, 13]. The probability of formation of such nuclei increases with increasing overpotential [21, 46] and η_i, can also be defined as the overpotential at which they are formed. Regardless of this, Eqs. (2.47), (2.49), and (2.53) illustrate well the effect of different parameters on the initiation of dendritic growth. It is obvious that the electrochemical conditions, as well as the crystallographic ones, under which dendritic deposits are formed can be precisely determined.

2.2 Electrodeposition on the Inert Substrate

2.2.1 *Cementation and Prevention of it by Deposition from the Complex Salt Solutions*

If zinc is immersed in a copper sulfate solution, the reaction

$$Cu^{2+} + Zn \rightarrow Zn^{2+} + Cu \tag{2.74}$$

occurs, whereby copper ions are converted to the metal, while the zinc dissolves. Such deposits are spongy and dendritic in a number of cases, and non-adherent as well. Hence, a good copper deposit on a zinc substrate cannot be formed in this manner. The basic characteristics of plating baths which produce good deposits are use of inert substrates and simple or complex salt solutions from which the metal is deposited at a sufficiently negative cathode potential [13].

In a cyanide-containing bath, the copper potential is sufficiently negative so cementation does not occur and copper can be successfully deposited onto zinc. This is due to the fact that the cyanide complexes of copper are very strong so the potential of copper in such a solution is much more negative than in simple salt solutions. On the other hand, the zinc cyanide complex is relatively weak and the potentials of two metals become comparable so an external power supply is required to deposit copper on the zinc from cyanide.

Analogously, from a copper sulfate solution, cementation of copper on immersed iron occurs according to the reaction

$$Cu^{2+} + Fe \rightarrow Fe^{2+} + Cu \tag{2.75}$$

but from a cyanide solution, copper can be successfully deposited using an external power supply. In this case the complexes of both copper and iron are very strong and cementation is theoretically possible if only the reversible potentials only are taken into consideration. The fact that copper can be successfully deposited onto steal from cyanide-containing solutions is explained by the fact that the reaction between iron and cyanide ions is very slow, and so the reversible potential is never reached.

2.2.2 *Surface Film Formation*

2.2.2.1 Crystallization Overpotential

The formation of the first crystals during galvanostatic metal electrodeposition on an inert substrate is sometimes accompanied by a pronounced increase in overpotential [47, 48]. The dependence of the overpotential on time in such situations is shown in Fig. 2.16.

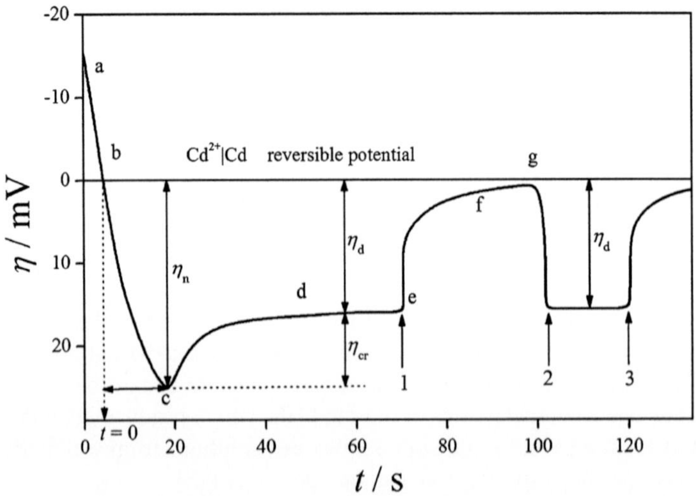

Fig. 2.16 The dependence of the overpotential on time during cadmium electrodeposition on a spiral platinum cathode (electrode surface area: 1.5 cm^2) from 0.50 M $CdSO_4$. The deposition current was 65 μA, η_n is the nucleation, η_{cr} is the crystallization, and η_d is the deposition overpotential (Reprinted from Ref. [47] with permission from Bulgarian Chemical Communications and Ref. [13] with kind permission from Springer)

The overpotential η_t changes with deposition time from point b to point c according to the Eq. (2.76) [47]:

$$\eta_t = \frac{RT}{nF} \ln \frac{C_{t,a}}{C_{0,a}} \tag{2.76}$$

where $C_{t,a}$ is the concentration of adatoms at time t and $C_{0,a}$ is the concentration of adatoms at $t = 0$.

On the other hand, the surface concentration of adatoms changes according to:

$$C_{t,a} = C_{0,a} + \frac{it}{nF} \tag{2.77}$$

Substitution of the Eq. (2.77) into the Eq. (2.76) produces

$$\eta_t = \frac{RT}{nF} \ln\left(1 + \frac{it}{nFC_{0,a}}\right) \tag{2.78}$$

which becomes

$$\eta_n = \frac{RT}{nF} \ln\left(\frac{C_{cr,a}}{C_{0,a}}\right) \tag{2.79}$$

at the moment of nucleation.

The Eq. (2.78) describes the dependence of the overpotential on the deposition time from point *b* to point *c*. The overpotential changes due to the change of the surface concentration of adatoms from $C_{0,a}$ at the equilibrium potential to some critical value $C_{cr,a}$ at the critical overpotential, η_n, at which the new phase is formed. Hence, the concentration of adatoms increases above the equilibrium concentration during the cathode reaction, meaning that at potentials from point *b* to point *c* there is some supersaturation. The concentration of adatoms increases to the extent to which the boundary of the equilibrium existence of adatoms and crystals has been assumed to enable the formation of crystal nuclei. On the other hand, the polarization curve can be expressed by the equation of the charge transfer reaction, modified in relation to the crystallization process, if diffusion and the reaction overpotential are negligible, as given by Klapka [48]:

$$i = i_0 \left\{ \frac{C_a}{C_{0,a}} \exp\left(\frac{\alpha n F \eta}{RT}\right) - \exp\left[\frac{(\alpha - 1) n F \eta}{RT}\right] \right\} \tag{2.80}$$

because the partial anode current density depends on the concentration of adatoms, which for $i/i_0 = 0$ becomes equal to the Eq. (2.79).

Obviously, the Eq. (2.80) becomes valid at the moment of the formation of the new phase, and it can be used for estimating the overpotential, η_n, at which the nucleation takes place. In order to calculate this overpotential, the supersaturation must be known. According to Pangarov et al. [22, 49, 50], the work of formation of differently oriented particles can be estimated using supersaturations of four to seven. Considering the nucleation overpotential (for different supersaturations), Klapka [48] assumed ten as the upper limit of supersaturation. The lower limit is obviously one and the Eq. (2.80) in this case becomes identical to the equation of the charge transfer reaction.

The difference in overpotential between the curves for a given supersaturation (nucleation on an inert substrate) and the curve for a supersaturation equal to unity (deposition on a native substrate) gives the value of the crystallization overpotential, η_{cr} [48]. It is equal to the difference in the overpotential at point *c* and at point *e* in Fig. 2.16. If the current is switched off at point *e*, the electrode potential will approach to the reversible potential of the deposited metal (point *g*); after switching on the current again at point *g*, the overpotential returns to the same value as at point *e*, *i.e.*, the deposition overpotential, η_d, meaning that a new phase is formed. On the contrary, if current is switched off before point *c*, the electrode potential will approach the initial stationary potential of the inert electrode, meaning that new phase has not been formed [47].

Using temperature of 25 °C, $\alpha = 0.5$, $n = 1$ or 2, and the Eq. (2.80), Klapka [48] calculated the dependencies of the nucleation overpotential on the i/i_0 ratio for $C_{cr,a}/C_{0,a} = 1$, 2, 5, and 10. The calculated curves are shown in Fig. 2.17a. From these curves, the dependencies of the crystallization overpotentials on the i/i_0 ratio, shown in Fig. 2.17b, can be derived.

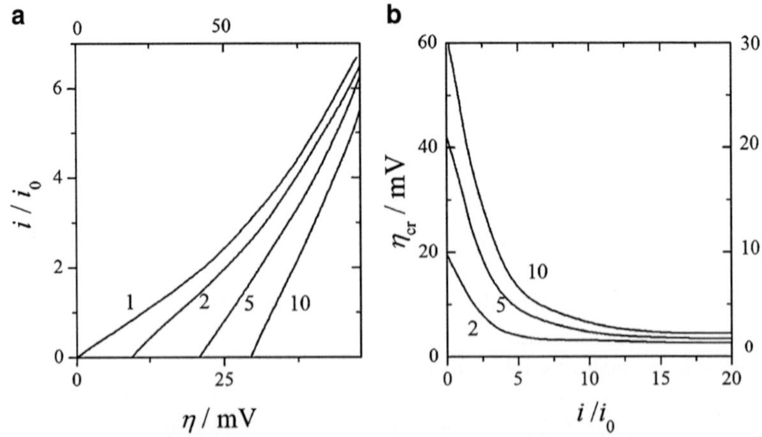

Fig. 2.17 (**a**) Dependence of the i/i_0 ratio on the overall overpotential for different $C_{cr,a}/C_{0,a}$ ratios, indicated on the curves (*upper scale* for $n = 1$, *lower scale* for $n = 2$) and (**b**) dependence of the crystallization overpotential, η_{cr} (in mV), for the cathodic reaction on i/i_0 for different $C_{cr,a}/C_{0,a}$ ratios, indicated on the curves (*left-hand scale* for $n = 1$, *right-hand scale* for $n = 2$) (Reprinted from Ref. [48] with permission from the Collection of Czechoslovak Chemical Communications and Ref. [13] with kind permission from Springer)

The crystallization overpotential strongly decreases with increasing i/i_0 ratio. As a result of this, it can be measured only in the case of a metal deposition which is characterized by very high values of the exchange current density [48].

2.2.2.2 The Nucleation Exclusion Zones

Metal electrodeposition on inert electrodes begins with the formation of separate growth centers until a continuous or disperse deposit is produced. Once a nucleus of the depositing metal has been formed, the current flowing causes a local deformation of the electric field in the vicinity of the growing center. As a result, an ohmic potential drop occurs along the nucleus–anode direction. Considering the high dependence of the nucleation rate on the overpotential, new nuclei would be expected to form only outside the spatial region around the initial nucleus. In that region, the potential difference between the cathode and the electrolyte surpasses some critical value η_{crit}. Using simple mathematics, one obtains for the radius of the screening zone, r_{sz}, in an ohmic-controlled deposition:

$$r_{sz} = f \frac{U_\Omega}{U_\Omega - \eta_{crit}} r_N \tag{2.81}$$

where η_{crit} is the critical overpotential for nucleation to occur, U_Ω is the ohmic drop between the anode and cathode, f is a numerical factor, and r_N is the radius of the nucleus. The radius of the screening zone depends on the value of both U_Ω and η_{crit}.

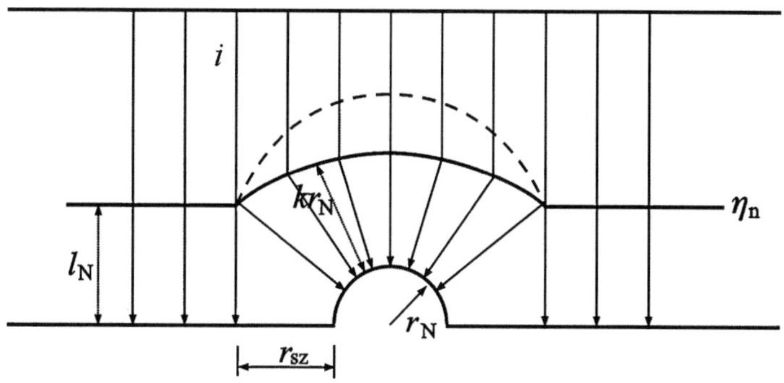

Fig. 2.18 A schematic representation of the deformation of the current field around a defect or a grain grown on a foreign substrate. For an explanation of the symbols, see Ref. [52] (Reprinted from Ref. [52] with permission from the Serbian Chemical Society and Ref. [13] with kind permission from Springer)

At a constant η_{crit}, an increase in U_Ω leads to a decrease in the radius of the screening zone, the same is true if η_{crit} decreases at constant U_Ω [51].

The radius of a nucleation exclusion zone can be calculated on the basis of the following discussion, taking into account the charge transfer overpotential also. If there is a half-spherical nucleus on a flat electrode, the extent of the deviation in the shape of the equipotential surfaces which occurs around it depends on the crystallization overpotential, current density, a resistivity of the solution and on the radius of the nucleus, r_N. If the distance from the flat part of the substrate surface to the equipotential surface which corresponds to the critical nucleation overpotential, η_n, is l_N, then this changes around defect to the extent $k{\cdot}r_N$, where k is a number, as is presented in Fig. 2.18.

Therefore, in this region, the current lines deviate from straight lines toward the defect, thus causing an increase in the deposition rate, while in the surrounding region nucleation does not occur, *i.e.*, a nucleation exclusion zone is formed. The voltage drop between the point from which the deviation occurs and the nucleus surface consists of the ohmic drop between these points and the charge transfer overpotential at the nucleus solution interface. The nucleation overpotential includes both the crystallization and charge transfer (deposition) overpotential:

$$\eta_n = \eta_{cr} + \eta_d \tag{2.82}$$

Hence, at the moment when $k{\cdot}r_N$ become equal to l_N

$$k\, r_N i \rho = \eta_{cr} \tag{2.83}$$

where i is the current density along the current lines and ρ is the specific ohmic resistivity of the solution. Hence, when the ohmic drop between the deviation point and nucleus surface becomes equal to the crystallization overpotential, a new nucleation becomes possible on inert substrate assuming in both cases the same charge transfer overpotential, and the same value of the current density between the two symmetrical points on the anode and inert cathode surface and between the same point on the anode and the point at the surface of the earlier formed nucleus.

The radius of the nucleation exclusion zone, r_{sz}, corresponds to the distance between the edge of a nucleus and the first current line which not deviates (when $k{\cdot}r_N$ becomes equal to l_N). Accordingly, nucleation will occur at distances from the edge of a nucleus equal or larger than r_{sz}, which can be calculated as:

$$r_{sz} = r_N\left(\sqrt{2k+1} - 1\right) \tag{2.84}$$

If the Eq. (2.83) is taken into account, one obtains:

$$r_{sz} = r_N\left(\sqrt{\frac{2\eta_{cr}}{r_N \rho i} + 1} - 1\right) \tag{2.85}$$

According to the Eq. (2.85), a new nucleation is possible in the vicinity of a nucleus if $\eta_{cr} \to 0$ or $i \to \infty$ or $\rho \to \infty$

The analysis of the nucleation rate around a growing grain can also be treated in a more rigorous way [53]. Regardless of this, the above model is sufficient to explain the role of nucleation–exclusion zones in the first stage of electrocrystallization. This is because the nuclei formed are extremely small, and the spherical diffusion control around them can be established after relatively large induction times [54]. During this induction time, the nucleation exclusion zones are due to the ohmic drop in the vicinity of a growing center. Simultaneously, the nucleation process is practically terminated, because it is very fast [55]. On the other hand, the rigorous treatment of this problem is very complicated while the effect of the kinetic parameters of the deposition process in the first stage of electrocrystallization can be qualitatively explained in a simple way using the described model, *i.e.*, Klapka's concept of crystallization overpotential and the classical nucleation theory.

During the cathodic process at low i/i_0, the crystallization overpotential is considerably high; with increasing i/i_0, however, it decreases rapidly [48]. Hence, for $i_0 \to 0$, it follows that $r_{sz} \to 0$.

2.2.2.3 Nucleation Rate and Deposition Overpotential

It has been established experimentally that the number of nuclei deposited electrolytically onto an inert electrode increases linearly with time after an induction period. After a sufficient length of time, it reaches a saturated value that is independent of time. The density of the saturation value increases with the increasing applied overpotential and is strongly dependent on the concentration of the electrolyte and the state of the electrode surface [55].

Kaischew and Mutaftchiew [56] explained the phenomenon of saturation on the basis of energetic inhomogeneity of the substrate surface. They assumed that the active centers have different activity, or different critical overpotential with respect to the formation of nuclei. Nuclei can be formed on those centers whose critical overpotential is lower or equal to the overpotential externally applied to the electrolytic cell. The higher the applied overpotential, the greater the number of weaker active sites taking part in the nucleation process and, hence, the greater the saturation nucleus density is. The formation and growth of nuclei is necessarily followed by the formation and growth of nucleation exclusion zones. After some time, the zones overlap to cover the substrate surface exposed for nucleation, thus terminating the nucleation process [55]. This is well illustrated in Fig. 2.19. It can be seen that the deposit obtained at low current densities consists of a small number of nuclei, but with increasing overpotential or current density, the number of growth sites increases and the grain size of the deposit decreases.

The simultaneous action of both active centers and nucleation exclusion zones must be taken into consideration when discussing the dependence of the number of nuclei on time. In the limiting case for active centers, when screening zones are not formed, the saturation nucleus density is exactly equal to the integral number of active centers. In the limiting case for nucleation exclusion zones, the saturation nucleus density is directly proportional to the nucleation rate and inversely to the zone growth rate [55]. It is obvious that the saturation nucleus density is larger in the first than in the second case, because of the deactivation of active centers by overlapping nucleation exclusion zones.

The classical expression for the steady-state nucleation rate, J, is given by [47, 57, 58]:

$$J = K_1 \exp\left(-\frac{K_2}{\eta^2}\right) \tag{2.86}$$

where K_1 and K_2 are practically overpotential-independent constants. Equation (2.86) is valid for a number of systems regardless of the value of the exchange current density for the deposition process [47, 58]. At one and the same deposition current density, i, decreasing i_0 leads to an increasing nucleation rate and decreasing nucleation exclusion zones radii. Hence, the limiting case for nucleation exclusion zones can be expected when $i/i_0 \rightarrow 0$, and the limiting case for active centers when $i/i_0 \rightarrow \infty$.

The saturation nucleus density and the exchange current density of the deposition process strongly affect the morphology of metal deposits. At high exchange current densities, the radii of the screening zones are large and the saturation nucleus density is low. This permits the formation of large, well-defined crystal grains and granular growth of the deposit. At low exchange current densities, the screening zones radii are low, or equal to zero, the nucleation rate is large, and a thin surface film can be easily formed. The saturation nucleus density depends also on the deposition overpotential.

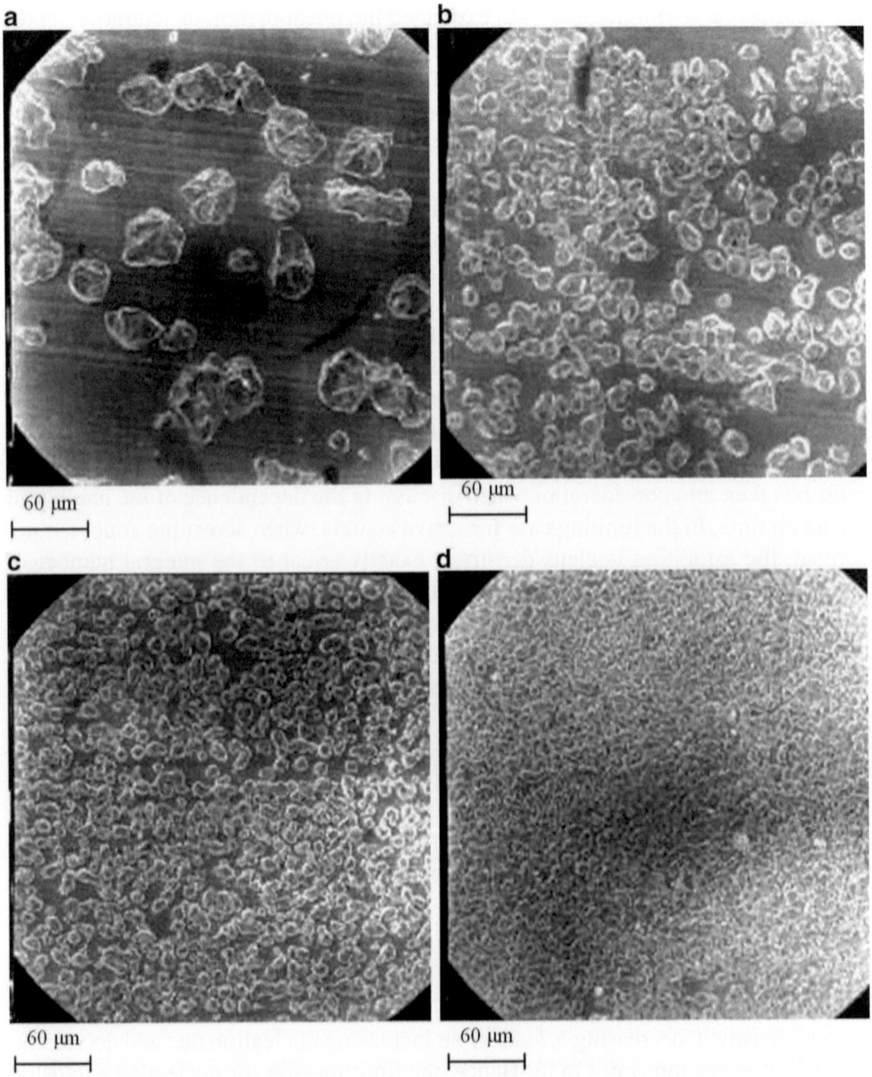

Fig. 2.19 Cadmium deposits obtained from 1.0 M $CdSO_4$ in 0.50 M H_2SO_4 solution onto a copper plane electrode: (**a**) deposition overpotential: 10 mV; deposition time: 24 min, (**b**) deposition overpotential: 40 mV; deposition time: 4 min, (**c**) deposition overpotential: 60 mV; deposition time: 2 min, and (**d**) deposition overpotential: 110 mV; deposition time: 80 s (Reprinted from Refs. [13, 54] with kind permission from Springer)

The nucleation law can be written as [59]:

$$N = N_0[1 - \exp(-At)] \tag{2.87}$$

where

$$A = K_1 i_0 \exp\left(-\frac{K_2}{\eta^2}\right) \tag{2.88}$$

and N_0 is the saturation nucleus surface density (nuclei cm^{-2}), being dependent on the exchange current density of deposition process and the deposition overpotential.

The overpotential and the current density in activation-controlled deposition inside the Tafel region are related by:

$$\eta_{\text{act}} = \frac{b_c}{2.3} \ln \frac{i}{i_0} \tag{2.89}$$

Therefore, increasing b_c and decreasing i_0 leads to an increase in the deposition overpotential. According to Eq. (2.88), the value of A increases with increasing overpotential and decreases with decreasing the exchange current density. It follows from all available data that the former effect is more pronounced resulting in deposits with a finer grain size with decreasing value of the exchange current density.

Nucleation does not occur simultaneously over the entire cathode surface, but it is a process extended in time so that crystals generated earlier may be considerably larger than those generated later. This causes periodicity in the surface structure of polycrystalline electrolytic deposits, as well as coarseness of the obtained thin metal film even when they are formed on ideally smooth substrate. Hence, the larger the nucleation rate, the more homogeneous the crystal grain size distribution is obtained leading to formation of smoother deposit. Obviously, periodicity in the surface structure is a more complicated problem, as shown by Kovarskii et al. [60–62], but for the purpose of this analysis, the above conclusion is sufficient.

The confirmation of the above theories with accent on the analysis of the effect of exchange current density on the thin metal film formation on inert substrates can be obtained in the following way. In Chap. 1 (see Fig. 1.13 and Table 1.2) are given the polarization curves, the exchange current density values, and i_L/i_0 ratios for three different electrodeposition systems (Cd, Cu, and Ni) characterized by the substantially different value exchange current densities [13, 63].

The deposits of Cd, Cu, and Ni obtained by the processes of electrochemical deposition on the inert substrates are shown in Fig. 2.20. In Cd electrodeposition, boulders or large grains are formed by the independent growth of formed nuclei inside zones of zero nucleation (Fig. 2.20a). Due to the high value of i_0 for Cd, the deposition overpotential is low, the crystallization overpotential is relatively large, and the screening zones are, according to Eq. (2.81), relatively large. On the other hand, the nucleation rate is low.

In the case of Cu, due to the lower exchange current density value, a surface film is practically formed by a smaller quantity of electricity (Fig. 2.20b). The value of the deposition overpotential is larger than in the case of Cd, and the crystallization overpotential is lower, resulting in a decrease of the zero nucleation zone radiuses. In the case of Cu, it is clear that a considerably larger nucleation rate is observed.

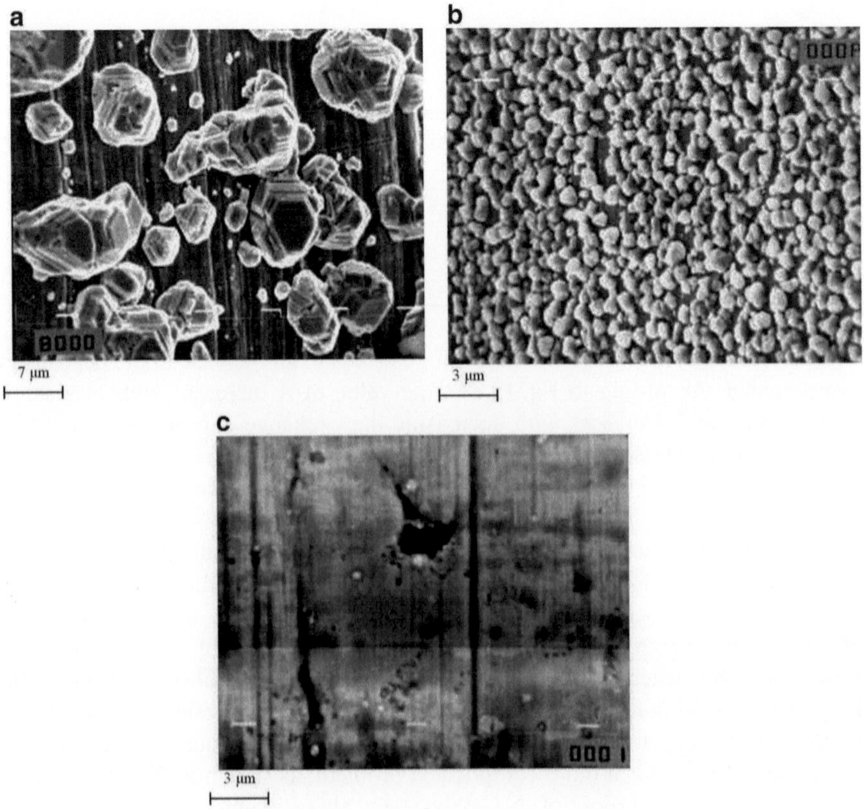

Fig. 2.20 (**a**) Cd deposit on Cu substrate obtained at a current density of 1 mA cm^{-2} from 0.070 M $CdSO_4$ in 0.50 M H_2SO_4. Deposition overpotential: 15 mV. Deposition time: 1200 s, (**b**) Cu deposit on Ag substrate obtained at a current density of 1 mA cm^{-2} from 0.070 M $CuSO_4$ in 0.50 M H_2SO_4. Deposition overpotential: 60 mV. Deposition time: 300 s, and (**c**) Ni deposit on Cu substrate obtained at a current density of 1 mA cm^{-2} from 0.070 M $NiSO_4$ in 0.50 M $Na_2SO_4 + 30$ g dm^{-3} H_3BO_3, pH = 4. Deposition overpotential: 715 mV. Deposition time: 120 s (Reprinted from Ref. [63] with permission from the Serbian Chemical Society and Refs. [8, 13] with kind permission from Springer)

A further decrease in the exchange current density value, as in the case of Ni, leads to the surface morphology shown in Fig. 2.20c. A surface film is formed, but it is porous, probably due to hydrogen evolution.

Hence, a decrease in the value of the exchange current density of the deposition process enhances thin surface metal film formation on inert substrates due to an increase in the nucleation rate and a decrease in the radius of the zero nucleation zones. As a result of this, a compact surface metal film is formed with a smaller quantity of electrodeposited metal, and its coarseness and porosity decrease with a decreasing exchange current density. On the other hand, at sufficiently negative equilibrium potentials and low hydrogen overpotential for an inert substrate,

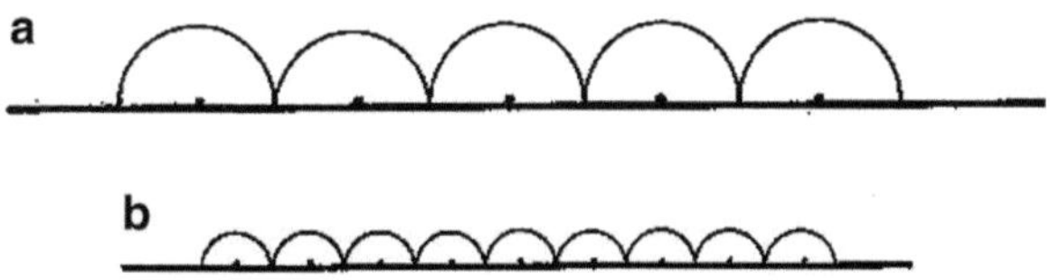

Fig. 2.21 Schematic representation of the formation of the surface film: (**a**) low nucleation rate and (**b**) high nucleation rate (Reprinted from Ref. [13] with permission from Springer)

decreasing the exchange current density of the deposition process can produce a porous deposit due to hydrogen evolution.

2.2.2.4 Compact Surface Film Formation

A surface metal film on an inert substrate is formed by the coalescence of growing grains developed from corresponding nuclei, as is illustrated in Fig. 2.21, whereby the surface properties of the inert substrate are changed to those of the electrodeposited metal.

It is obvious that the larger nucleus density, the thinner is the thickness of the metal film required to isolate the substrate from the solution. At the same time, a thinner surface film will be less coarse than a thicker one. This means that a smoother and thinner surface film will be obtained at larger deposition overpotentials and nucleation rates, *i.e.*, by electrodeposition processes characterized by high cathodic Tafel slopes and low exchange current densities.

The main ways of the decrease of the exchange current density and the increase of the cathodic Tafel slope can be considered as follows.

2.2.2.5 Deposition from Complex Salt Solutions

The silver deposits obtained from nitrate electrolyte-containing 0.10 M $AgNO_3$ in 0.20 M HNO_3 (the basic electrolyte) at the overpotential corresponding to an initial current density of $0.5i_L$ and from the ammonium complex electrolyte (0.10 M $AgNO_3$ in 0.50 M $(NH_4)_2SO_4$ solution to which ammonium hydroxide had been added to dissolve the silver sulfate precipitate) at the overpotential corresponding to an initial current density of $0.3i_L$ are presented in Fig. 2.22 [64].

From Fig. 2.22a, it can be seen that the silver deposit obtained from the nitrate electrolyte was constructed from a small number of nuclei. Irregular large grains were formed even with an initial current density of $0.5i_L$, causing formation of a non-compact deposit. On the other hand, the microcrystalline deposit was obtained by electrodeposition from the ammonium electrolyte (Fig. 2.22b).

The poor microthrowing power of the Ag deposit obtained from the nitrate electrolytes at the small current densities can be explained in the following way:

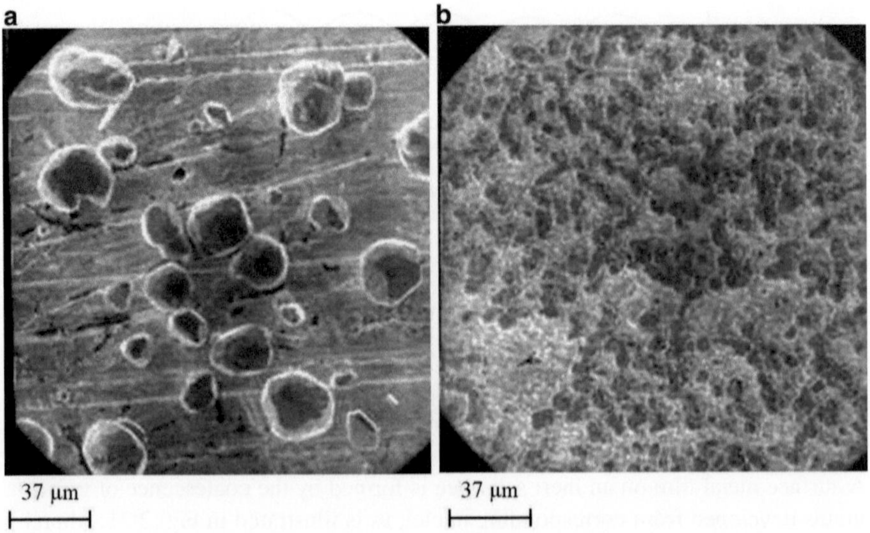

Fig. 2.22 Silver deposits obtained from: (**a**) the nitrate solution at an overpotential of 18 mV (corresponding to an initial current density of $0.5i_L$) (90 mA min cm^{-2}) and (**b**) an ammonium complex solution at an overpotential of 65 mV (corresponding to an initial current density of $0.3i_L$) (90 mA min cm^{-2}) (Reprinted from Ref. [64] with permission from Elsevier and Ref. [13] with kind permission from Springer)

For $i_0 >> i_L$, the deposition overpotential according to Eq. (1.33) is given by [13, 64]:

$$\eta = \frac{RT}{nF} \ln\left(1 - \frac{i}{i_L}\right) \tag{2.90}$$

It follows from Eqs. (2.87) and (2.88) that for $i << i_L$, $\eta \rightarrow 0$ and $N \rightarrow 0$. Thus, at low current densities, poor microthrowing power is expected.

On the other hand, for the ammonium complex solution, it is valid $i_0 << i_L$ [13]. For $i_0 << i$ and $i << i_L$, Eqs. (2.87), (2.88), and (1.29) are valid meaning that $N > 0$. Hence, for deposition at low current densities, the decreasing i_0 leads to the increasing coverage of an inert substrate for a given quantity of deposited metal.

2.2.2.6 Deposition in the Presence of Adsorbed Additives

Inorganic Compounds

The effect of addition of inorganic compounds to the electroplating baths on electrodeposition processes was illustrated by the analysis of Ag deposits obtained from the nitrate electrolytes (0.50 M $AgNO_3$ in 100 g dm^{-3} $NaNO_3$) without (Fig. 2.23a) and with the addition of 6 g dm^{-3} H_3PO_4 (Fig. 2.23b) [65–67].

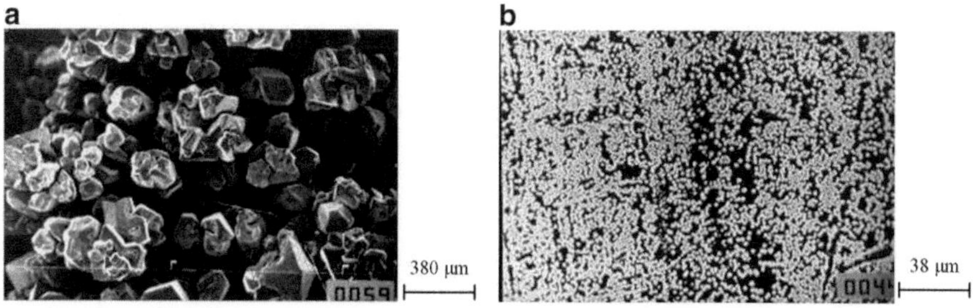

Fig. 2.23 Silver deposits electrodeposited from 0.50 M $AgNO_3$ in 100 g dm^{-3} $NaNO_3$: (**a**) in the potentiostatic regime of electrolysis onto Ag wire electrode without the addition of H_3PO_4. Quantity of electricity: 100 mA h cm^{-2}. Deposition overpotential: 120 mV. The exchange current density: 26 mA cm^{-2} and (**b**) in the galvanostatic regime of electrolysis onto Pt wire electrodes with the addition of 6 g dm^{-3} H_3PO_4. Current: 30 mA. Time: 2 s. The exchange current density: 5 mA cm^{-2} (Reprinted from Refs. [13, 65, 66] with kind permission from Springer)

In the case of electrodeposition from the solution containing PO_4^{3-}-ions, almost complete surface coverage was achieved even with a charge quantity of 2 mA h cm^{-2} at a current density of 30 mA cm^{-2} (*i.e.*, under the optimal film deposition conditions, as determined in Ref. [65]). This is probably due to the possibility of further nucleation occurring immediately next to the already existing nuclei, as a result of the smaller values of the radii of the nucleation exclusion zones caused by the decrease of the exchange current density for the deposition process by the addition of PO_4^{3-}-ions. For comparison, in phosphate-free nitrate solution, a compact Ag film had not been deposited even after 100 mA h cm^{-2} had been passed through the cell, as can be seen from Fig. 2.23a [65].

Organic Compounds

The effect of organic compounds on morphology of electrodeposited metal has been examined by the analysis of cadmium electrodeposition processes from solution containing 0.25 M $CdSO_4$ in 0.50 M H_2SO_4 to which was added 3.3 g dm^{-3} poly(oxyethylene alkylphenol) (9.5 mol ethylene oxide) [68].

The overpotential-log (current density) plot is given in Fig. 2.24. A well-defined Tafel line characterized by $i_0 \approx 10^{-6}$ A cm^{-2} and $b_c = 160$ mV dec^{-1} was observed at higher potentials also. This phenomenon is explained by the formation of a film of the organic additive which completely covers the cathode at sufficiently negative potentials [69, 70]. Tafel linearity was also observed over a short overpotential range at low overpotentials. The values of $i_0 \approx 10^{-3}$ A cm^{-2} and $b_c = 60$ mV dec^{-1} obtained in this case are close to the values expected for deposition from a pure solution [71].

It has been shown [16] that the optimum plating overpotential and current density are determined by the upper limit of the validity of the Tafel equation for

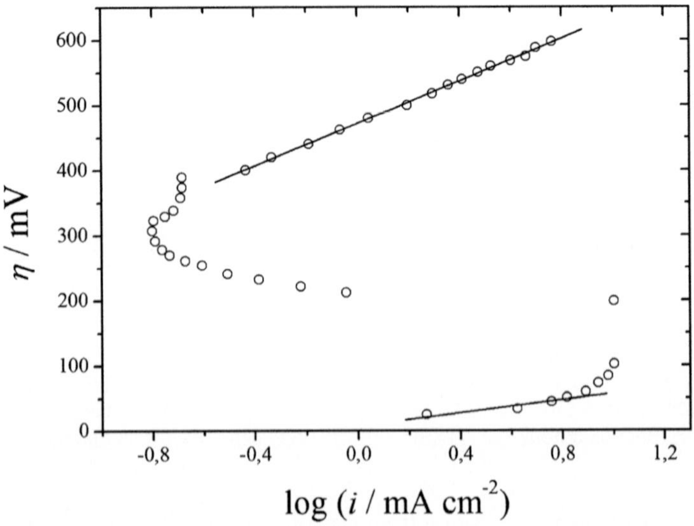

Fig. 2.24 Overpotential *vs.* log (current density) for cadmium deposition in the presence of additive (Reprinted from Ref. [68] with permission from Elsevier and Ref. [13] with kind permission from Springer)

the deposition process. In this case, as can be seen in Fig. 2.24, the optimum deposition overpotentials are about 530 mV and 40 mV in the presence and in the absence of adsorption of this additive, respectively. Cadmium deposits 3 μm thick obtained at 40 mV and 530 mV are shown in Fig. 2.25. It can be seen that the deposit obtained at 40 mV is constructed from large grains (Fig. 2.25a), whereas the one obtained at 530 mV is fine grained due to the larger overpotential (Fig. 2.25b).

Apart from the nucleation density, the preferential orientation of the nuclei is important in surface metal film formation. As the deposit becomes free of the influence of the substrate structure on thickening, instead of the formation of a randomly oriented grain structure, a preferred crystal orientation can develop, which gives a definite texture to the cross section of the deposit [72]. Texture can be expressed in terms of degree of orientation of the grains constituting the deposit.

It is to note that the theoretical approach to the problems of deposit orientation was successfully developed by Pangarov et al. [21, 49, 50]. Using this theory, it was possible to determine the preferred orientation as a function of overpotential from silver single crystals [50] to nickel and iron thin films [21, 50].

Finally, it can be concluded that increasing the overpotential at one and the same current density leads to less coarse deposits with prolonged deposition, because of lower grains formed during the formation of the surface metal film. This conclusion is valid if $i_0 < i_L$. In the opposite case, a rough deposit is often not formed because of disperse or granular deposit formation. Hence, the first condition that must be satisfied in thick non-rough metal film deposition is that the exchange current density must be considerably lower than the limiting diffusion current density in the system under consideration. The second condition is that the deposition current

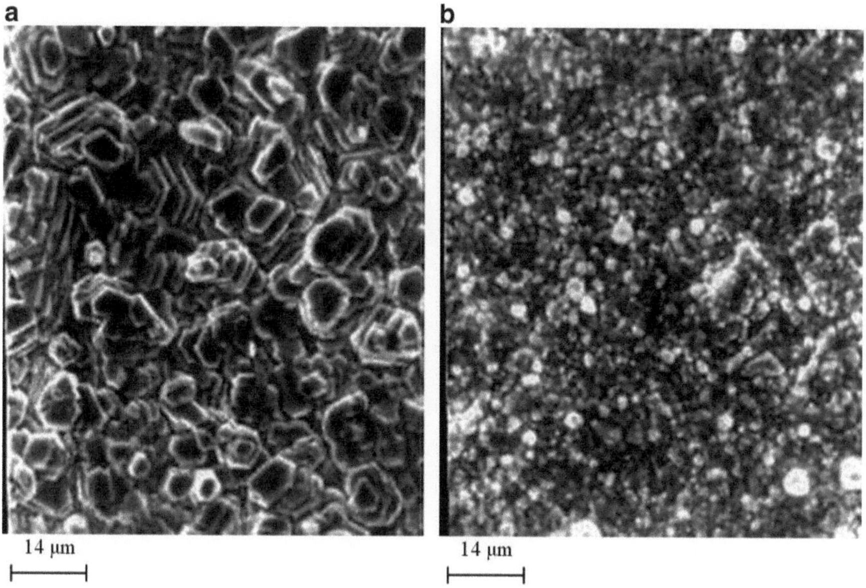

Fig. 2.25 Cadmium deposits formed in the middle of Cu electrodes at overpotentials of (**a**) 40 mV; in the absence of adsorption of the additive and (**b**) 530 mV; in the presence of adsorption of the additive. The deposit thickness is 3 μm (Reprinted from Ref. [68] with permission from Elsevier and Ref. [13] with kind permission from Springer)

density must be a little larger than the one corresponding to the end of the Tafel linearity. In this way, the formation of large and well-defined crystal grains due to deposition under activation control will be prevented. Simultaneously, the increase of the surface coarseness due to the deposition in mixed activation–diffusion control will be minimum, and the formation of cauliflower-like and carrot-like protrusions will be avoided.

On the other hand, smooth and bright deposits can be only obtained in the presence of specific substances denoted as additives. They are the most often organic compounds that either are incorporated into the deposits or undergo electrochemical reaction. Also, they can be adsorbed on the flat part of the surfaces permitting the electrodeposition at the steps [73, 74].

2.2.2.7 Smooth Surfaces

Basic Facts

It is well known [11] that in electrolytes containing specific substances as additives, a phenomenon opposite to the ones described so far can occur, *i.e.*, a more rapid metal deposition at recessed points of the surface than at elevated points. This causes leveling of the surface irregularities as is illustrated in Fig. 2.26.

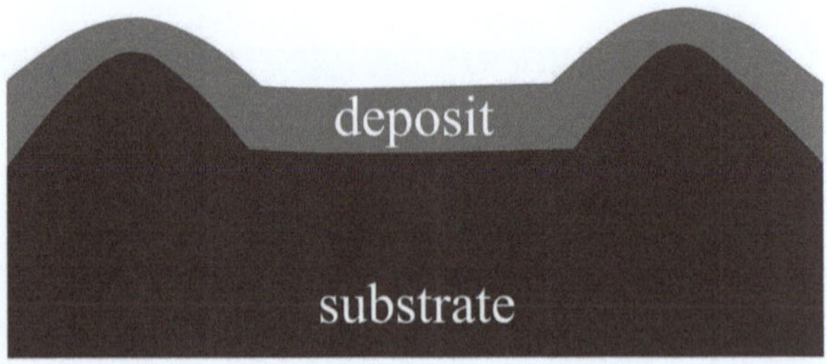

Fig. 2.26 Illustration of leveling action of an additive during metal electrodeposition

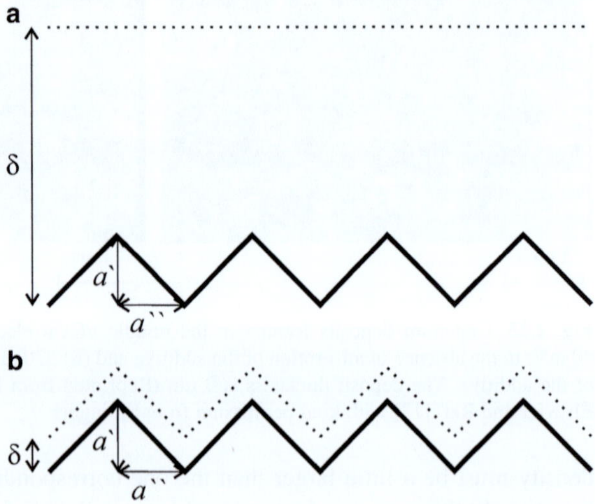

Fig. 2.27 Serrated surface profile with diffusion layer, showing the difference between (**a**) microprofile ($a' << \delta$) and (**b**) macroprofile ($a' >> \delta$) (Reprinted from Refs. [13, 76] with kind permission from Springer)

The fact that this phenomenon is only observed at microprofiles not exceeding 100 μm in amplitude necessitated the introduction of the concept of "microthrowing power" as a category different from ordinary throwing power [11]. The latter is used in technical literature to describe the quality of electrolytes in plating on macroprofiles, at which a similar effect is never observed. The difference between a microprofile and a macroprofile can be seen from Fig. 2.27.

Detailed surveys of the literature on leveling are available in Refs [11, 73–75, 77, 78].

Model of Leveling

All the experimental evidence points to the conclusion that leveling takes place under conditions when the supply of the substance causing inhibition of the electrode process is under diffusion control. It was already clear to early investigators that the explanation should be sought in the local variations in the supply of the leveling agent over the surface profile. Peaks at the surface receive larger amounts of an additive than the recesses. This results in an increase in inhibition and a decrease in the local current density of deposition at the protrusion relative to less-exposed parts of the surface. Thus, leveling is directly related to differences in the surface concentration of the additive which leads to differences in the local current density of deposition [11].

The deposition current density of metal ions must be close to the end of the Tafel linearity, *i.e.*, it can be treated as the current density in activation-controlled deposition, being independent on the geometry of the system. Hence, the current density at the tip of a protrusion will be equal to the current density on the flat surface in the absence of additive. It should be noted that under such condition "geometric leveling" occurs, but true leveling requires the presence of an additive. If the additive is consumed at the electrode by the reaction

$$\text{Additive} + n^{\bullet}\text{e}^{-} \rightarrow \text{Product} \tag{2.91}$$

the limiting diffusion current density of the additive, $i^{\bullet}_{\text{L,tip}}$, to the tip of the protrusion from Fig. 2.1, if the spherical diffusion can be neglected, is given by:

$$i^{\bullet}_{\text{L,tip}} = \frac{n^{\bullet}FD^{\bullet}C^{\bullet}_{0}}{\delta - h_{\text{p}}} \tag{2.92}$$

and that to the flat part of the electrode, $i^{\bullet}_{\text{L,f}}$, by:

$$i^{\bullet}_{\text{L,f}} = \frac{n^{\bullet}FD^{\bullet}C^{\bullet}_{0}}{\delta} \tag{2.93}$$

where $n^{\bullet}$ is the number of electrons in Eq. (2.91), $D^{\bullet}$ and C_0 are the diffusion coefficient and concentration of the additive, respectively.

Assuming that the overall current density, i, at each point of the electrode surface is equal, the effective current density, i_{eff}, of metal deposition at the tip of a protrusion is given by:

$$i_{\text{eff},1} = i - \frac{n^{\bullet}FD^{\bullet}C^{\bullet}_{0}}{\delta - h_{\text{p}}} \tag{2.94}$$

and on the flat part of the surface by:

$$i_{\text{eff},2} = i - \frac{n^{\bullet}FD^{\bullet}C^{\bullet}_{0}}{\delta} \tag{2.95}$$

and following the same procedure as in the treatment of the increase of coarseness (see section "Non-dendritic amplification of surface coarseness"), one can write:

$$\frac{\text{d}h_{\text{p}}}{\text{d}t} = \frac{V}{nF}\left(i_{\text{eff},1} - i_{\text{eff},2}\right) \tag{2.96}$$

or in the integral form:

$$h_{\mathrm{p}} = h_{\mathrm{p},0} \exp\left(-\frac{t}{\tau_1}\right) \tag{2.97}$$

where

$$\tau_1 = \frac{n\delta^2}{n^\bullet V D^\bullet C_0^\bullet} \tag{2.98}$$

where V (molar volume of metal) and n correspond to the electrodeposited metal. This is the simplest mathematical model of the leveling process [79]. Despite this, it elucidates the physical essence of the phenomenon under consideration well.

Obviously, for this model to be operative, two conditions must be satisfied: (a) the additive must be consumed in some manner at the electrode, so that it must be continuously supplied in order to maintain a certain surface concentration, and (b) the diffusion layer must not follow the microprofile but must have a smoother outer boundary, so that variations in its thickness arise, which cause variations of the diffusion flux of the additive [11].

The first condition is fulfilled with all good leveling agents. Most of them undergo sufficiently strong adsorption to remain long enough at the metal surface to be surrounded by depositing atoms and be incorporated into the deposit. It is the balance between the rate of incorporation and that of diffusion of the substance from bulk of solution which maintains a given surface concentration of the additive. The larger the diffusion flux, the higher is the steady-state surface concentration of the additive. Conversely, higher rates of metal deposition cause a lowering of the latter [11].

There is an optimal range of additive concentration and current density of deposition at which the differences in inhibition of deposition between the peaks and recesses, and hence the effect of leveling, are maximum. At too low surface concentrations of the additive, *i.e.*, low bulk concentration and high current density of deposition, the process is practically uninhibited and little difference in the local current density of deposition can arise. This explains the decrease in the leveling effect with increasing current density [11].

At somewhat higher bulk concentrations and lower current densities, linearity exists between the bulk and the surface concentration. This is the range of maximum difference in inhibition [11].

However, at still higher concentrations, an adsorption/desorption equilibrium tends to be approached leading to a Langmuir-type relationship. Eventually, in spite of incorporation, saturation of the surface is reached and the surface concentration is no longer sensitive to local changes in the diffusion flux of the additives. Hence, differences in inhibition vanish and leveling is lost [11].

One should appreciate that some time is needed for the diffusion layer to develop to the extent that it separates from the surface microprofile and provides for local differences in the diffusion flux of the additives. Hence, an induction time should be expected before the leveling effect appears. This is demonstrated by the observed sensitivity of the process to current interruptions [77].

Quantitative Treatment

An attempt at a comprehensive quantitative consideration of the problem was made by Krichmar [80]. He proposed that additives adsorbed to the surface of an electrode are incorporated into the deposit at a rate proportional to the surface coverage and current density.

For a sinusoidal profile of the electrode surface, Krichmar [80] obtained an exponential decrease of the amplitude, $H_0(t)$ with time:

$$H_0(t) = H_0 \exp\left(-\frac{t}{\tau_2}\right) \tag{2.99}$$

where τ_2 is a time constant given by:

$$\tau_2 = \frac{a_1}{A_1\left[B_1\left(C_0^\bullet/\delta\right) - i\right]} \coth\left(\frac{2\pi\delta}{a_1}\right) \tag{2.100}$$

where

$$A_1 = \frac{2\pi V}{nF} \tag{2.101}$$

and

$$B_1 = \frac{K_{\text{lev}}\,\alpha_{\text{lev}}\,F\,n D^\bullet C_{0,\text{a}}^\bullet}{RT\,\delta V} \tag{2.102}$$

K_{lev} and α_{lev} in the above equations are constants.

An exponential decrease of the amplitude of the surface profile was found experimentally by Krichmar and Pronskaya [81].

Regardless, a model of the current distribution and numerical procedure for the calculation of the change in shape of an electrode, for electrodeposition followed by diffusion-controlled reduction or incorporation of a leveling agent, has been developed [78, 82–84]; the approach of Krichmar [80] seems to be the most important one for understanding the leveling process.

2.2.2.8 Bright Surfaces

Basic Facts

In order for metal coatings to be used for decorative purposes, it is necessary that their surfaces must be smooth and bright. In everyday experience, the optical properties of the metals are associated with high reflectivity and low transmission. For this reason, the terms "mirror bright," "semi bright," and "high bright" are often used when considering electrodeposited metal coatings.

However, there are no precise definitions of these terms. Brightness of a metal surface is usually defined as its reflecting power, measured by the amount of light specularly reflected off the surface, *i.e.*, at an angle equal and opposite to that of the incidence light with respect to the normal to the geometrical surface. A more precise definition, not involving the actual reflectivity of the surface, would be the ratio between specularly and diffusely reflected light [32, 85–87]. The ratio of specular to total reflectivity can also be used for an estimation of the brightness of a surface [88].

The development of the techniques of scanning tunneling microscopy (STM) and atomic force microscopy (AFM) enabled the determination and systematization of structural characteristics which metal surfaces should fulfill in order for light from metal surfaces to be to a high degree specularly reflected [89, 90]. The value of these techniques lies in their ability to provide *local* information concerning the metal deposit, which is unsurpassed by any other technique owing to the resolution. These techniques enable a precise analysis of the topography of a metal surface, *i.e.*, examination of the submicrostructures of metal surfaces to be made. Also, the digital images obtained by these techniques can easily be analyzed by powerful software packages.

The (total) reflected light from an ideally flat (geometrical) surface is exclusively mirror-reflected light, *i.e.*, the light is reflected at an angle equal and opposite to that of the incident light with respect to the normal to the geometrical surface. On the other hand, any (every) real surface possesses surface irregularities, which lead to scattering of the light. Then, reflected light from a metal surface includes diffusely reflected, not only specularly reflected light.

According to Bockris and Razumney [32], if the brightness of a surface is defined as the ability to reflect light only in a direction making an angle with the normal equal to that of the incident beam (*i.e.*, approach to an "ideal mirror"), then the surface will be increasingly bright as the scattering introduced in the reflected light by surface irregularities decreases, that is, as deviations of the actual surface from the ideal, geometrical boundaries are eliminated.

The system which is the nearest approximation to an ideally flat surface, from which reflection of light is exclusively mirror reflection, is the surface of a silver mirror or of mercury. For that reason, reflection and structural characteristics of a silver mirror surface can be taken as the reference standard for bright surfaces [87, 91, 92].

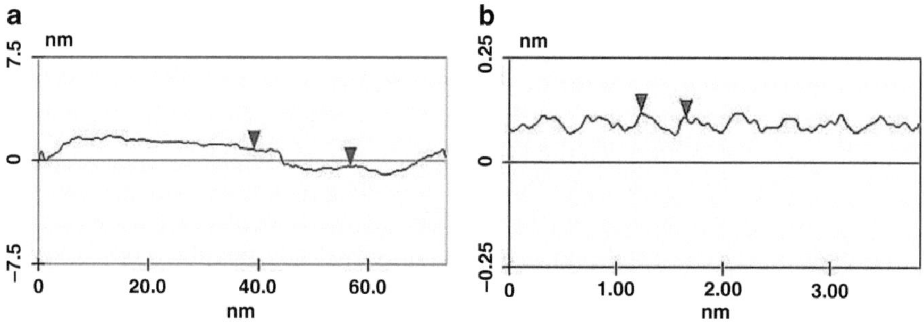

Fig. 2.28 Typical surface profiles of a silver mirror: (**a**) the vertical distance between the markers represents 1.246 nm, and (**b**) line section analysis of the flat part of the surface shown in Fig. 2.28a (Reprinted from Ref. [92] with permission from the Serbian Chemical Society and from Refs. [13, 87] with kind permission from Springer)

Silver Mirror

The reflection of light from a silver mirror surface is mostly mirror reflection and the degree of diffuse reflection is very small (up to 2 %). The degree of mirror reflection from this surface is also very close to the ideal reflectance of silver [87, 91, 92]. The silver mirror surface consists of parts parallel to the base and flat on the atomic level with low-step heights between them, as is shown in Fig. 2.28. It is clear that bright metal surfaces must be similar to the surface of the mirror.

Electropolished Surfaces

The phenomenon of decreasing the surface coarseness of a metal upon anodic dissolution under certain conditions is defined as electropolishing. In cases when polishing occurs, the current–voltage curve was found to exhibit a plateau characteristic for diffusion control of the dissolution process. Some facts point to the complex nature of the phenomenon of electropolishing.

It was also found that systems undergoing electropolishing exhibit a significant photoelectrochemical effect. This corresponds to the region of limiting diffusion current densities and also to the maximum polishing effect.

This suggests to the existence of a photosensitive semiconducting film at the surface and to a possible role of this film in the electropolishing process. Subsequent measurements of the capacitance and resistance of the double layer as functions of potential have shown that this film must be a very thin and a well-conducting one.

In spite of all this, considerable evidence has accumulated justifying the treatment of the electropolishing process as an essentially transport-controlled phenomenon; this film could be related to the effect of brightening [11].

A quantitative model was suggested by Edwards [93] and elaborated by Wagner [94]. According to this model, the metal ions produced are complexed by a component of the electropolishing solution (*e.g.*, phosphate ions or water molecules). Hence, for the reaction to be completed, not only must ions be formed, but also acceptor species have to diffuse to the surface from the bulk of the solution in order to form a complex reaction product. Diffusion of the acceptor from the bulk of the solution to the surface determines the overall rate of reaction.

Hence, differences in acceptor fluxes of different points at the surface arise. The slower diffusion of the acceptor to the recessed parts could cause an increased concentration of free metal ions. This would have many possible consequences such as increasing the cathodic partial current which reduces the net dissolution current, producing changes in the reaction layer such as the formation of an oxide film by hydrolysis, making room for an additional phenomenon observed in electropolishing–brightening. This could be related to the dissolution of facets and other crystallites. Brightening seems to occur when the surface becomes covered by a protective film, which controls the rate of dissolution and makes it a random process, the energetic advantages of atoms at facets and dislocations being lost. Thus, it could be concluded that the Edwards–Wagner model seems to provide a reasonable basis for development of a comprehensive theory of the electropolishing process [11].

The reaction between a metal ion and a ligand is usually sufficiently fast, but it is the insufficient supply of the latter, which could cause the inhibition of this step and make it the rate-determining step.

The anodic partial current is independent of the presence of the ligand. However, if the ligand becomes scarce, the concentration of the free metal ions increases, and, as a result, the cathodic partial current is increased as well. In an ideal case, when diffusion of free metal ions away from the electrode is negligible, the difference between the two partial currents, *i.e.*, the net dissolution current, must be proportional to the flux of the ligand. This case was considered in detail by Wagner [94] who assumed a molecular diffusion mechanism of supply through a fixed hydrodynamic boundary layer of thickness, δ, much larger than wavelength and amplitude of the surface profile.

For a sinusoidal profile of the electrode surface, Wagner [94] gives:

$$H_0(t) = H_0 \exp\left(-\frac{t}{\tau_3}\right) \tag{2.103}$$

where H_0 is the initial surface amplitude, and τ_3 is given by:

$$\tau_3 = \frac{\omega\delta}{2\pi D V C^0} \tag{2.104}$$

ω is the wavelength of the profile and C^0 is the concentration of the acceptor.

Equation (2.104) shows that the electropolishing process is faster if the thickness of the diffusion layer is smaller and if the bulk concentration of the ligand is larger. Also, it is faster, the smaller the wavelength of the surface profile is. The latter reflects the radius of curvature of the elevations and recesses, and the result indicates that "microroughness" will disappear more rapidly than "macroroughness," which is in accordance with experimental experience (see also Fig. 2.29c–g). The exponential time dependence of the height of the elevations given by Eq. (2.104) is in agreement with the findings of Krichmar and Pronskaya [95]. The electropolished metal surfaces are characterized by a specular reflection of light.

The 3D STM images of the Cu surface polished only mechanically and the Cu surface polished both mechanically and electrochemically are shown in Fig. 2.29a, b, respectively. In order to determine which structural features determine brightness, the flat parts of the profile, which were parallel to the base, were examined under different magnifications. The line sections of parts of the flat surfaces were investigated first at lower magnifications and then at increasingly higher ones. Following this procedure, Fig. 2.29c–f were obtained.

In the case of the mechanically polished surface, the amplitude of roughness was several atomic diameters of Cu (Fig. 2.29c, d), whereas the mechanically and electrochemically polished surface exhibited smoothness on the atomic level (Fig. 2.29e, f). The increase in the specular reflection of 20–25 % is due to this fact. It is interesting to note (Fig. 2.29g) that the structure of the bright surface was not oriented. This is in accordance with the assumption that the dissolution process under polishing conditions is a random process.

Hence, it can be concluded that the condition for mirror brightness of a metal surface is smoothness on the atomic level of a suitably oriented flat part of the metal surface with low-step heights between, as in the case of a silver mirror.

Electrodeposited Surfaces

The STM images of the copper coating electrodeposited in the presence of the brightening additives that exhibit high degree of mirror reflection which approached very nearly the ideal reflectance of copper are shown in Fig. 2.30 [87, 91, 96].

The high degree of mirror reflection of this coating was determined by flat and mutually parallel parts of the surface (Fig. 2.30a, b). The distances between adjacent flat parts were several atomic diameters of copper (Fig. 2.30c) [87]. The flat parts of the surfaces were smooth on the atomic scale (Fig. 2.30d). The mean size of these atomically flat parts was estimated by STM software measurements, using option of the determination of the autocovariance (ACVF) and power spectral density (PSD) functions, and they were: $(160 \times 170)\ \mathrm{nm}^2$ – determined by ACVF function and 198 nm – determined by PSD function [87].

The flat and mutually parts of the surface appear because the growth of the deposit in the vertical direction is suppressed by the adsorption of the additive [89]. This happens if the organic additive adsorbs strongly on the top of the copper crystallites and inhibits the formation of new growth centers. Deposition then occurs dominantly at the edges of the crystallite, resulting in lateral expansion of the layers and the formation of smooth terraces on the atomic level. The growth of a

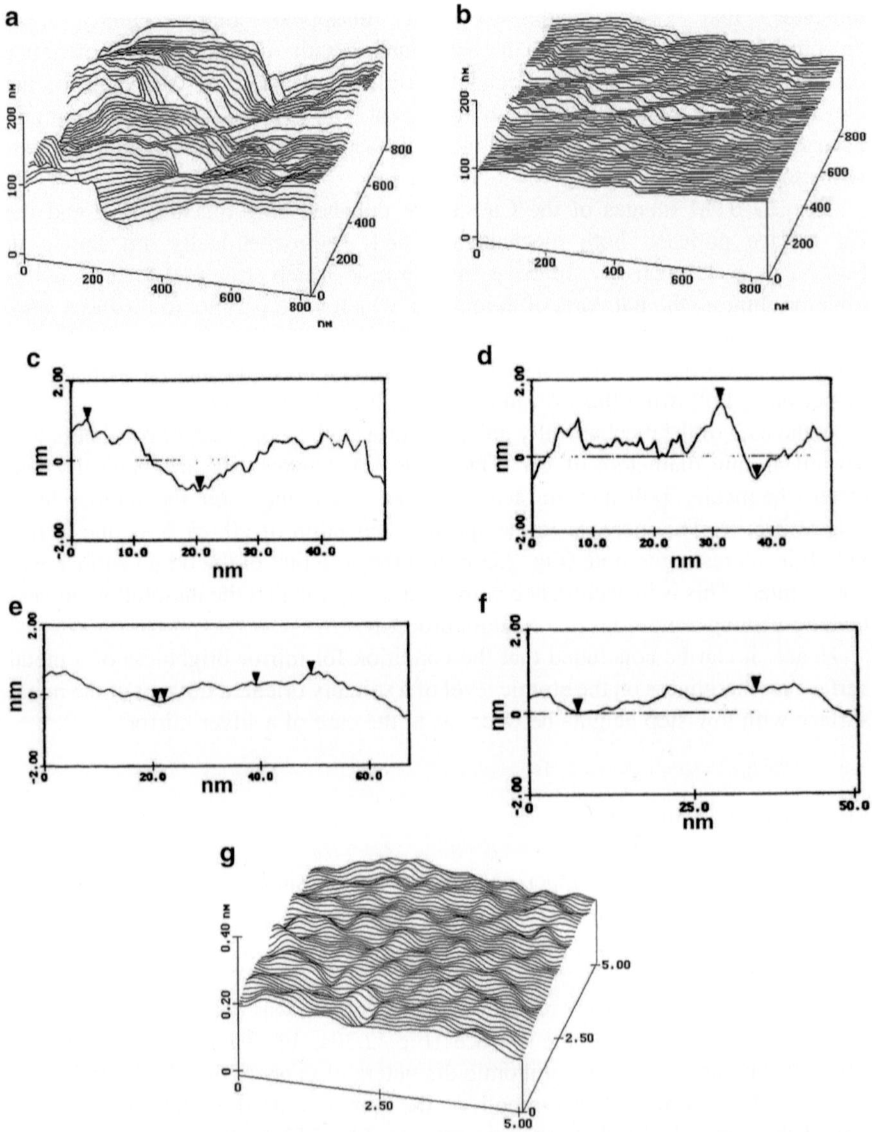

Fig. 2.29 The 3D STM images of (**a**) only mechanically polished Cu surfaces, (**b**, **g**) both mechanically and electrochemically polished, and the surface profiles of the flat parts from (50 × 50) nm^2 samples areas: mechanically (**c–d**) and both, mechanically and electrochemically (**e–f**) polished Cu surface. The vertical distances between the labels of the maximum and minimum points are (**c–d**) $\Delta l = 2.0$ nm, (**e–f**) $\Delta l = 0.51$ nm (Reprinted from Ref. [87] with kind permission from Springer, Ref. [91] with permission from Elsevier, and Ref. [96] with permission from the Serbian Chemical Society)

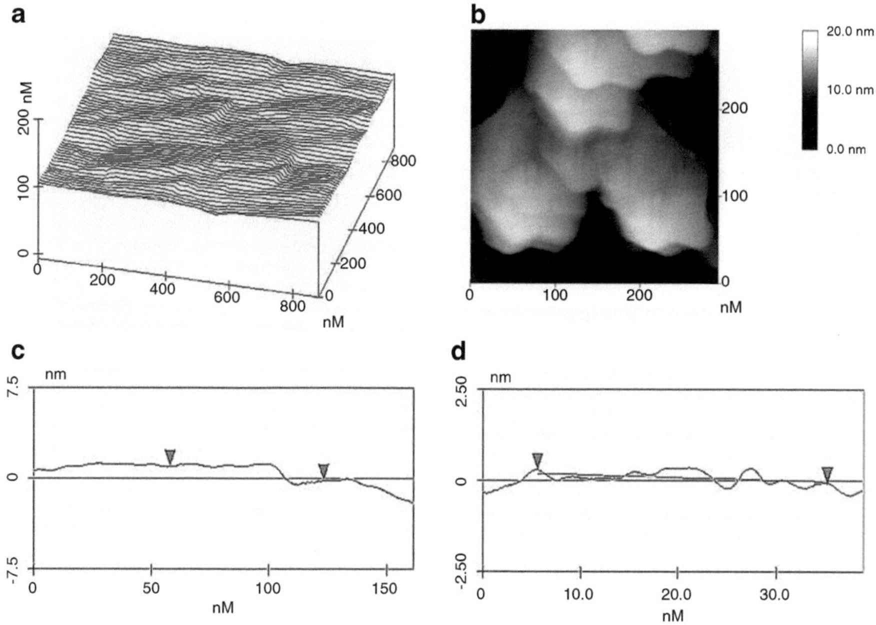

Fig. 2.30 The STM images of 25 μm thick the copper coating electrodeposited from solution: 240 g dm^{-3} $CuSO_4 \cdot 5\ H_2O$ + 60 g dm^{-3} H_2SO_4 + 0.124 g dm^{-3} NaCl + 1.0 g dm^{-3} modified polyglycol ether (Lutron HF 1) + 1.0 g dm^{-3} poly(ethylene glycol) $M_n = 6000$ (PEG 6000) + 1.5 mg dm^{-3} 3 -mercapto propane sulfonate. Scan size: (**a**) (880 × 880) nm^2, and (**b**) (300 × 300) nm^2, (**c**) the line section analysis from the portion of the STM surface of the copper coating shown in Fig. 2.30a. The distance between markers is 1.177 nm, and (**d**) the line section analysis of the flat part of surface of the copper coating shown in Fig. 2.30c (Reprinted from Ref. [91] with permission from Elsevier and Ref. [87] with kind permission from Springer)

new layer starts from defect arising at the interface between two adjacent crystallites, and surfaces like those in Fig. 2.30a are formed.

Mirror brightness of copper surfaces is not associated with their preferred orientation. This copper coating showed (200) preferred orientation [87]. On the other hand, the copper coating electrodeposited from the copper solution containing thiourea as the brightening additive showed the (111) preferred orientation, while the mirror bright copper surface polished both mechanically and electrochemically was relatively disordered with an increased ratio of copper crystallites oriented in (200), (220), and (311) planes [87, 91, 97].

The 3D STM image of the zinc coating having the high degree of mirror reflection is shown in Fig. 2.31a. This zinc coating obtained with the addition of dextrin/salicylaldehyde mixture as the brightening additives were relatively smooth, but without large flat and mutually parallel parts of the surface, which were characteristic of the previously observed copper surfaces [98]. Analysis of the zinc coating at even higher magnification (Fig. 2.31b) showed that the zinc coating was constructed from hexagonal zinc crystals. XRD (X-ray diffraction) analysis of

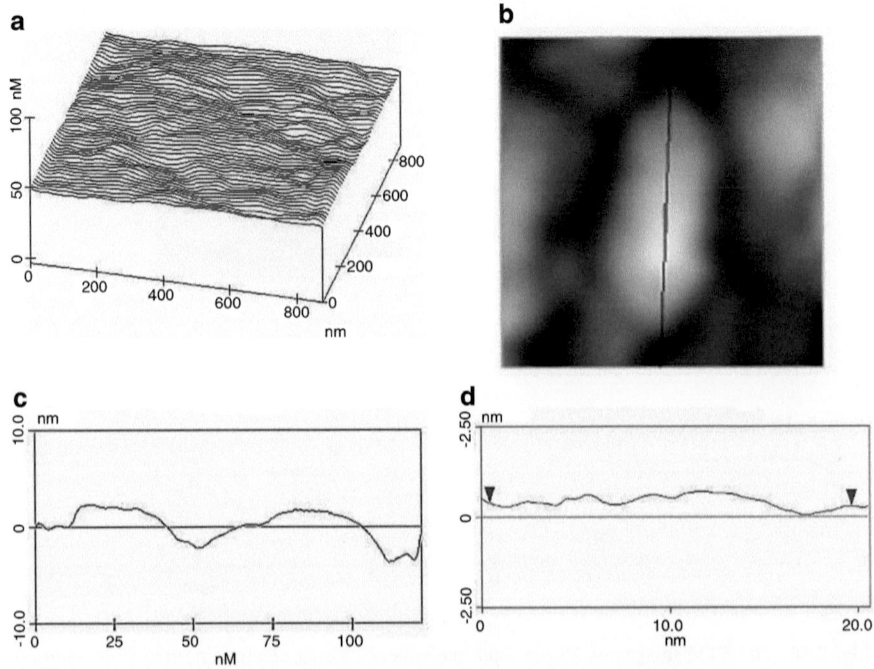

Fig. 2.31 The STM images of 25 μm thick the zinc coating electrodeposited from solution: 300 g dm^{-3} $ZnSO_4 \cdot 7\,H_2O$ + 30 g dm^{-3} $Al_2(SO_4)_3 \cdot 18\,H_2O$ + 15 g dm^{-3} NaCl + 30 g dm^{-3} H_3BO_3 + 3.0 g dm^{-3} dextrin + 2.8 ml dm^{-3} salicylaldehyde. Scan size: (**a**) (880 × 880) nm^2, and (**b**) (50 × 50) nm^2, (**c**) the line section analysis from the portion of the STM surface of the zinc coating shown in Fig. 2.31a, and (**d**) the line section analysis of the hexagonal zinc crystal shown in Fig. 2.31c (Reprinted from Ref. [98] with permission from the Serbian Chemical Society and Refs. [87, 97] with kind permission from Springer)

the zinc coating showed that the zinc crystallites were oriented in the (110) plane only. The (110) plane of the h.c.p. crystal lattice is the plane where the most densely packed atom rows in the most densely packed lattice plane (basal plane) stand upright to the cathode surface and when the crystallites of the deposits are oriented so that the most densely populated atom plane is perpendicular to the substrate, then the deposited metal adopts an outward type of growth [87].

The line section analysis of the zinc coating shown in Fig. 2.31a is given in Fig. 2.31c, from which it can be observed that the top surfaces of the hexagonal zinc crystals are relatively flat and mutually parallel. The distance between the top plane of a hexagonal zinc crystal and the boundary with the adjacent zinc crystal was several atomic diameters of zinc [87]. The roughness of the hexagonal zinc crystals was less than the atomic diameter of zinc, *i.e.*, these zinc crystals were smooth at the atomic level (Fig. 2.31d). The mean size of these hexagonal zinc crystals (estimated by the STM software data processing) was by approximately (18 × 20) nm^2 by the ACVF and about 20 nm by the PSD functions.

The similar reflection and structural characteristics also showed the zinc coatings electrodeposited onto the zinc substrate [99], as well as those electrodeposited from acid sulfate solution with the addition of dextrin/furfural mixture [100].

The similarity of reflection and structural characteristics of the copper surfaces polished both mechanically and electrochemically, and the copper and zinc coatings electrodeposited with the corresponding brightening additives with those of silver mirror as the reference standard just enabled systematization of the conditions which metal surfaces must fulfill to be mirror bright. These conditions are:

I. *Reflection condition*: high degree of mirror reflection which approaches to the ideal reflectance of the corresponding metal
II. *Structural conditions*:
   (a) Flat parts of the surface parallel to the base which are smooth on the atomic level
   (b) Distances between adjacent flat parts which are comparable with the distances between the adjacent flat parts of a silver mirror

Mathematic model that treats brightness of metal surfaces from the point of view of geometrical optics only is proposed [87, 91]. The proposed model is valid if the following assumptions were fulfilled:

- Metal surfaces are divided into equal elementary parts for which the surface area is n_s.
- The flat parts are smooth.
- The light falls onto the surface at a determined angle and is reflected from the surface only in the direction making an angle with the normal equal to the angle of incidence.
- *The upper flat parts of a surface area* ($k_u \cdot n_s$ parts) reflect light completely.
- The parts of *the lower flat parts of a surface area* ($k_d \cdot n_s$ parts) do not reflect light completely because they are screened by the height of *the lateral parts of a surface area* ($k_l \cdot n$ parts). The screened zones are made by both incident and reflected lights.

In total, the brightness of a surface is determined by the ratio of flat parts to the surface between adjacent flat parts. Increasing the distance between adjacent flat parts leads to a decreasing brightness of the surface [91].

The parts of the surface which are not able to reflect light, e_i, depend on the angle of incidence and the height of the lateral part of the surface.

All metal surfaces can be represented by one equivalent surface which consists of row elementary surface units. One such surface unit is represented in Fig. 2.32.

The part of the surface which is screened, e_i, and so cannot reflect light is given, for an angle of incidence, θ_i, by Eq. (2.105):

$$e_i = k_{l,i} \cdot n_s \cdot \text{tg}\,\theta_i \tag{2.105}$$

where $i = 0, 1, 2, 3.......$

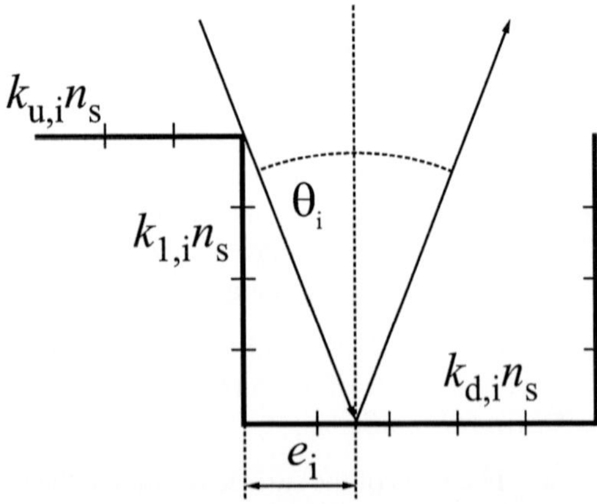

Fig. 2.32 Line section simulation of one surface unit a metal surface (Reprinted from Ref. [91] with permission from Elsevier and Ref. [87] with kind permission from Springer)

Then, having in mind the fact that the screened zones are made by both incident and reflected lights, the brightness of a surface, defined as the ratio of the geometrical minus the screened surface and the geometrical surface, can be given by Eq. (2.106):

$$\text{brightness} = \frac{\sum_1^\infty k_{u,i} \cdot n_s + \left(\sum_1^\infty k_{d,i} \cdot n_s - 2\sum_1^\infty e_i\right)}{\sum_1^\infty k_{u,i} \cdot n_s + \sum_1^\infty k_{d,i} \cdot n_s}$$

$$= \frac{k_u \cdot n_s + (k_d \cdot n_s - 2 \cdot e)}{k_u \cdot n_s + k_d \cdot n_s} = \frac{k_u \cdot n_s + \left(k_d \cdot n_s - 2 \cdot \sum_1^\infty k_{l,i} \cdot n_s \cdot \mathrm{tg}\theta_i\right)}{k_u \cdot n_s + k_d \cdot n_s}$$

$$= \frac{k_u + (k_d - 2 \cdot k_l \cdot \mathrm{tg}\theta_i)}{k_u + k_d} = 1 - 2\frac{k_l}{k_u + k_d}\mathrm{tg}\theta_i \tag{2.106}$$

where

$$k_u = \sum_1^\infty k_{u,i},\ k_d = \sum_1^\infty k_{d,i},\ k_l = \sum_1^\infty k_{l,i},\ e = \sum_1^\infty e_i$$

Hence, according to Eq. (2.106), the brightness of a surface is a function of the incident angle, θ_i, and the ratio between the flat parts, $(k_u + k_d)$ and the lateral parts, k_l. Increasing the lateral parts of a surface, k_l, leads to a decrease in the brightness of the surface. Therefore, when $k_l \rightarrow 0$, then the brightness of the surface approaches that of a mirror.

The boundary conditions of Eq. (2.106) depend on the expression in the bracket [87]. *The lower flat parts of a surface area* will reflect light only if tangent of an incident angle of light is smaller than the ratio ($k_{d,i}/2\ k_{l,i}$). On the other hand, for larger incident angles or an incident angle equal to this ratio, *the lower flat parts of a surface area* will not be able to reflect light.

It is necessary to note that an increase of the lateral parts of a surface, k_l, is accompanied by an increase of the "degree of development of the surface." The later quantity, S', defined as the ratio of the real (actual) to geometrical surface, is given by Eq. (2.107):

$$S' = \frac{k_u \cdot n_s + k_d \cdot n_s + 2k_l \cdot n_s}{k_u \cdot n_s + k_d \cdot n_s} = 1 + 2\frac{k_l}{k_u + k_d} \tag{2.107}$$

Then, according to Eqs. (2.106) and (2.107), the brightness of a surface can be defined by Eq. (2.108):

$$\text{brightness} = 1 - S_s \text{tg}\theta_i \tag{2.108}$$

or, in %, by Eq. (2.108a):

$$\text{brightness (in\%)} = 100 - S_s(\text{in\%})\text{tg}\theta_i \tag{2.108a}$$

where $S_s = (S' - 1)$ – a quantity experimentally measurably by STM data processing as *the surface area diff.*

The excellent agreement within the limits of the experimental error, between experimentally obtained values of mirror-reflected light for incident angles of 5° and 30° and those of brightness obtained according to the proposed model using experimentally determined values of *the surface area diff* (Eq. 2.108a), was observed [87].

2.2.3 Active Microelectrodes Inside the Diffusion Layer of the Inert Macroelectrode

2.2.3.1 Granular Growth

Granular deposits consisting of independently growing grains with a highly developed surface area are obtained by deposition of metals in processes characterized by large values of the exchange current density [101–103]. In some cases, they consist of grains growing independent of each other until a compact film [54, 104] or spongy deposit on it is formed [54].

Granular Growth of Silver

It is shown in Fig. 2.33 that various crystallographic forms, some of them ideal, are obtained during silver deposition at low overpotentials from pure silver nitrate

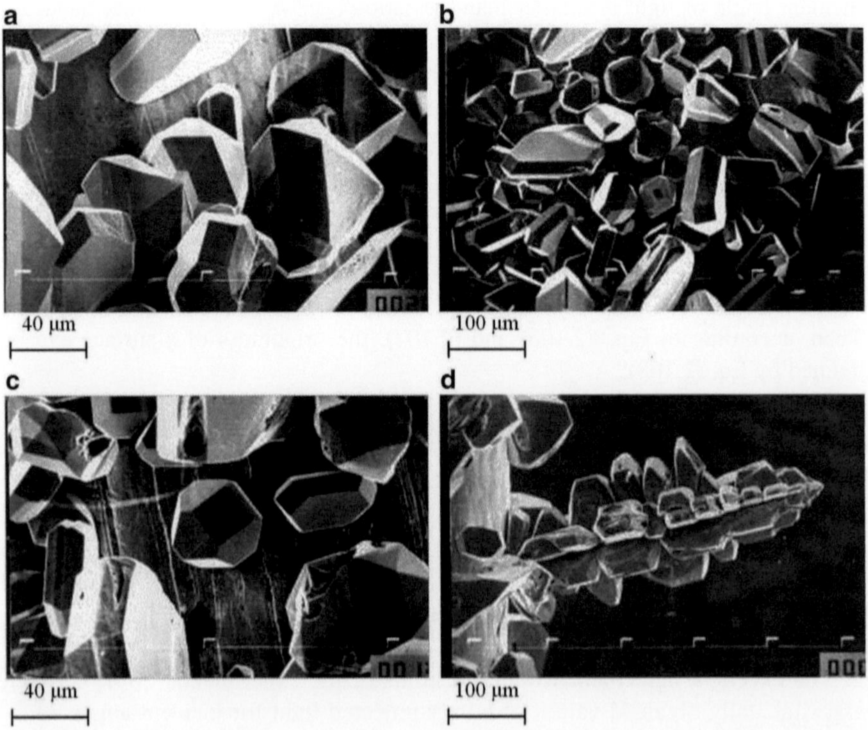

Fig. 2.33 Silver deposits obtained potentiostatically on platinum electrode from 0.10 M $AgNO_3$ in 100 g dm^{-3} $NaNO_3$ at the room temperature. Quantity of electricity: 2 mA h cm^{-2}. Deposition overpotential: (**a**) 60 mV, (**b**) 60 mV, (**c**) 70 mV, and (**d**) 70 mV (Reprinted from Refs. [13, 67] with kind permission from Springer)

solution, due to the independent grain growth inside the zones of zero nucleation, probably if 2D nucleation is the rate-determining step [105]. Dendritic growth starts at higher overpotentials. It can be seen from Fig. 2.33b, c that dendritic growth initiation is mainly related to the appearance of twinned forms in which indestructible reentrant groove is formed, being the precursors of dendrite. Hence, a granular deposit will be formed at overpotentials lower than those for the formation of dendrite precursors and dendritic growth initiation.

Granular Growth of Lead

The typical Pb granules obtained in the potentiostatic regime of electrolysis are shown in Fig. 2.34. Similar to silver granules, granules of various shapes were produced by the electrodeposition processes. The granules, such as octahedrons and hexagons, as well as many various types of twinned particles: single-twinned, multiply-twinned (MTPs), lamellar-twinned (LTPs), and many other complicated

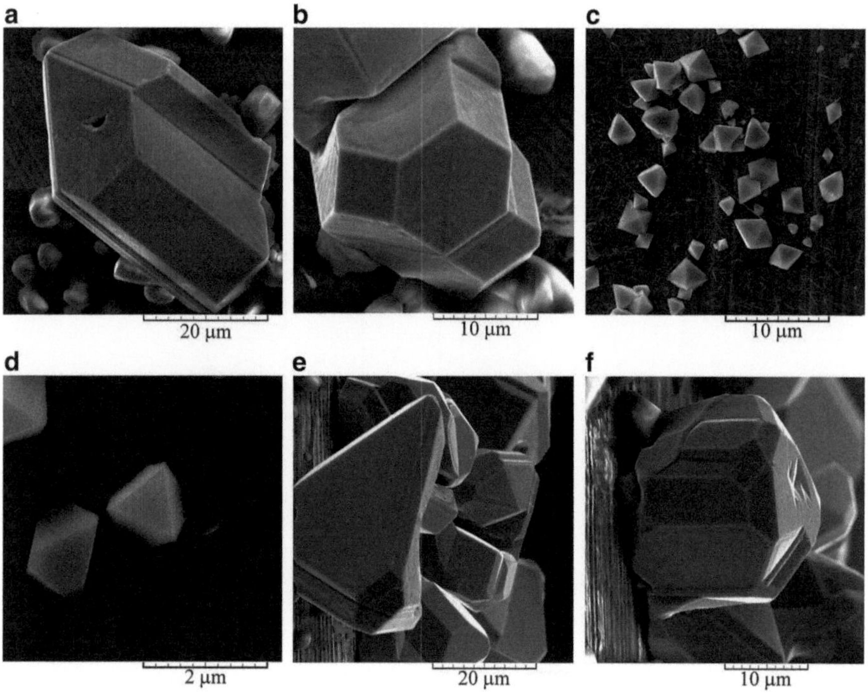

Fig. 2.34 Lead granules obtained by potentiostatic regime of electrolysis at an overpotential of 5 mV from: (**a–d**) 0.020 M $Pb(NO_3)_2$ in 2.0 M $NaNO_3$, and (**e** and **f**) 0.040 M $Pb(NO_3)_2$ in 2.0 M $NaNO_3$

shapes denoted as polyparticles, were synthesized through regulation of the parameters of electrolysis.

2.2.3.2 Mathematical Model and Polarization Curve Equation for the Metal Deposition on Partially Covered Inert Electrode

The ensemble of microelectrodes consists of microelectrodes placed on the inert surface at distances between their centers larger than their diameter. There are two conditions under which the particles of active metal placed on the surface of a macroelectrode can represent microelectrodes. The first condition is that the substrate is conducting, but inert relative to the process under consideration. The second one is that the grains are sufficiently small to permit activation control of the electrochemical process on them, making, thus, mixed overall control, as in the case of the tips of growing dendrites [3, 5, 20], or during the induction time of the formation of spongy deposits [54].

Assuming homogeneously distributed, equal to each other, hemispherical grains of the catalyst on an inert substrate, in a way similar to that used by Gileadi [10] to

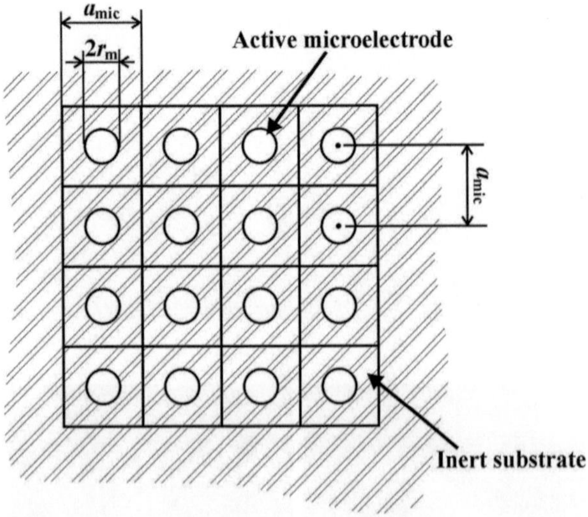

Fig. 2.35 A schematic presentation of the top view of the surface of an inert macroelectrode modified with microelectrodes of catalyst (Reprinted from Ref. [6] with permission from Elsevier and Ref. [1] with kind permission from Springer)

describe an ensemble of microelectrodes, the surface of the macroelectrode can be presented by an idealized model, as in Fig. 2.35, and the number of grains per cm^2 is given by:

$$N = \frac{1}{a_{mic}^2} \tag{2.109}$$

It is obvious from Fig. 2.35 that the edge size, a_{mic}, of a square of the surface of the inert macroelectrode belonging to one such particle is equal to the distance between the centers of the closest particles. If

$$a_{mic} = x \cdot 2r_m \tag{2.110}$$

where r_m is the radius of the microelectrode and x is the ratio of the distance between the centers of the particles and the particle diameter, Eq. (2.109), can be rewritten in the form:

$$N = \frac{1}{4r_m^2 \cdot x^2} \tag{2.111}$$

The surface of each hemispherical microelectrode, S_m, is:

$$S_m = 2r_m^2 \cdot \pi \tag{2.112}$$

and the working surface of catalyst per square centimeter of macroelectrode, S_w, is then:

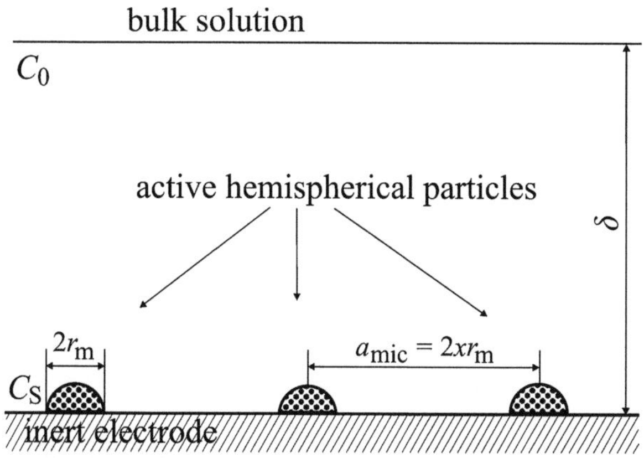

Fig. 2.36 Schematic presentation of the cross section of the diffusion layer of a partially covered inert electrode with hemispherical active particles, where r_m is the radius of the microelectrodes, δ is the diffusion layer thickness of the macroelectrode, C_0 and C_s are the bulk and the surface concentrations of reacting ions, respectively, x is the ratio of the distance between the centers of neighboring particles and the particle diameter, and $\delta >> r_m$ (Reprinted from Ref. [6] with permission from Elsevier and Ref. [1] with kind permission from Springer)

$$S_w = N \cdot S_m = \frac{\pi}{2x^2} \tag{2.113}$$

or

$$S_w = \frac{2r_m^2 \pi}{a_{mic}^2} \tag{2.114}$$

if Eq. (2.110) is taken into account, being dependent on the ratio r_m/a_{mic}.

A schematic presentation of the cross section of the diffusion layer of an inert macroelectrode partially covered with small active hemispherical particles is shown in Fig. 2.36, taking into account that $\delta >> r_m$.

A mathematical model can be derived under the assumption that the electrochemical process on the microelectrodes inside the diffusion layer of a partially covered inert macroelectrode is under activation control, despite the overall rate being controlled by the diffusion layer of the macroelectrode. The process on the microelectrodes decreases the concentration of the electrochemically active ions on the surfaces of the microelectrodes inside the diffusion layer of the macroelectrode, and the zones of decreased concentration around them overlap, giving way to linear mass transfer to an effectively planar surface [42]. Assuming that the surface concentration is the same on the total area of the electrode surface, under steady-state conditions, the current density on the whole electrode surface, i, is given by:

$$i = \frac{nFD\,(C_0 - C_s)}{\delta} \tag{2.115}$$

Obviously, the current density from Eq. (2.115) is due to the difference in the bulk, C_0, and surface concentration, C_s, of the reactive ions. Considering the concentration dependence of the exchange current density [37] (see Eqs. (1.16) and (1.17)), the current density on the macroelectrode can also be written as:

$$i = S_w i_0 \left(\frac{C_s}{C_0}\right)^{\zeta} (f_c - f_a) \tag{2.116}$$

assuming a reversible activation-controlled electrode process on the hemispherical active microelectrodes on an inert substrate, where S_w is the active surface per cm^2 of the macroelectrode and i_0 is the exchange current density on the massive active electrode, standardized to the apparent electrode surface.

The current densities given by Eqs. (2.115) and (2.116) are mutually equal and substitution of C_s/C_0 from Eq. (1.18) into Eq. (2.116) produces

$$i = S_w i_0 \left(1 - \frac{i}{i_L}\right)^{\zeta} (f_c - f_a) \tag{2.117}$$

or, for $\zeta = 1$, after rearranging

$$i = \frac{S_w i_0 (f_c - f_a)}{1 + \frac{S_w i_0 (f_c - f_a)}{i_L}} \tag{2.118}$$

where i_L is the limiting diffusion current density on the macroelectrode standardized to the apparent electrode surface.

It is obvious that

$$i_{0,\mathrm{eff}} = S_w i_0 \tag{2.119}$$

where $i_{0,\mathrm{eff}}$ is the effective value of the exchange current density relative to the total surface of a partially covered electrode.

If Eq. (2.120)

$$\frac{S_w i_0 (f_c - f_a)}{i_L} >> 1 \quad \text{and} \quad f_c > f_a \tag{2.120}$$

is valid even for small S_w and overpotentials, all the surface behaves as an active one if $i_0/i_L \to \infty$ meaning the ohmic control of electrodeposition.

Figure 2.37 shows the polarization curves for Ag electrodeposition from 0.50 M $AgNO_3$ in 0.20 M HNO_3 recorded on both a graphite electrode and graphite electrode previously plated with silver from the ammonium electrolyte (electrolyte containing 0.10 M $AgNO_3 + 0.50$ M $(NH_4)_2SO_4 + NH_4OH$ to dissolve the silver sulfate precipitate that was used for this purpose) [1, 6].

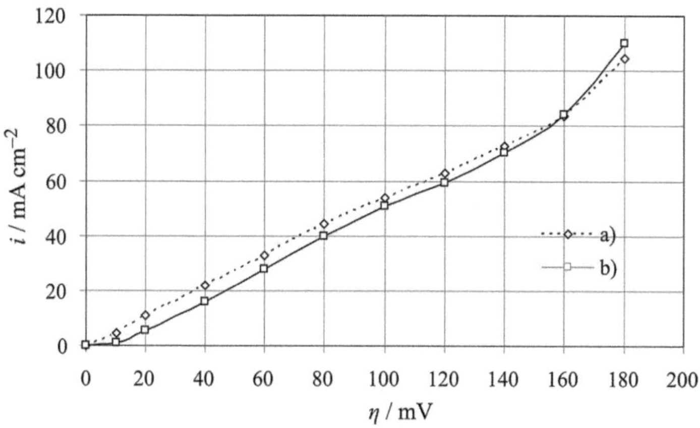

Fig. 2.37 Polarization curves for silver electrodeposition from the nitrate solution on (*a*) a graphite electrode previously plated with silver from the ammonium solution and (*b*) an uncovered graphite electrode (Reprinted from Ref. [1] with kind permission from Springer and Ref. [6] with permission from Elsevier)

From Fig. 2.37, it can be noticed that the polarization curves for deposition on the compact silver layer and on the uncovered graphite electrode are practically the same (please note that there is small difference in initial parts of the polarization curves due to the different nucleations on silver and graphite electrodes). In both cases, the ohmic-controlled deposition can be noticed up to about 100 mV. In both cases, an overpotential of 120 mV belongs to the region in which a slight decline in the slope of the polarization curve indicates an increased degree of diffusion control. At overpotentials larger than 140 mV, a strong increase in the current density with increasing overpotential occurs because of the initiation of dendritic growth [3, 5]. The polarization curves on platinum electrodes were very similar to those obtained on graphite ones [6].

The surface of completely covered graphite electrode by deposition from ammonium bath is shown in Fig. 2.38a, while the silver deposit obtained after polarization measurement up to an overpotential of 120 mV on an uncovered graphite electrode is shown in Fig. 2.38b. This electrode surface is partially covered because of the overlapping of the nucleation exclusion zones, being the ideal physical model of a partially covered inert electrode. The regular crystal form of the grains in Fig. 2.38b confirms that the deposition on the microelectrodes is not under diffusion control [16, 106], despite the overall deposition rate that is determined by diffusion to the macroelectrode.

The current density on the electrode from Fig. 2.38b, with coverage of about 20 %, is practically the same as on a completely covered graphite electrode, as can be seen from Fig. 2.37 at an overpotential of 120 mV. This is because the exchange current density for the silver electrodeposition process from nitrate baths is large [58, 107]. A similar situation appears in silver electrodeposition on platinum [6].

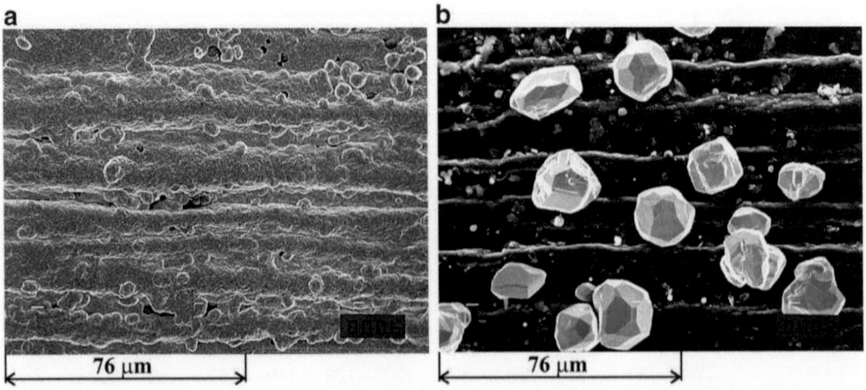

Fig. 2.38 The physical model of a partially covered inert electrode with active grains and a completely covered inert electrode: (**a**) a graphite electrode completely covered by deposition from the ammonium electrolyte; current density on the electrode completely covered with silver was 62.5 mA cm^{-2} at an overpotential of 120 mV in the nitrate electrolyte and (**b**) the silver deposit on the graphite electrode after the polarization measurements ended at an overpotential of 120 mV in the nitrate electrolyte; current density on such electrode was 59.4 mA cm^{-2} at the same overpotential in the nitrate electrolyte (Reprinted from Ref. [1] with kind permission from Springer and Ref. [6] with permission from Elsevier)

2.2.4 *Dendritic Growth Initiation Inside Diffusion Layer of the Macroelectrode in the Case of Very Fast Electrodeposition Processes*

The initiation of dendritic growth in the case of very fast electrodeposition processes also will be followed by an increase of the deposition current density, and the overall current density will be larger than the limiting diffusion current on a flat active electrode. Based on the above discussion, the polarization curve equation in the mixed ohmic diffusion-controlled electrodeposition of metals can be determined as [108]:

$$i = \frac{\kappa\eta}{L_c} \text{ for } 0 \le \eta < i_L \frac{L_c}{\kappa} = \eta_i \tag{2.121}$$

$$i = i_L \text{ for } \eta_i \le \eta < \eta_c \tag{2.122}$$

$$i = i_L\theta + (1-\theta)i_0 \frac{(f_c - f_a)}{N} \sum_{i=1}^{i=N} \left(\frac{h_{i,e}}{\delta}\right)^{\xi} \text{ for } \eta_c \le \eta \tag{2.123}$$

where $N = N(t)$ is the number of dendrites and $\theta = \theta(t) \le 1$, where θ is the part of the electrode surface uncovered by dendrites.

Equation (2.121) describes the linear part of the polarization curves for tin [109], silver [1, 6], and lead [34, 35, 110, 111] deposition, and Eq. (2.122) foresees the inflection point in the cases when η_c is low and the resistance of electrolyte is large. Finally, Eq. (2.123) describes the part of the polarization curve after the inflection point characterized by the fast increase of current density with increasing the overpotential (see Chap. 1; Sect. 1.3.1.1).

In the case of the fast electrochemical processes, the diffusion control follows after the linear dependence of the current density on the overpotential, *i.e.*, after the ohmic part at the polarization curve. The overpotential which initiates dendritic growth is denoted as the critical overpotential for dendritic growth initiation, η_i, and it is overpotential at which the system enters diffusion control of the electrodeposition. The critical overpotential for instantaneous dendritic growth, η_c, is overpotential at which diffusion control becomes complete, and this overpotential corresponds to the inflection point at the polarization curve [111]. After the inflection point, the electrodeposition systems remain diffusion controlled ones, and the rapid increase of the current density after the inflection point is result of the fast growth of dendrites and strong increase of the surface area. Then, using the electrochemical definition of a dendrite, it follows that this sudden and rapid increase of the current density can be mainly ascribed to the activation-controlled electrodeposition at the tips of the formed dendrites. Hence, for this part of the polarization curve, it can be concluded that there is elimination of mass transfer limitations due to instantaneous dendritic growth and the activation control at the tips of growing dendrites.

The polarization curve for Ag electrodeposition from 0.50 M $AgNO_3$ in 0.20 M HNO_3, onto a graphite electrode, is shown in Fig. 2.39. The polarization curve in Fig. 2.39 means that mass transfer limitations were decreased or even eliminated. The SEM photomicrographs of the deposit corresponding to the points from Fig. 2.39 are shown in Fig. 2.40.

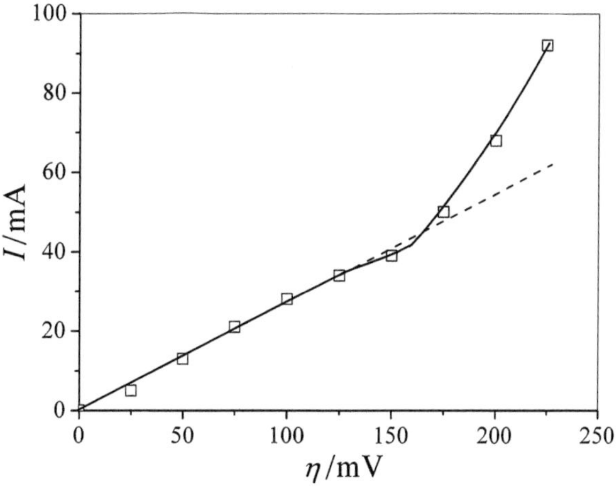

Fig. 2.39 The polarization curve for silver electrodeposition from nitrate solution on a graphite electrode (Reprinted from Ref. [108] with permission from Elsevier and Ref. [1] with kind permission from Springer)

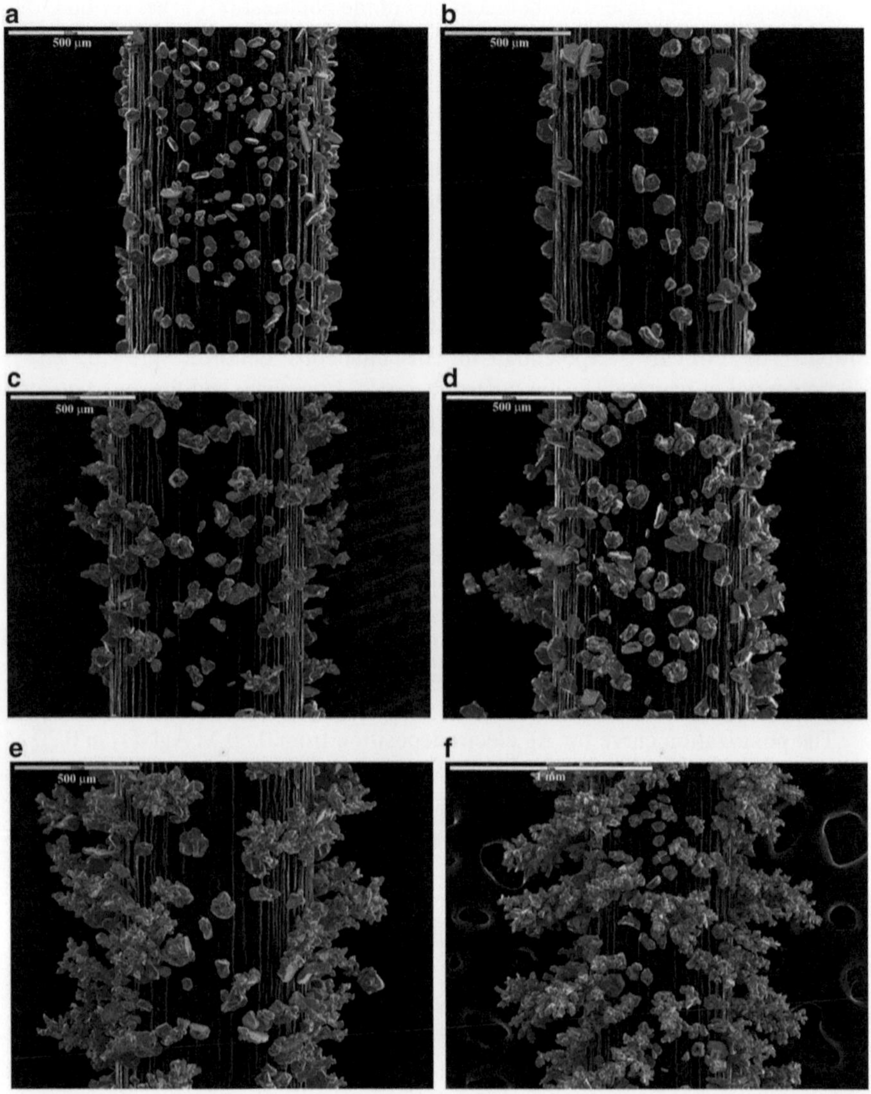

Fig. 2.40 The SEM photomicrographs of the silver deposit obtained on a graphite electrode obtained after the recording of the current at different overpotentials in polarization measurements: (**a**) 100 mV, (**b**) 125 mV, (**c**) 150 mV, (**d**) 175 mV, (**e**) 200 mV, and (**f**) 225 mV (Reprinted from Ref. [108] with permission from Elsevier and Ref. [1] with kind permission from Springer)

It can be seen from Figs. 2.39 and 2.40a that only grains are obtained at an overpotential of 100 mV indicating that electrodeposition process was not under diffusion control [108]. From Figs. 2.39 and 2.40b, it follows that electrodeposition at 125 mV is still out of the diffusion control. At 150 mV, the current density is somewhat lower than that which could be expected from the linear dependence of the current on the overpotential. This indicates on the initiation of diffusion control of the deposition process, but also to the initiation of dendrite growth, which compensates the mass transfer limitations (Figs. 2.39 and 2.40c). The point corresponding to an overpotential of 150 mV can be considered as the inflection point at the polarization curve in Fig. 2.39. At overpotentials larger than 175 mV, the current density is considerably larger than the one expected from the linear dependence of current on overpotential. The formation of dendritic deposits (Fig. 2.40d–f) confirms that the electrodeposition process was dominantly under activation control at the tips of dendrites. It is necessary to note that the silver deposits shown in Fig. 2.40d–f are not similar to ideal silver dendrites [26], but they behave as dendritic ones in regard to their electrochemical properties. Hence, they can be considered as degenerate dendritic deposits. Occasionally, the needle-like dendrites can also be formed.

For metals characterized by $i_0 < i_L$ (electrodeposition under mixed activation–diffusion control; *e.g.*, Cu), both η_i and η_c increase with increasing concentration of the depositing ions, indicating a decrease of the i_0/i_L ratios with the increasing concentration of metal ions [111]. The difference between η_c and η_i (see Eqs. (2.46) and (2.47)) is given by:

$$\eta_c - \eta_i = \frac{\zeta b_c}{2.3} \ln \frac{\delta}{h_p} \tag{2.124}$$

It is very clear from Eq. (2.124) that this difference does not depend on the i_L/i_0 ratio and, hence, on the concentration of the depositing ions. This difference will always have a finite value which will never tend to be zero.

For metals characterized by $i_0 >> j_L$ (electrodeposition in mixed ohmic–diffusion control of the electrodeposition; *e.g.*, Pb and Ag), increasing concentration of metal ions causes a decrease in both η_i and η_c [111]. Simultaneously, opposite to electrodeposition of metals in mixed activation–diffusion control, increasing the concentration of depositing ions leads to a strong increase in the i_0/i_L ratio.

For this group of metals (see Eqs. (2.48) and (2.49)), the difference between η_c and η_i is given by Eq. (2.125):

$$\eta_c - \eta_i = \frac{RT}{nF}\frac{i_L}{i_0}\left[\left(\frac{\delta}{h_p}\right)^{\zeta} - 1\right] \tag{2.125}$$

This difference decreases with the decreasing i_L/i_0 value (or increasing i_0/i_L value) and becomes negligible at the sufficiently large concentrations of depositing ions with the increasing concentration. Due to the decrease i_L/i_0 ratio with increasing concentration of depositing ions and at the sufficiently high concentration of depositing ions, this ratio will tend to be zero ($i_L/i_0 \rightarrow 0$). Then, the right side of Eq. (2.125) will also tend to be zero and hence $\eta_c - \eta_i \rightarrow 0$ (or $\eta_c \approx \eta_i$), which corresponds to full ohmic control of the electrodeposition process (see Chap. 1; Sect. 1.3.1.1). In this way, the limiting diffusion current density plateau transforms into the inflection point.

2.2.5 Spongy Deposit Formation

2.2.5.1 Low Overpotentials

Mathematical Model

In systems characterized with low exchange current density values, electrodeposition process enters the full diffusion control at sufficiently large overpotentials [13]. On the other hand, if condition $0.1 < i_0/i_L \leq 1$ is fulfilled, deposition will be under complete diffusion control at all overpotentials if some other kind of control does not take place (*e.g.*, for silver deposition on a well-defined silver crystal grains at a silver electrode at low overpotentials, two-dimensional nucleation is the rate-determining step [105]).

At low overpotentials, a small number of nuclei are formed, and they can grow independently. The limiting diffusion current density to the growing nucleus $i_{L,N}$ is given by:

$$i_{L,N} = \frac{nFDC_0}{r_N} \tag{2.126}$$

or

$$i_{L,N} = \frac{i_L \delta}{r_N} \tag{2.127}$$

where r_N is the tip radius of the nucleus. Hence, if $r_N \rightarrow 0$, condition $0.1 < i_0/i_L \leq 1$ is not satisfied and electrodeposition is under activation or mixed control. Pure activation-controlled deposition is, thus, possible even at $i_0 >> i_L$ on very small electrodes such as nuclei on an inert substrate.

An increase in r_N leads to a decrease of $i_{L,N}$, and, at sufficiently large r_N, the electrodeposition comes under mixed activation–diffusion control, *i.e.*, when:

$$r_N > r_{mix} \tag{2.128}$$

where r_{mix} is the radius of a growing nucleus where the process comes under mixed control [7, 54].

Under mixed control of the deposition, amplification of the surface irregularities on the growing nucleus occurs, leading to the formation of a spherical agglomerate of filaments. Thereby a spongy deposit is formed. The above reasoning is valid if spherical diffusion control can occur around growing grains, as in the case of cauliflower-like deposit growth. Assuming that around each grain with radius r_N, growing under spherical diffusion control, a diffusion layer of the same thickness is formed, then the initiation of spongy growth is possible if the number of nuclei per cm^2, N, satisfies the condition:

$$N \leq \frac{1}{(4r_{\text{mix}})^2} \tag{2.129}$$

On the basis of all the above facts, it can be concluded that the formation of a spongy deposit on an inert substrate may be caused by mass-transport limitations when the nucleation rate is low. Hence, suitable conditions for the formation of spongy deposits arise at low overpotentials in systems where $i_L < i_0$ [7, 54].

Physical Model

As was mentioned before, at a fixed value of the overpotential, the growth of a spongy deposit is possible if condition given by Eq. (2.128) is fulfilled.

The situation in which spongy deposits can start to grow can easily be demonstrated [112]. Grains of the desired size and distribution can be grown at low overpotentials under conditions of activation-controlled deposition. This corresponds to growth of grains when $r < r_{\text{mix}}$. The situation in which $r > r_{\text{mix}}$ can be simulated by increasing the overpotential to a sufficiently high value to result in diffusion control around the growing grains and the amplification of surface irregularities. With increasing overpotential, r_{mix} decreases. This permits the simulation of the initial stage of spongy growth, as illustrated in Fig. 2.41a, b.

Spongy growth can start on the growing nucleus if the conditions given by Eqs. (2.128) and (2.129) are both satisfied simultaneously.

In the first stage of deposition, the formation of nuclei having a regular crystal shape can be expected because the deposition is activation controlled. After r_{mix} is reached, the system comes under mixed control, producing polycrystalline grains like those shown in Fig. 2.41c, just as in the case of mixed control of copper deposition (see Fig. 2.4c) [12]. In this situation, amplification of the surface irregularities on the growing grains occurs, and spongy growth is initiated.

An ideal spongy nucleus obtained in a real system is shown in Fig. 2.41c which illustrates the above discussion and physical simulation well [113]. The agglomerate of filaments in Fig. 2.42b is obviously formed by further growth of nuclei like that in Fig. 2.41d. Hence, it can be concluded that at low overpotentials the initiation of spongy growth is due to the amplification of surface protrusions directly inside the spherical diffusion layer formed around each independently growing grain, as in the case of the formation of cauliflower-like deposits. The growth of protrusions in all directions is good proof that the initial stage of deposition on the grain is under spherical diffusion control, while further growth takes place in the diffusion layer of the macroelectrode. In less ideal situations, non-ideal spongy nuclei are formed, which, however, after further deposition, results in a macroelectrode with the same appearance.

The growth of protrusions in all directions is a good proof that the deposition on the grain is under spherical diffusion control. At longer deposition times, the protrusions branch and interweave causing the macroelectrode to have a spongy appearance.

Fig. 2.41 Cadmium deposits obtained from (**a**) 1.0 M $CdSO_4$ in 0.50 M H_2SO_4 at 12 mV; Deposition time: 15 min, (**b**) 0.10 M $CdSO_4$ in 0.50 M H_2SO_4 at 120 mV. Deposition time: 45 s, and Zinc deposits obtained by deposition at 35 mV from 0.10 M zincate solution in 1.0 M KOH. Deposition time: (**c**) 7 min and (**d**) 15 min (Reprinted from Refs. [7, 13, 112] with kind permission from Springer and Ref. [113] with permission from the Serbian Chemical Society)

Real Systems

Typical spongy electrodeposits are formed during zinc and cadmium electrodeposition at low overpotentials [7, 54]. Scanning electron microscopy images of zinc deposited at an overpotential of 20 mV onto a Cu electrode from an alkaline zincate solution are shown in Fig. 2.42. The increase in the number of nuclei formed with increasing deposition time can be seen in Fig. 2.42a, b, and a spongy deposit is formed as can be seen in Fig. 2.42b. The spongy growth takes place on a relatively small number of nuclei, as is shown in Fig. 2.42b, c.

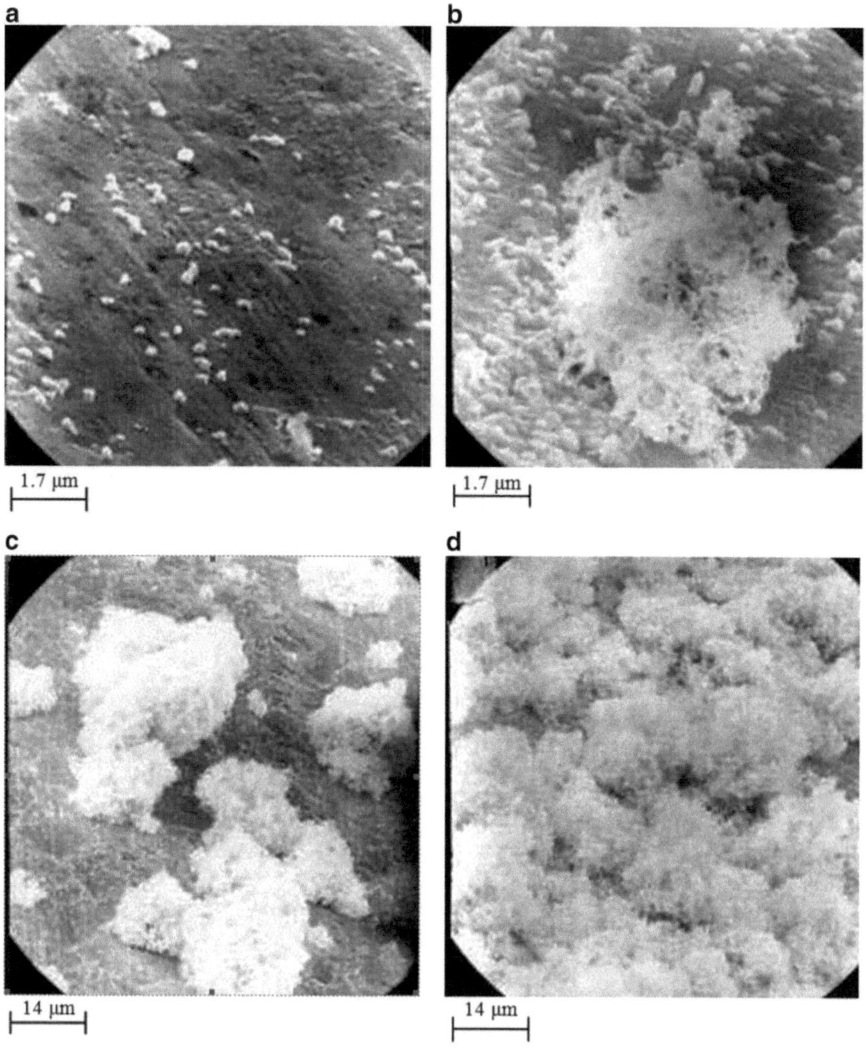

Fig. 2.42 Zinc deposits obtained by electrodeposition at 20 mV from 0.10 M zincate and 1.0 M KOH. Deposition time: (**a**) 10 min, (**b**) 20 min, (**c**) 30 min, and (**d**) 60 min (Reprinted from Refs. [7, 13, 54] with kind permission from Springer)

The initiation of spongy growth at a fixed overpotential is possible if the condition $r_N > r_{mix}$ (Eq. (2.128)) is satisfied, which is the case after some time. On the other hand, increasing the deposition time leads to the formation of a larger number of nuclei, and so the condition given by Eq. (2.129) is not satisfied over a large part of the electrode surface. Regardless of this, the coverage of the electrode surface by spongy deposits increases with increasing deposition time up to full coverage, as can be seen in Fig. 2.42d.

It should be noted that some other possible mechanisms of spongy deposit formation have been considered in a qualitative way, as reviewed in Refs. [114, 115], but the mechanism presented above seems to be the most probable [115]. However, the mechanism of formation of a spongy deposit over an initial coating, which is not seen in the case of cadmium but occurs in zinc deposition, requires clarification [54, 113]. For instance, the mechanism of spongy growth initiation in this case has not been elucidated.

2.2.5.2 Large Overpotentials

Mathematical Model

As already told, the limiting diffusion current to the growing nucleus, $i_{L,N}$, is related to i_L by Eq. (2.127).

Equation (2.4) for the growing nucleus can be rewritten in the form:

$$i_N = \frac{i_0 f_c}{1 + \frac{i_0 f_c}{i_L} \frac{r_N}{\delta}} \tag{2.130}$$

where i_N is the current density to the growing nucleus. It is obvious from Eqs. (2.4) and (2.130) that deposition process on the macroelectrode can be under complete diffusion control if the condition given by Eq. (2.3) is fulfilled.

At the same overpotential, process on the growing nucleus can be under pure activation control if conditions

$$\frac{i_0 f_c}{i_L} \frac{r_N}{\delta} << 1 \tag{2.131}$$

or

$$\frac{r_N}{\delta} \to 0 \tag{2.132}$$

are satisfied [116].

Equations (2.131) and (2.132) are fulfilled in the initial stage of electrodeposition to the nuclei of metal formed on the inert substrate [112]. In this case, the nuclei behave as microelectrodes, because of their complete independent growth well before the formation of the diffusion layer of the macroelectrode. The radius r_0 of the initial stable nucleus, at overpotential η, is given by [117]:

$$r_0 = \frac{2\gamma V}{nF} \tag{2.133}$$

where γ is the interfacial energy between metal and solution. The radius of the growing nucleus will vary with the time according to [101]:

$$r_{\mathrm{N}} = r_0 + \frac{V}{nF} i_0 f_{\mathrm{c}} t \tag{2.134}$$

or

$$r_{\mathrm{N}} \approx \frac{V}{nF} i_0 f_{\mathrm{c}} t \tag{2.135}$$

because r_0 is extremely low.

Obviously, Eqs. (2.130), (2.131), (2.134), and (2.135) are only the approximation on growth times, because the effect of surface energy has not been taken into consideration. At larger deposition times, they are valid, because the surface energy term at higher value can be neglected [5].

An increase in r_{N} leads to a decrease of $i_{\mathrm{L,N}}$, and at sufficiently large r_{N} deposition comes under mixed activation–diffusion control. It can be assumed that this happens at

$$i_{\mathrm{N}} > \upsilon\, i_{\mathrm{L,N}} \tag{2.136}$$

where $0 < \upsilon < 1$. By combining Eqs. (2.127), (2.130), and (2.136) one obtain:

$$r_{\mathrm{mix}} \approx \frac{i_{\mathrm{L}} \upsilon \delta}{i_0 f_{\mathrm{c}} (1 - \upsilon)} \tag{2.137}$$

where r_{mix} is the radius of the growing nucleus when the process comes under mixed or spherical diffusion control. According to Barton and Bockris [5], the diffusion layer around such grain forms very fast. The further combination of Eqs. (2.135) and (2.137) gives the corresponding induction time, t_{i}, given by:

$$t_{\mathrm{i}} = \frac{i_{\mathrm{L}}}{i_0^2} \frac{\upsilon nF\delta}{Vf_{\mathrm{c}}^2 (1 - \upsilon)} \tag{2.138}$$

For $\upsilon = 0.2$, Eqs. (2.137) and (2.138) can be rewritten in the forms [118]

$$r_{\mathrm{mix}} \approx \frac{i_{\mathrm{L}}}{i_0} \frac{\delta}{4 f_c} \tag{2.139}$$

and

$$t_{\mathrm{i}} = \frac{i_{\mathrm{L}}}{i_0^2} \frac{nF\delta}{4Vf_{\mathrm{c}}^2} \tag{2.140}$$

At $r_N < r_{mix}$ and $t < t_i$ the deposition on the growing grain is under activation control.

Hence, if $r_N > r_{mix}$, the spherical diffusion layer around microelectrode can be formed. This is the condition for deposition in spherical diffusion control.

The nucleus of spongy deposit, *i.e.*, hedgehog-like particle, appears when amplification of surface coarseness on the nucleus in spherical diffusion control growing starts. It was shown earlier [118] that this amplification is very fast so the induction time when growing nucleus enters mixed control can be taken also as induction time of spongy formation. It follows from Eqs. (2.139) and (2.140) that r_{mix} and t_i decrease with increasing overpotential.

On the other hand, it was also shown [54] that spongy deposit can be formed only if around each grain with radius r_{mix}, growing under spherical diffusion control, a diffusion layer of the same thickness is formed, as illustrated earlier (Eq. (2.129)).

Hence, deposition in spherical diffusion control on the growing grain is possible if both Eqs. (2.139) and (2.129) are satisfied in the same time. The nucleation law is already given by Eqs. (2.87) and (2.88).

Then, spongy deposits formation is possible if

$$N_0[1 - \exp(-At_i)] \leq \frac{1}{(4r_{mix})^2} \tag{2.141}$$

and $At_i \approx 0$ which happens at sufficiently high overpotentials where $K_2/\eta^2 \to 0$, $A \to K_1 i_0$ and $t_i \to 0$. Hence, the spongy deposit formation at high overpotentials starts at very low deposition times, when the spherical diffusion layer formed around grains does not overlap. The critical overpotential of spongy formation can be obtained by substitution of r_{mix} from Eq. (2.139) and t_i from Eq. (2.140) in Eq. (2.141) and further calculation if this overpotential is larger than critical one for instantaneous dendritic growth the dendrite spongy nuclei can be formed over inert substrate.

Real Systems

The experimental verification of the above discussion is given by the consideration of the morphology of electrodeposited silver from 0.50 M $AgNO_3$ in 0.20 M HNO_3 on the graphite electrode at different overpotentials of deposition and with different deposition times [116]. In Fig. 2.43a, the deposit obtained at an overpotential of 100 mV during 180 s is shown.

As expected, the large irregular grains are obtained. In Fig. 2.43b, c, the deposits obtained at 200 mV during 1 and 10 s are presented. At 1 s, the grains are formed, but at 10 s the needle-like deposit is obtained. This means that the spherical diffusion layer around the growing grains is not formed before the formation of the diffusion layer of the macroelectrode. The electrodeposition inside the diffusion layer of the macroelectrode is confirmed by the growth of needles toward the bulk of solution.

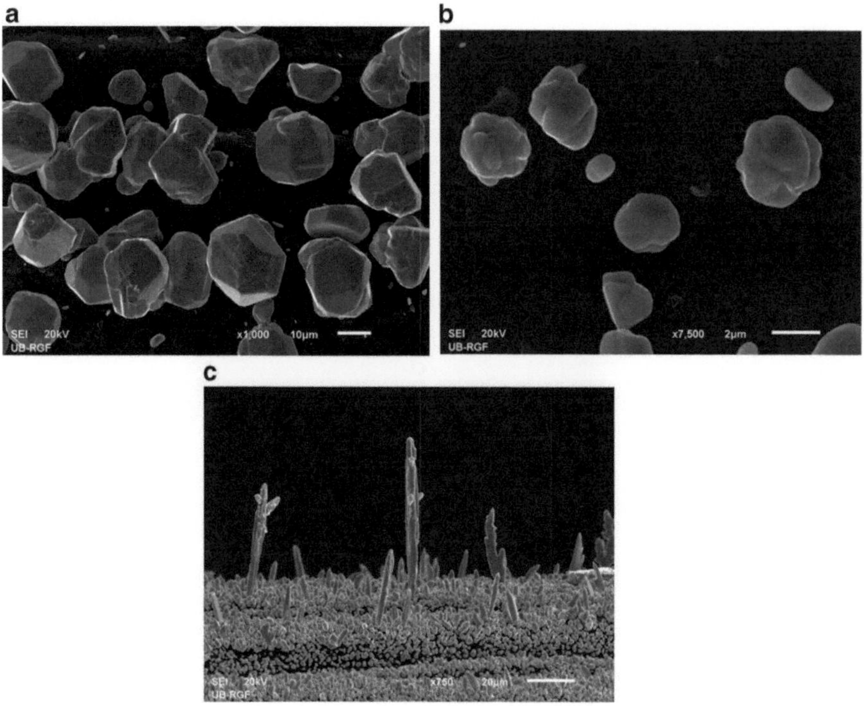

Fig. 2.43 Silver deposit obtained from 0.50 M $AgNO_3$ in 0.20 M HNO_3 at overpotentials of (**a**) 100 mV; time: 180 s, (**b**) 200 mV; time: 1 s, and (**c**) 200 mV; time: 10 s (Reprinted from Ref. [116] with permission from ESG)

At an overpotential of 300 mV, the conditions of the spherical diffusion control around the growing grains are fulfilled and dendritic-spongy deposit is formed, as can be seen from Fig. 2.44.

The growth of dendrites in all directions means that there is a spherical control to the growing grains in the initial stage of the electrodeposition.

Finally, the fact that r_{mix} and t_{i} (Eqs. (2.139) and (2.140)) decrease with the increasing overpotential can be verified by Figs. 2.44 and 2.45.

2.2.6 *Whisker Deposits*

This form of crystal growth differs from that of dendrites in that (a) it tends to have a still larger ratio between the longitudinal and the lateral dimensions with an almost perfect preservation of the latter during growth, and (b) it exhibits no tendency to side branching. Impurities or additives in the electrolyte seem to be a prerequisite for its appearance [11, 119].

Gorbunova et al. [120, 121] grew silver whiskers from fairly concentrated silver nitrate solutions (>0.30 M) containing oleic acid, gelatin, albumin, heptyl, octyl, and nonyl alcohols.

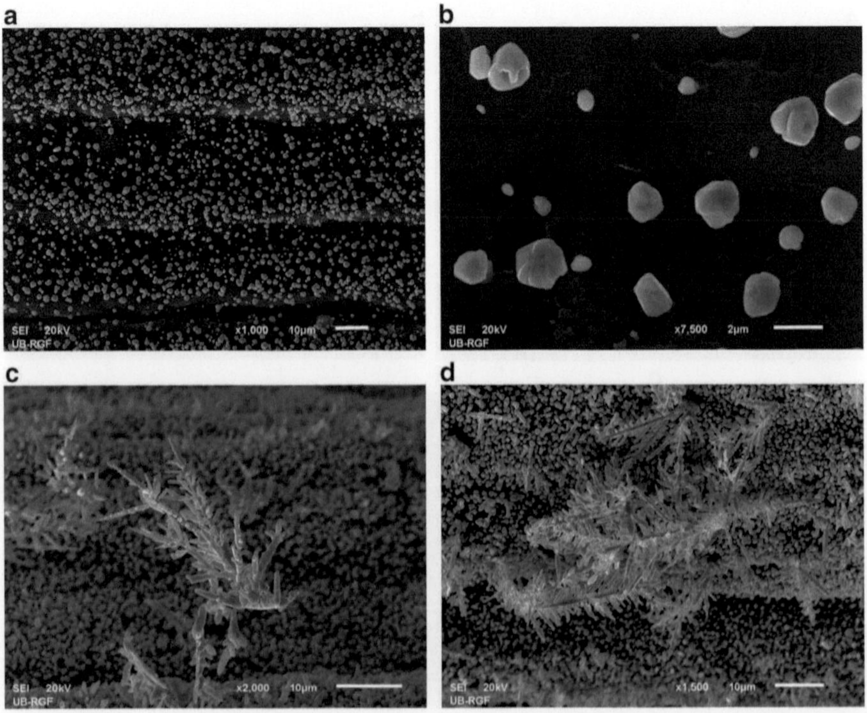

Fig. 2.44 Silver deposits obtained at an overpotential of 300 mV with a time of (**a**) 1 s, (**b**) 1 s, (**c**) 3 s, and (**d**) 5 s (Reprinted from Ref. [116] with permission from ESG)

A few more phenomena should be noted: (a) while growing exclusively in one direction only, whiskers dissolve anodically at a practically uniform rate from all sides [119] and at an overpotential much smaller than that needed for growth; (b) a higher overpotential is needed temporarily for the initiation of growth (or continuation after interruption) than for growth at a steady rate; (c) if the growth is interrupted for a longer period of time, then it may continue at the tip, but usually assuming a new direction, or else it may be completely prevented and a new whisker started elsewhere. The minimum time required for complete cessation of further growth was found to depend on the concentration of the additive; (d) if a constant rate of growth is maintained, by a constant current flow through the cell to the individual whisker tip, fluctuations of overpotential are observed.

Finally, it should be noted that whiskers differ from other crystals of the same metal in two respects at least: they have an increased electrical resistivity (two to three times that of crystals deposited in the absence of additives) and an increased tensile strength compared to large, pure silver single crystals [11].

A model of the growth mechanism was developed by Price et al. [119] which gives a good account of most of the phenomena observed. The basic assumption of the model is that molecules of impurities or additives are strongly adsorbed at all but one crystal plane and at such a concentration as to completely block the

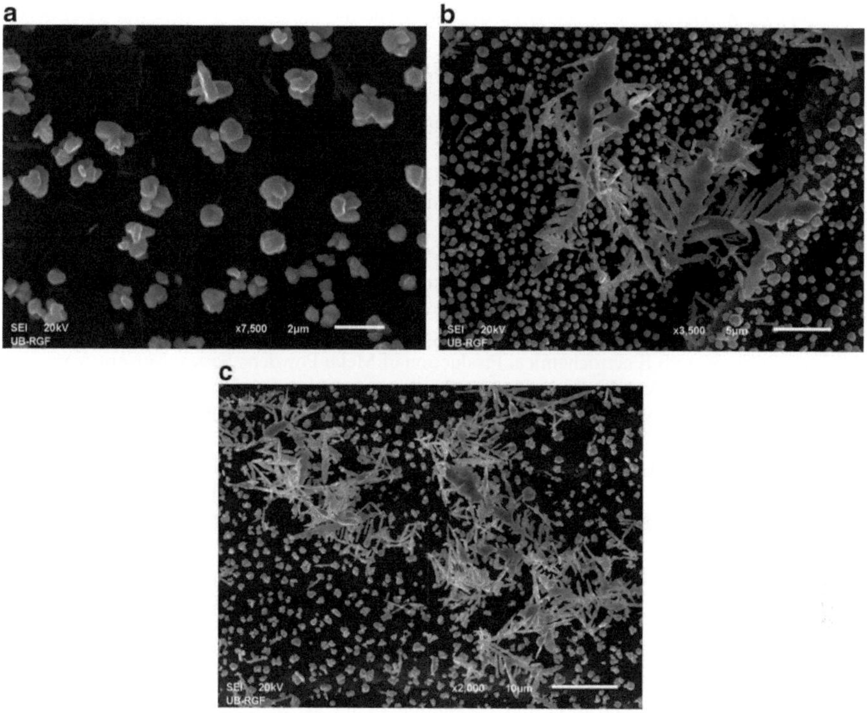

Fig. 2.45 Silver deposits obtained at an overpotential of 700 mV with a time of (**a**) 0.15 s, (**b**) 0.30 s, and (**c**) 0.50 s (Reprinted from Ref. [116] with permission from ESG)

deposition and extension of the lattice. On the one plane, however, the process of adsorption is competitive with that of metal deposition whereby the adsorbed molecules are buried, and, at a steady state, a sufficiently low surface coverage of foreign molecules is maintained for growth to be possible. The latter is assumed to occur by continuous nucleation and movement of steps over the close-packed surface. Indeed, the appearance of some whiskers suggests repeated one-dimensional nucleation of the type shown in Fig. 2.8a and the extension of the step in two directions to the edge of the crystal.

References

1. Popov KI, Živković PM, Nikolić ND (2010) The effect of morphology of activated electrodes on their electrochemical activity. In: Djokić SS (ed) Electrodeposition: theory and practice, vol 48, Series: modern aspects of electrochemistry. Springer, New York, pp 163–213
2. Bockris JO'M, Reddy AKN, Gamboa–Aldeco M (2000) Modern electrochemistry 2A, fundamentals of electrodics, 2nd edn. Kluwer Academic/Plenum Publishers, New York, p 1248

3. Diggle JW, Despić AR, Bockris JO'M (1969) The mechanism of the dendritic electrocrystallization of zinc. J Electrochem Soc 116:1503–1514
4. Živković PM, Grgur BN, Popov KI (2008) The validity of the general polarization curve equation approximation for the metal deposition process. J Serb Chem Soc 73:227–231
5. Barton JL, Bockris JO'M (1962) The electrolytic growth of dendrites from ionic solutions. Proc Roy Soc A268:485–505
6. Popov KI, Živković PM, Grgur BN (2007) Physical and mathematical models of an inert macroelectrode modified with active hemispherical microelectrodes. Electrochim Acta 52:4696–4707
7. Popov KI, Krstajić NV, Čekerevac MI (1996) The mechanism of formation of coarse and disperse electrodeposits. In: White RE, Conway BE, Bockris JO'M (eds) Modern aspects of electrochemistry, vol 30. Plenum Press, New York, pp 261–311
8. Popov KI, Nikolić ND (2012) General Theory of Disperse Metal Electrodeposits Formation. In: Djokić SS (ed) Electrochemical Production of Metal Powders. Series: Modern Aspects of Electrochemistry, vol 54. Springer, New York, pp 1–62
9. Despić AR, Diggle JW, Bockris JO'M (1968) Mechanism of formation of zinc dendrites. J Electrochem Soc 115:507–508
10. Gilleadi E (1993) Electrode kinetics. VCH Publishers Inc, New York, p 443
11. Despić AR, Popov KI (1972) Transport controlled deposition and dissolution of metals. In: Conway BE, Bockris JO'M (eds) Modern aspects of electrochemistry, vol 7. Plenum Press, New York, pp 199–313
12. Popov KI, Grgur BN, Pavlović MG, Radmilović V (1993) The morphology of copper electrodeposits: I. The mechanism of copper cauliflower-like electrodeposit formation. J Serb Chem Soc 58:1055–1062
13. Popov KI, Djokić SS, Grgur BN (2002) Fundamental aspects of electrometallurgy. Kluwer Academic/Plenum Publishers, New York
14. Popov KI (1971) Deposition and dissolution of metals in diffusion control conditions. University of Belgrade, Dissertation (in Serbian)
15. Popov KI, Pavlović LjJ, Pavlović MG, Čekerevac MI (1988) Electrode surface coarsening in potentiostatic copper electrodeposition. Surf Coat Technol 35:39–45
16. Popov KI, Pavlović MG, Pavlović LjJ, Čekerevac MI, Remović GŽ (1988) Electrode surface coarsening in pulsating overpotential copper electrodeposition. Surf Coat Technol 34:355–363
17. Popov KI, Maksimović MD, Pavlović MG, Lukić DT (1980) The mechanism of copper powder formation in potentiostatic deposition. J Appl Electrochem 10:299–308
18. Damjanović A (1965) On the mechanism of metal electrocrystallization. Plating 52:1017–1026
19. Popov KI, Radmilović V, Grgur BN, Pavlović MG (1994) The morphology of copper electrodeposits. III. The disperse deposits formation. J Serb Chem Soc 59:119–125
20. Popov KI, Radmilović V, Grgur BN, Pavlović MG (1994) The morphology of copper electrodeposits. II. The mechanism of carrot-like electrodeposit formations. J Serb Chem Soc 59:47–52
21. Pangarov NA (1964) On the crystal orientation of electrodeposited metals. Electrochim Acta 9:721–726
22. Pangarov NA, Vitkova SD (1966) Preferred orientation of electrodeposited cobalt crystallites. Electrochim Acta 11:1733–1745
23. Maksimović MD, Popov KI, Jović LjJ, Pavlović MG (1979) Mechanism of electrode surface roughening in galvanostatic metal deposition at natural convection. Bull Soc Chim 44:547–554
24. Maksimović MD, Popov KI, Pavlović MG (1979) Roughening in galvanostatic metal deposition on a rotating disc electrode. Bull Soc Chim 44:687–693

25. Popov KI, Maksimović MD (1989) Theory of the effect of electrodeposition at a periodically changing rate on the morphology of metal deposits. In: Conway BE, Bockris JO'M, White RE (eds) Modern aspects of electrochemistry, vol 19. Plenum Press, New York, pp 193–250
26. Wranglen G (1960) Dendrites and growth layers in the electrocrystallization of metals. Electrochim Acta 2:130–146
27. Bechtoldt CJ, Ogburn F, Smith J (1968) Structure and morphology of electrodeposited molybdenum dendrites. J Electrochem Soc 115:813–816
28. Faust JW, John HF (1963) Growth twins in F.C.C metals. J Electrochem Soc 110:463–464
29. Faust JW, John HF (1961) Germanium dendritic studies. I. Studies of thin structures and the seeding mechanism. J Electrochem Soc 108:855–860
30. Smith J, Ogburn F, Bechtoldt CJ (1968) Multiple twin structures in electrodeposited silver dendrites. J Electrochem Soc 115:371–374
31. Justinijanović IN, Despić AR (1973) Some observation on the properties of zinc electrodeposited from alkaline zincate solutions. Electrochim Acta 18:709–712
32. Bockris JO'M, Razumney GA (1967) Fundamental aspects of electrocrystallization. Plenum Press, New York
33. Popov KI, Čekerevac MI (1989) Dendritic electrocrystallization of cadmium from acid sulphate solution II: the effect of the geometry of dendrite precursors on the shape of dendrites. Surf Coat Technol 37:435–440
34. Nikolić ND, Popov KI, Ivanović ER, Branković G, Stevanović SI, Živković PM (2015) The potentiostatic current transients and the role of local diffusion fields in formation of the 2D lead dendrites from the concentrated electrolyte. J Electroanal Chem 739:137–148
35. Nikolić ND, Popov KI (2014) A new approach to the understanding of the mechanism of lead electrodeposition. In: Djokić SS (ed) Electrodeposition and surface finishing, vol 57, Series: modern aspects of electrochemistry. Springer, New York, pp 85–132
36. Nikolić ND, Popov KI, Ivanović ER, Branković G (2014) Effect of orientation of initially formed grains on the final morphology of electrodeposited lead. J Serb Chem Soc 79:993–1005
37. Newman JS (1973) Electrochemical systems. Prentice-Hall Inc, Engelwood Cliffs, p 177
38. Popov KI, Maksimović MD, Trnjavčev JD, Pavlović MG (1981) Dendritic electrocrystallization and the mechanism of powder formation in the potentiostatic electrodeposition of metals. J Appl Electrochem 11:239–246
39. Popov KI, Krstajić NV, Pantelić RM, Popov SR (1985) Dendritic electrocrystallisation of lead from lead nitrate solution. Surf Technol 26:177–183
40. Popov KI, Pavlović MG, Jovićević JN (1989) Morphology of tin powder particles obtained in electrodeposition on copper cathode by constant and square – wave pulsating overpotential from Sn(II) alkaline solution. Hydrometallurgy 23:127–137
41. Epstein IM (1966) Priblizhennaya formula polubeskonechnoi cilindricheskoi diffuzii. Elektrokhimiya 2:734–736 (in Russian)
42. Scharifker B, Hills G (1983) Theoretical and experimental studies of multiple nucleation. Electrochim Acta 28:879–889
43. Popov KI, Pavlović MG (1993) Electrodeposition of metal powders with controlled grain size and morphology. In: White RE, Bockris JO'M, Conway BE (eds) Modern aspects of electrochemistry, vol 24. Plenum Press, New York, pp 299–391
44. Calusaru A (1979) Electrodeposition of metal powders. Material science monography, vol 3. Elsevier, Amsterdam
45. Ibl N (1962) The formation of powdered metal deposits. In: Delahay P, Tobias CW (eds) Advances in electrochemistry and electrochemical engineering, vol 2. Intersience, New York
46. Pangarov NA (1967) Twining processes in the electrocrystallization of face-centred cubic metals. Phys Stat Sol 20:371–377
47. Gutzov I (1964) Kinetics of electrolytic phase-formation under galvanostatic conditions. Izv Inst Fiz Chim Bulgar Acad Nauk 4:69–88 (in Bulgarian)

48. Klapka V (1970) To the problem of crystallization overvoltage during electrocrystallization of metals. Coll Czechoslov Chem Comun 35:899–906
49. Pangarov NA, Velinov V (1966) The orientation of silver nuclei on a platinum substrate. Electrochim Acta 11:1753–1758
50. Pangarov NA, Vitkova SD, Uzunova I (1966) Electronographic investigation of the degree of preferred orientation of nickel electrodeposits. Electrochim Acta 11:1747–1751
51. Markov I, Boynov A, Toshev S (1973) Screening action and growth kinetics of electrodeposited mercury droplets. Electrochim Acta 18:377–384
52. Štrbac S, Rakočević Z, Popov KI, Pavlović MG, Petrović R (1999) The role of surface defects in HOPG on the electrochemical and physical deposition of Ag. J Serb Chem Soc 64:483–493
53. Milchev A, Kruijt WS, Sluyters-Rehbach M, Sluyters JH (1993) Distribution of the nucleation rate in the vicinity of a groving spherical cluster. Part 1. Theory and simulation results. J Electroanal Chem 362:21–31; Kruijt WS, Sluyters-Rehbach M, Sluyters JH, Milchev A (1994) Distribution of the nucleation rate in the vicinity of a growing spherical cluster. Part 2. Theory of some special cases and experimaental results. J Electroanal Chem 371:13–26
54. Popov KI, Krstajić NV (1983) The mechanism of spongy electrodeposits formation on inert substrate at low overpotentials. J Appl Electrochem 13:775–782
55. Markov I (1976) Saturation nucleus density in the electrodeposition of metal onto inert electrodes. I. Theory. Thin Solid Films 35:11–20; Markov I, Stoycheva E (1976) Saturation nucleus density in the electrodeposition of metal onto inert electrodes. II. Experimental. Thin Solid Films 35:21–35
56. Kaishew R, Mutafctschiew B (1965) Electrolytic nucleation of mercury. Electrochim Acta 10:643–650 (in German)
57. Erdey-Grúz T, Volmer Z (1931) Overvoltage of metals. Z Phys Chem 157A:165–181 (in German)
58. Fetter K (1967) Electrochemical kinetics. Khimiya, Moscow (in Russian)
59. Fleischmann M, Thirsk HR (1959) The potentiostatic study of the growth of deposits on electrodes. Electrochim Acta 1:146–160
60. Kovarskii NY, Lisov AV (1984) Periodicities in the surface structure of polycrystalline electrolytic deposits. Elektrokhimiya 20:221–225 (in Russian)
61. Kovarskii NY, Lisov AV (1984) The reasons for the structure periodicity in the surfaces of electrolyte copper deposits. Elektrokhimiya 20:833–837 (in Russian)
62. Kovarskii NY, Arzhanova TA (1986) On the nature of the "no nucleation" zones in the electrocrystallization process. Elektrokhimiya 20:452–458 (in Russian)
63. Popov KI, Grgur BN, Stojilković ER, Pavlović MG, Nikolić ND (1997) The effect of deposition process exchange current density on the thin metal films formation on inert substrate. J Serb Chem Soc 62:433–442
64. Popov KI, Krstajić NV, Popov SR (1983) Fundamental aspects of plating technology. II: morphological aspects of metal electrodeposition from complex salt solutions. Surf Technol 20:203–208
65. Dimitrov AT, Hadži-Jordanov S, Popov KI, Pavlović MG, Radmilović V (1998) Electrodeposition of silver from nitrate solutions: part I. Effect of phosphate ions on morphology. J Appl Electrochem 28:791–796
66. Popov KI, Pavlović MG, Grgur BN, Dimitrov AT, Hadži-Jordanov S (1998) Electrodeposition of silver from nitrate solutions: part II. Mechanism of the effect of phospate ions. J Appl Electrochem 28:797–801
67. Radmilović V, Popov KI, Pavlović MG, Dimitrov AT, Hadži-Jordanov S (1998) The mechanism of silver granular electrodeposits formation. J Solid State Electrochem 2:162–169
68. Popov KI, Rodaljević ZP, Krstajić NV, Novaković SD (1985) Fundamental aspects of plating technology V: the effect of strongly adsorbed species on the morphology of metal deposit. Surf Technol 25:217–222

69. Meibuhr S, Yeger E, Kozawa A, Hovorka F (1963) The electrochemistry of tin I. Effect of nonionic addition agents on electrodeposition from stannous sulfate solutions. J Electrochem Soc 110:190–202
70. Kabanov BN (1966) Electrochemistry of metal and adsorption. Nauka, Moscow (in Russian)
71. Lorenz W (1954) Oscillographic overvoltage measurements. Z Electrochem 58:912–918 (in German)
72. Despić AR (1983) Deposition and dissolution of metals and alloys. In: Bockris JO'M, Conway BE, Yeger E, White RE (eds) Comprehensive treatise of electrochemistry, 2nd edn. Plenum Press, New York
73. Oniciu L, Muresan LM (1991) Some fundamental aspects of levelling and brightening in metal electrodeposition. J Appl Electrochem 21:565–574
74. Muresan LM, Varvara SC (2005) Levelling and brightening mechanisms in metal electrodeposition. In: Nunez M (ed) Metal electrodeposition. Nova Science Publishers Inc, ew York, pp 1–45
75. Kardos O, Foulke G (1962) Applications of mass transfer theory. In: Delahay P, Tobias CW (eds) Electrodeposition on small-scale profiles. Advances in electrochemistry and electrochemical engineering, 2nd edn. Interscience, New York
76. Ibl N (1983) Current distribution. In: Yeger E, Bockris JO'M, Conway BE, Sarangapani S (eds) Comprehensive treatise of electrochemistry, 6th edn. Plenum Press, New York
77. Kruglikov SS, Kudriavtsev NT, Vorobiova GF, Antonov AY (1965) On the mechanism of levelling by addition agents in electrodeposition of metals. Electrochim Acta 10:253–262
78. Dukovic JO, Tobias CW (1990) Simulation of levelling in electrodeposition. J Electrochem Soc 137:3748–3755
79. Đorđević S, Maksimović MD, Pavlović MG, Popov KI (1997) Electroplating. Tehnička knjiga, Beograd (in Serbian)
80. Krichmar SI (1965) On the theory of the levelling action in the electrochemical behaviour of metals. Elektrokhimiya 1:858–861 (in Russian); Krichmar SI (1965) Levelling mechanism in the cathodic deposition of nickel. Zh Fiz Khim 39:602–603 (in Russian)
81. Krichmar SI, Pronskaya AY (1965) Experimental investigation of the levelling effect in the cathodic deposition of nickel from coumarine containing electrolytes. Zh Fiz Khim 39:741–744 (in Russian)
82. Jordan KG, Tobias CW (1991) The effect of inhibitor transport on levelling in electrodeposition. J Electrochem Soc 138:1251–1259
83. Dukovic JO (1993) Feature-scale simulation of resist-pattered electrodposition. IBM J Res Develop 37:125–141
84. Andricacos PC, Uzoh C, Dukovic JO, Horkans J, Deligianni H (1998) Damascene copper electroplating for chip interconnection. IBM J Res Develop 42:567–574
85. Schlesinger M, Paunovic M (eds) (2000) Modern electroplating. Wiley-interscience publication, New York
86. Weil R, Paquin R (1960) The relationship between brightness and structure in electroplated nickel. J Elec-trochem Soc 107:87–91
87. Nikolić ND, Rakočević Z, Popov KI (2005) Nanostructural analysis of bright metal surfaces in relation to their reflectivities. In: Conway BE, Vayenas CG, White RE, Gamboa-Adelco M (eds) Modern aspects of electrochemistry, vol 38. Kluwer Academic/Plenum Publishers, New York, pp 425–474
88. Dennis JK, Such TE (1993) Nickel and chromium plating. Wood head publ. ltd, Cambridge
89. Nichols RJ, Bach CE, Meyer H (1993) The effect of three organic additives on the structure and growth of electrodeposited copper: an in-sity scaning probe microscopy study. Ber Bunsenges Phys Chem 97:1012–1020
90. Czerwinski F, Kondo K, Szpunar JA (1997) Atomic force microscopy study of surface morphology of zinc-iron electrodeposits. J Electrochem Soc 144:481–484
91. Nikolić ND, Rakočević Z, Popov KI (2001) Structural characteristics of bright copper surfaces. J Electroanal Chem 514:56–66

92. Nikolić ND, Rakočević Z, Popov KI (2001) The STM analysis of a silver mirror surface. J Serb Chem Soc 66:723–727
93. Edwards J (1953) The mechanism of electropolishing of copper in phosphoric acid solutions. I. Processes preceding the establishment of polishing conditions. J Electrochem Soc 100:189c–194c
94. Wagner C (1954) Contribution to the theory of electropolishing. J Electrochem Soc 101:225–228
95. Krichmar SI, Pronskaya AY (1966) Study of the levelling effect in electrochemical polishing of metals. Elektrokhimiya 2:69–73 (in Russian)
96. Popov KI, Pavlović MG, Rakočević Z, Škorić DM (1995) The structure of bright copper surfaces. J Serb Chem Soc 60:873–878
97. Nikolić ND, Rakočević Z, Popov KI (2004) Reflection and structural analyses of mirror bright metal coatings. J Solid State Electrochem 8:526–531
98. Nikolić ND, Popov KI, Rakočević Z, Đurović DR, Pavlović MG, Stojanović M (2000) The structure of bright zinc coatings. J Serb Chem Soc 65:819–827
99. Nikolić ND, Rakočević Z, Đurović DR, Popov KI (2006) Nanostructural analysis of mirror – bright zinc coatings. Russ J Electrochem 42:1121–1126
100. Nikolić ND, Novaković G, Rakočević Z, Đurović DR, Popov KI (2002) Comparative reflection and structural analyses of copper and zinc coatings electrodeposited from acid sulfate solutions without and with additives. Surf Coat Technol 161:188–194
101. Bockris JO'M, Nagy Z, Dražić DM (1973) On the morphology of zinc electrodeposited from alkaline solutions. J Electrochem Soc 120:30–41
102. Jovićević JN, Despić AR, Dražić DM (1977) Studies of the deposition of cadmium on foreign substrates. Electrochim Acta 22:577–587
103. Despić AR, Dražić MD, Mirjanić MD (1978) Granular growth of electrochemically deposited metals. Faraday Discuss Chem Soc 12:126–135
104. Popov KI, Čekerevac MI, Nikolić LjN (1988) The dendritic electrocrystallization of cadmium from acid sulphate solutions I: Granular cadmium substrate. Surf Coat Technol 34:219–229
105. Popov KI, Krstajić NV, Jerotijević ZD, Marinković SR (1985) Electrocrystallization of silver from silver nitrate solutions at low overpotentials. Surf Technol 26:185–188
106. Jović VD, Nikolić ND, Lačnjevac UČ, Jović BM, Popov KI (2012) Morphology of different electrodeposited pure metal powders. In: Djokić SS (ed) Electrochemical production of metal powders, vol 54, Series: modern aspects of electrochemistry. Springer, New York, pp 63–123
107. Price PB, Vermilyea DA (1958) Kinetics of electrodeposition of silver. J Chem Phys 28:720–721
108. Popov KI, Živković PM, Krstić SB, Nikolić ND (2009) Polarization curves in the Ohmic controlled electrodeposition of metals. Electrochim Acta 54:2924–2931
109. Meibhur S, Yeager E, Kozawa A, Hovorka F (1963) The electrochemistry of tin: I. Effects of nonionic addition agents on electrodeposition from stannous sulfate solutions. J Electrochem Soc 110:190–202
110. Popov KI, Pavlović MG, Stojilković ER, Stevanović ZŽ (1997) The current density distribution on stationary wire electrodes during copper and lead electro-deposition. Hydrometallurgy 46:321–336
111. Nikolić ND, Popov KI, Živković PM, Branković G (2013) A new insight into the mechanism of lead electrodeposition: Ohmic–diffusion control of the electrodeposition process. J Electroanal Chem 691:66–76
112. Popov KI, Krstajić NV, Popov SR, Čekerevac MI (1986) Spongy electrodeposit formation. J Appl Electrochem 16:771–774
113. Popov KI, Krstajić NV, Simičić MV, Bibić NM (1992) The initial stage of spongy electrodeposit formation on inert substrate. J Serb Chem Soc 57:927–933
114. Jakšić MM (1985) Impurity effects on the macromorphology of electrodeposited zinc I: theoretical consideration and a review of existing knowledge. Surf Technol 24:193–217

115. Murashova I, Pomosov B (1989) Electrodeposition of metals in dendritic shapes. In: Polukarov YM (ed) In Itogi nauki i tehniki, Seria Elektrokhimiya, vol 30. Acad Sci, Moscow (in Russian)
116. Popov KI, Živković PM, Nikolić ND (2012) Formation of disperse silver deposits by the electrodeposition processes at high overpotentials. Int J Electrochem Sci 7:686–696
117. Toshev S, Markov I (1967) Electrolytic nucleation of cadmium. Electrochim Acta 12:281–286
118. Popov KI, Krstajić NV, Popov SR (1985) The morphology of cadmium deposits obtained on foreign substrates at high overpotentials. J Appl Electrochem 15:151–154
119. Price PB, Vermilyea DA, Webb MB (1958) The growth and properties of electrolytic whiskers. Acta Mat 6:524–531
120. Gorbunova KM, Zhukova AJ (1949) Crystallochemical and diffusion mechanism of electrocrystallization. Zh Fiz Khim 23:605–615 (in Russian)
121. Gorbunova KM, Pankov PD (1949) Regularities in the crystallization of thin silver filaments. Zh Fiz Khim 23:616–624 (in Russian)

116. [illegible]

Chapter 3
Current Distribution in Electrochemical Cells

3.1 Introduction

The current distribution on a macroprofile is very important in technical metal electrodeposition. In electroplating, the current distribution determines the local variations in the thickness of the coating. In electrowinning and electrorefining of metals, a non-homogeneous current distribution can cause a short circuit with the counter electrode and the corner weakness effect in electroforming. This is very important in the three-dimensional electrodes, as well as in some storage batteries. In all the cited cases, a uniform current density distribution over the macroprofile is required.

The aim of this chapter is to present the procedure, based on simple equations of electrode kinetics, by the condition in which a desired current density distribution can be obtained or an undesired one avoided, under the assumption that the limiting diffusion current density does not vary over the whole electrode surface, including the edges of flat and the tips of wire electrodes.

The simplest representation of an electrochemical cell is presented in Fig. 3.1 [1].

If electrodes are placed in the cell as shown in Fig. 3.1, three ways for the propagation of the current lines between the anode and cathode are possible.

I. There is only one current line between two symmetrically positioned points on the cathode and anode in the homogeneous field (A in Fig. 3.1; see Sect. 3.2).
II. There are many current lines between two symmetrically positioned points at the edges of both the cathode and anode (B in Fig. 3.1; see Sect. 3.3).
III. There is a division of a current line into two parts at the corner of the cathode (C in Fig. 3.1; see Sect. 3.4.7).

All these cases will be treated separately.

K.I. Popov et al., *Morphology of Electrochemically and Chemically Deposited Metals*, DOI 10.1007/978-3-319-26073-0_3

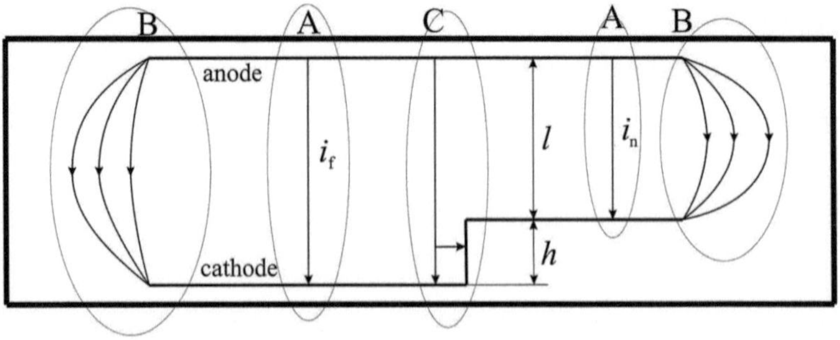

Fig. 3.1 Schematic representation of an electrochemical cell (Reprinted from Ref. [1] with permission from the Serbian Chemical Society)

3.2 The Current Density Distribution in Homogeneous Fields

The current density distribution between parts of the electrode surface with smaller and larger distance between cathode and anode can be calculated using the following procedure [1]:

The voltage, U, imposed to a cell shown in Fig. 3.1 is given by

$$U - E = b_a \log \frac{i_f}{i_{0,a}} + b_c \log \frac{i_f}{i_{0,c}} \frac{i_L}{i_L - i_f} + \rho(l + h) i_f \quad (3.1)$$

as well as by

$$U - E = b_a \log \frac{i_n}{i_{0,a}} + b_c \log \frac{i_n}{i_{0,c}} \frac{i_L}{i_L - i_n} + \rho l i_n \quad (3.2)$$

being valid within the Tafel region and at larger overpotentials, where E is a equilibrium potential difference; b_a and b_c and $i_{0,a}$ and $i_{0,c}$ are the anodic and cathodic Tafel slopes and exchange current densities, respectively; i_L is the limiting diffusion current density for the cathodic process; and ρ is the specific ohmic resistivity of the solution and the current densities for the part of the cathode at larger, i_f, and smaller, i_n, distances from the anode, respectively. For soluble anode considered in this case, it will be $i_{0,a} = i_{0,c} = i_0$ and $E = 0$. In these calculations, absolute values of both the cathodic and anodic current densities as well as the cathodic and anodic overpotentials are taken. The meanings of l and h are seen from Fig. 3.1. It is obvious from Eqs. (3.1) and (3.2) that

$$i_n = i_f \quad (3.3)$$

if $\rho = 0$ and if the system is under complete diffusion control, *i.e.*, if

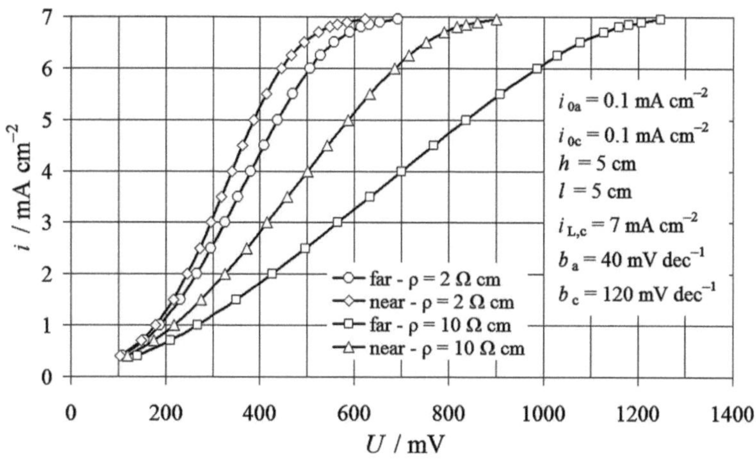

Fig. 3.2 The dependencies of the current densities at parts of the cathode closer to and further from the anode on the cell voltage for different solution resistivity (Reprinted from Ref. [1] with permission from the Serbian Chemical Society)

$$i_n = i_f = i_L \tag{3.4}$$

It follows from Eq. (3.3) that the dominant effect on current distribution is the effect of the resistivity of the solution.

On the other hand, if the electrochemical part of Eqs. (3.1) and (3.2) can be neglected relative to the ohmic voltage drop, complete ohmic control appears and the relation:

$$i_n = i_f \frac{l + h}{l} = i_f \left(1 + \frac{h}{l}\right) \tag{3.5}$$

is valid, corresponding to the primary current distribution.

An estimation of the effect of different parameters on the current density distribution can be made from Fig. 3.2, which shows the dependencies of the current densities at the closer, i_n, and further, i_f, part of the cathode from the anode on the cell voltage, U, for different solution resistivity. As can be seen, the increase of the conductivity of the electrolyte leads to the more uniform current density distribution of deposits at the electrode surface. A similar but less pronounced effect of the increase of the cathodic Tafel slope can be seen, while the change of i_0 does not affect the current density distribution. It is necessary to note that a soluble anode is considered in this case and, hence, the anodic and cathodic exchange current densities are the same.

Finally, the effect of the h/l ratio will be discussed. As expected, for $h/l \rightarrow 0$, the current density distribution approaches to a uniform one, while for $h/l >> 1$, it is similar to the primary current distribution. Based on the above results, it follows that the ohmic resistance of the solution is the most important parameter producing

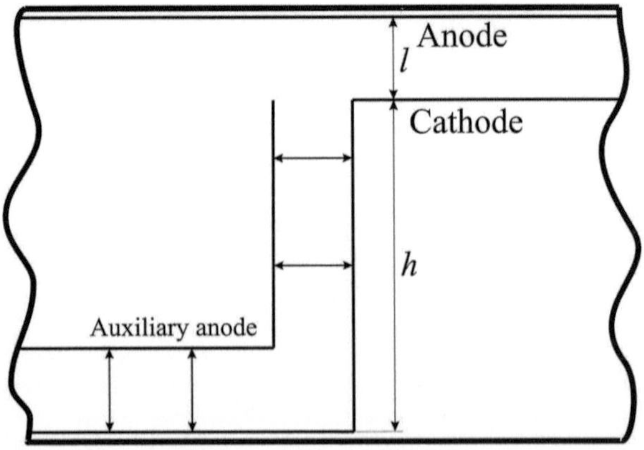

Fig. 3.3 The auxiliary anode adjusted to the cathode shape in order to improve the current density distribution (Reprinted from Ref. [1] with permission from the Serbian Chemical Society)

an effect on the current density distribution in electrochemical cells. If $h/l >> 1$, the effect of the geometry of the system can also be important, but an auxiliary anode adjusted to the cathode shape should be used in this case, as illustrated in Fig. 3.3.

3.3 The Edge Effect

The current density distribution in a cell with plane parallel electrodes, with edges not touching the cell wall, is illustrated in Fig. 3.4 [1].

It can be assumed that a homogeneous electric field and, consequently, a uniform current density distribution are present over the entire electrode surface up to the very edge of the electrode, where the current density increases abruptly. This problem has been studied in detail in Refs. [2, 3, 5–7], and a relatively rough approximation will be used here. This approximation is based on the following assumptions:

(a) There is only one current line between two symmetrically positioned points on the anode and cathode in a homogeneous field, and current density in the homogeneous field should be lower than current density at the edges.
(b) There is an infinitely large number of the current lines between two symmetrically positioned points at the edges of the electrodes. Then, it can be taken that the overall resistance between these two points will be equal to an infinitely large number of resistances connected in parallel, being lower than in the homogeneous field.

This approximation is more appropriate for the tip of a stationary wire electrode due to the dissipation of current lines occurring through the space [6]. On the other hand, in the case of the edges of plane parallel electrodes, the dissipation of current lines occurs in one plane normal to the electrodes to which the two symmetrically positioned points belong.

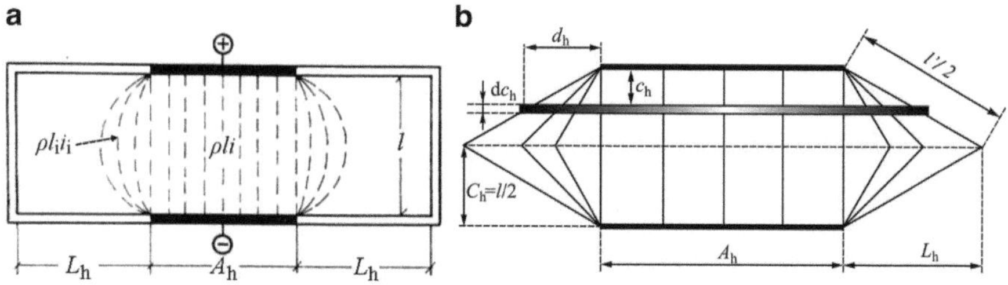

Fig. 3.4 (**a**) Current distribution in parallel plate electrode geometry (A_h is the electrode width, L_h is the distance between the edge of the electrode and the side walls, and l is the distance between the electrodes) and (**b**) the linear approximation model showing the current flow passing around the space between the plane parallel electrodes (Reprinted from Ref. [2] with permission from the Serbian Chemical Society, Ref. [3] with kind permission from Springer, and Ref. [4] with permission from Elsevier)

3.4 Two Equal Plane Parallel Electrode Arrangement

The cell with two equal plane parallel electrodes represents the elementary cell of electrode arrangement in electrochemical refining and winning processes [3].

In a cell with parallel electrodes (if the electrode edges do not touch the side walls of the cell), the current density is higher at the edges than in the center of the electrode [2]. This is because the current flow passes partially around the rectangular space between the electrodes. The increased current density at the edges of the electrodes can be easily checked by the analysis of the quality of the metal electrodeposit at the cathode. In some cases, the deposit in the central part of the cathode may be compact and flat, whereas the occurrence of dendrites is observed at the edges. The appearance of dendrites at the edges of the cathodes is undesirable because the growing dendrites could cause short circuits followed by a decrease in the current efficiency, or even to damage the power supply.

The aim of this section is also to show in which way dendritic growth at the cathode edges can be avoided in electrowinning and refining processes.

3.4.1 Ohmic Resistance of the Cell

The current density distribution in a rectangular electrolytic cell in which parallel electrodes cover only part of the wall of the cell is illustrated in Fig. 3.4. The linear approximation of the current density distribution in the cell with plane parallel electrodes shown in Fig. 3.4a is presented schematically in Fig. 3.4b.

The analysis performed here for the current density distribution between the edges of electrode and the side walls of cell is also valid for the case where there is the distance between the upper edges of the electrodes and free surface of solution, as well as between the lower edges of the electrode and the bottom of the cell. In the case under consideration, these two distances are taken to be zero.

The resistance dR of a section of the electrolyte of thickness dc is given by

$$\mathrm{d}R = \frac{\rho}{B_\mathrm{h}} \frac{\mathrm{d}c_\mathrm{h}}{A_\mathrm{h} + 2d_\mathrm{h}} \tag{3.6}$$

where B_h is the height of the electrode and ρ is the specific resistance of the electrolyte. From the linear approximation of the current lines, it follows:

$$d_\mathrm{h} = \frac{L_\mathrm{h}}{C_\mathrm{h}} c_\mathrm{h} \tag{3.7}$$

The parameters d_h and c_h are indicated in Fig. 3.4b.

The resistance of the whole electrolyte is then given by Popov et al. [4]:

$$R = \frac{\rho\, C_\mathrm{h}}{B_\mathrm{h}\, L_\mathrm{h}} \ln\left(\frac{A_\mathrm{h} + 2L_\mathrm{h}}{A_\mathrm{h}}\right) \tag{3.8}$$

and for $L_\mathrm{h} \rightarrow 0$, by

$$\lim_{L_\mathrm{h}\rightarrow 0} R = \frac{2\rho\, C_\mathrm{h}}{B_\mathrm{h} A_\mathrm{h}} = \frac{\rho\, l}{B_\mathrm{h} A_\mathrm{h}} = R_\mathrm{h} \tag{3.9}$$

where R_h corresponds to the resistance of a system with a homogeneous current density distribution (the side walls touch the edges of the electrodes). For $0 \leq L_\mathrm{h} << \infty$, L_h can be related to A_h by a linear coefficient k as follows:

$$L_\mathrm{h} = k\, A_\mathrm{h} \tag{3.10}$$

which transforms Eq. (3.8) to

$$R = \frac{R_\mathrm{h}}{2k} \ln(1 + 2k) \tag{3.11}$$

and

$$l_\mathrm{eff} = \frac{l}{2k} \ln(1 + 2k) \tag{3.12}$$

taking into account Eq. (3.9), where l_eff represents the inter-electrode distance in a cell with $L_\mathrm{h} = 0$ which the resistance is equal to the resistance of a cell with the inter-electrode distance l and $L_\mathrm{h} > 0$.

The cell for the determination of the total resistance of the cell is shown in Fig. 3.5.

Fig. 3.5 Schematic representation of the cell used for the determination of the total resistance of the cell (Reprinted from Ref. [3] with kind permission from Springer and Ref. [4] with permission from Elsevier)

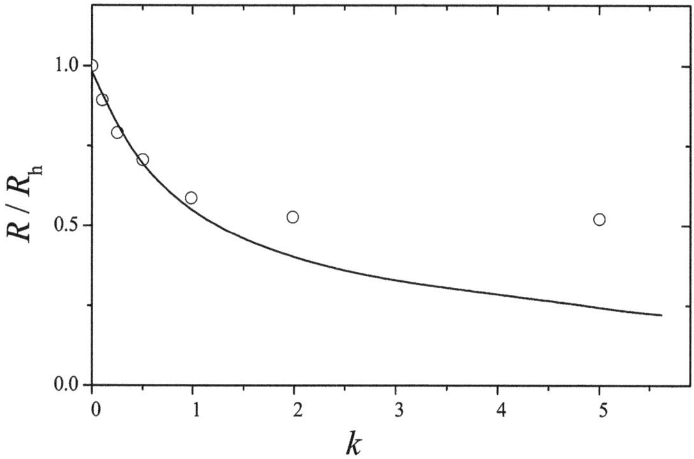

Fig. 3.6 The dependence of the total resistance of the system on k: (o) experimentally determined values; (—) calculated values (Reprinted from Ref. [3] with kind permission from Springer and Ref. [4] with permission from Elsevier)

The side screens enabled that the distance L_h between the edges of the electrodes and the side walls of the cell to be varied for given values A_h and $2C_h$. The resistance of the system for various adjusted values L_h was measured by the bridge method using the platinum electrodes in a 0.020 M KCl solution. The electrodes were 2 cm long and 1 cm wide. The inter-electrode distance was 2 cm. Hence, in this case $A_h = C_h$. The back sides of the electrodes were insulated. The upper edges of the electrode touch the free surface of the solution and the lower edges of the electrode touch the bottom of the cell.

The dependence of the total resistance of a system with plane parallel electrodes on the distance between the electrode edges and the cell side walls is shown in Fig. 3.6.

The open circles represent the experimental values, and the curve is obtained using Eq. (3.11). The good agreement between the experimental results and the values predicted by Eq. (3.11) extends to $k \approx 1$. It can be concluded that for this system, Eq. (3.11) is valid for $k < 1$. This means that the maximum penetration of the current lines occurs when $L_h = C_h = A_h$ in this case and that the maximum length of the current line l' is $l\sqrt{2}$.

3.4.2 The Very Edge Ohmic Resistance

This consideration of the very edge current density can be elaborated mathematically in the following way [5]: assuming the total ohmic control, the voltage drop in the solution between the electrodes inside the homogeneous field is given by

$$U - E = \rho l i \tag{3.13}$$

and outside of the homogeneous field by

$$U - E = \rho l_{\mathrm{i}} i_{\mathrm{i}} \tag{3.14}$$

where U is the cell voltage, E is the equilibrium potential difference, l is the inter-electrode distance, l_{i} is the length of the ith current line, and i_{i} is the current density corresponding to the ith current line, as can be seen from Fig. 3.4a.

The difference in the current lines outside the homogeneous field is given by

$$i_{\mathrm{i}} + \Delta i_{\mathrm{i}} - i_{\mathrm{i}} = \frac{U - E}{\rho}\left(\frac{1}{l_{\mathrm{i}}} - \frac{1}{l_{\mathrm{i}} + \Delta l_{\mathrm{i}}}\right) \tag{3.15}$$

or in the differential form

$$\frac{\mathrm{d}i_{\mathrm{i}}}{\mathrm{d}l_{\mathrm{i}}} = \frac{U - E}{\rho}\frac{1}{l_{\mathrm{i}}^2} \tag{3.16}$$

When Eq. (3.16) is integrated from the inter-electrode distance l to the maximum length of the current line, l', the maximum contribution to the edge current density due to current line propagation between the electrode edges and the side walls of the cell, i', is obtained as

$$i' = \frac{U - E}{\rho}\left(\frac{1}{l} - \frac{1}{l'}\right) = \frac{U - E}{\rho l}\frac{l' - l}{l'} \tag{3.17}$$

Taking into account Eq. (3.13), one obtains

$$i' = i\frac{l' - l}{l'} \tag{3.18}$$

The edge current density, i_{e}, can be written as

$$i_{\mathrm{e}} = i + i' \tag{3.19}$$

The maximum value of i' is obtained from Eq. (3.18) as

$$i'_{\mathrm{max}} = i\frac{2 - \sqrt{2}}{2} \tag{3.20}$$

Combining Eqs. (3.19) and (3.20), the maximum edge current density can be given as

$$i_{\mathrm{e,max}} = \left(2 - \frac{\sqrt{2}}{2}\right) i \approx 1.3i \tag{3.21}$$

for $l' = l\sqrt{2}$, as followed from Figs. 3.4 and 3.6.

This means that the very edge resistance is lower than in the homogeneous field and that the minimum effective inter-electrode distance, $l_{\mathrm{eff,e,min}}$, between the edges of the anode and cathode will be

$$l_{\mathrm{eff,e,min}} = \frac{2}{4 - \sqrt{2}} l \tag{3.22}$$

because of

$$\rho l\, i = \rho l_{\mathrm{eff,e,min}} i_{\mathrm{e,max}} \tag{3.23}$$

3.4.3 *The Quantitative Consideration of the Edge Effect*

In a cell with parallel plate electrodes, if the electrode edges do not touch the cell side walls, the potential difference between two points in the homogeneous field symmetrically positioned on the electrodes is given by [3]

$$U = E + \eta_{\mathrm{a}} + \eta_{\mathrm{c}} + \rho l\, i \tag{3.24}$$

Analogously, the cell voltage at the edges of the electrode can be expressed as

$$U = E + \eta_{\mathrm{a,e}} + \eta_{\mathrm{c,e}} + \rho l_{\mathrm{eff,e}}\, i_{\mathrm{e}} \tag{3.25}$$

where η_{a} and η_{c} are the anodic and cathodic overpotentials corresponding to the homogeneous field and $\eta_{\mathrm{a,e}}$ and $\eta_{\mathrm{c,e}}$ are the anodic and cathodic overpotentials corresponding to the edges of the electrode, respectively.

Elimination of U from Eqs. (3.24) and (3.25) gives

$$\eta_{\mathrm{a,e}} + \eta_{\mathrm{c,e}} = \eta_{\mathrm{a}} + \eta_{\mathrm{c}} + \rho l i - \rho l_{\mathrm{eff,e}}\, i_{\mathrm{e}} \tag{3.26}$$

In this case, $\rho \cdot l \cdot i > \rho \cdot l_{\mathrm{eff,e}} \cdot i_{\mathrm{e}}$, because increasing the current density leads to the increase of both the cathodic and anodic overpotentials at the edges.

In this way, a part of the ohmic potential drop in a homogeneous field transforms into electrochemical overpotential for points at the plane electrode edges, meaning the edge current density is larger than in the homogeneous field. In this way, it is possible to explain the change in the quality of the metal deposit near the edge and at the very edge of an electrode. It should be noted, however, that according to the proposed model, the entire edge current is located at the very edge of the electrode. In other words, a homogeneous electric field and, consequently, a uniform current distribution are assumed over the entire electrode surface up to the very edge of the electrode, where the current density increases abruptly, which is quite close to the real state described in Refs. [8–12].

3.4.4 The Depth of the Penetration of a Current Line Between the Electrode Edges and the Cell Side Walls

3.4.4.1 Mathematical Model

Equation (3.17) can be rewritten in the form:

$$l' = \frac{l}{1 - \frac{\rho l i'}{U-E}} \tag{3.27}$$

and i' can be substituted by $i\,\frac{2-\sqrt{2}}{2}$ giving

$$l' = \frac{l}{1 - \frac{\rho l i}{U-E}\frac{2-\sqrt{2}}{2}} \tag{3.28}$$

as the maximum length of a current line. $U - E$ in Eqs. (3.27) and (3.28) is the ohmic potential drop, but it can be substituted by the cell potential due to the following facts.

The current along each line should be very low, and because of this, the electrochemical overpotentials at the edges of electrodes at the ends of one current line can be neglected relative to the ohmic potential drop. Hence, the cell potential transforms into the ohmic potential drop along each current line and $U - E$ in Eq. (3.28) can be substituted by the cell potential from Eq. (3.24).

Substitution of $U - E$ in Eq. (3.28) by the cell voltage from Eq. (3.24) gives

$$l' = \frac{l}{1 - \frac{\rho l i}{\eta_a + \eta_c + \rho l i}\frac{2-\sqrt{2}}{2}} \tag{3.29}$$

or after rearrangement

$$l' = l\frac{\eta_a + \eta_c + \rho l i}{\eta_a + \eta_c + \frac{\sqrt{2}}{2}\rho l i} \tag{3.30}$$

Assuming a linear approximation of the propagation of a current line the relation between L' (which is the maximum depth of the propagation of a current line penetration in the space among the edges of the electrodes and the cell side walls), l, and l' is given by

$$L' = \frac{1}{2}\sqrt{l'^2 - l^2} \tag{3.31}$$

Substituting l' from Eq. (3.30) into Eq. (3.31) and rearranging gives

$$L' = \frac{l}{2}\left[\left(\frac{\eta_a + \eta_c + \rho l i}{\eta_a + \eta_c + \frac{\sqrt{2}}{2}\rho l i}\right)^2 - 1\right]^{1/2} \tag{3.32}$$

It can be shown that if

$$\eta_a + \eta_c >> \rho l i \tag{3.33}$$

then $l' \rightarrow l$ and $L' \rightarrow 0$.

In the opposite case, if $\eta_a + \eta_c << \rho l i$ is valid, then $l' \rightarrow l\sqrt{2}$ and $L' \rightarrow l/2$.

This shows that the ability of an electrolyte to distribute the current density uniformly increases with decreasing $\rho{\cdot}l{\cdot}i$ product, *i.e.*, with decreasing the ohmic polarization. Furthermore, it is to be expected that the increase of spacing (the distance between the polarization curves, $i - U$ for $L_h = 0$ and $L_h > 0$) leads to worse current density distribution [13].

3.4.4.2 Cell Voltage–Current Density Dependencies

The Effect of Geometry of the Cell

Taking into account Eqs. (3.10) and (3.12) for $L_h > 0$ (assuming that the current density in the homogeneous field is equal to the overall one, *i.e.*, that the current density distribution is uniform over the entire electrode, except for the very edge), Eq. (3.24) can be rewritten in the form:

$$U = E + \eta_a + \eta_c + \frac{\rho l i A_h}{2L_h}\ln\left(1 + \frac{2L_h}{A_h}\right) \tag{3.34}$$

If $L' < L_h$, L_h in Eq. (3.34) should be substituted by L':

$$U = E + \eta_a + \eta_c + \frac{\rho l i A_h}{2L'}\ln\left(1 + \frac{2L'}{A_h}\right) \tag{3.35}$$

Hence, calculations of the current density–cell voltage dependencies must begin with the determination of the relation between L and L'.

It is obvious that if

$$\eta_a + \eta_c >> \frac{\rho l i A_h}{2L_h} \ln\left(1 + \frac{2L_h}{A_h}\right) \tag{3.36}$$

electrolysis is predominantly under electrochemical control, and if

$$\eta_a + \eta_c << \frac{\rho l i A_h}{2L_h} \ln\left(1 + \frac{2L_h}{A_h}\right) \tag{3.37}$$

the electrolysis is predominantly under ohmic control.

In the first case, an "S"-shaped polarization curve can be expected, and a straight line in the second one. Hence, the shape of the polarization curve is a good indicator of the nature of the electrolysis process. On the other hand, the degree of current line penetration into the solution between the edges of the electrodes and the side walls of the cell is an indicator of the ability of an electrolyte to distribute the current density uniformly over the entire electrode. The experiments were carried out in the cell presented in Fig. 3.5.

The dependencies of the current density on the cell voltage for different inter-electrode distances and different distances between the edge of the electrode and the side wall of the cell, for the system (−) Cu | $CuSO_4$, H_2SO_4, H_2O | Cu (+) at a temperature of 20 °C is shown in Fig. 3.7.

From Fig. 3.7, the change in the shape of the $i-U$ dependencies is observed with the increase of the inter-electrode distance. In the region of lower current densities, at the shortest inter-electrode distance, the system is under mixed activation–ohmic control. At higher current densities, although a concentration overvoltage appears,

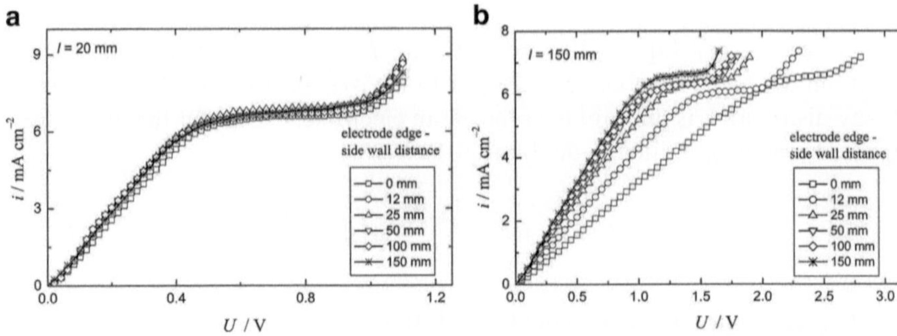

Fig. 3.7 Current density–cell voltage dependencies for the system (−) Cu | 0.10 M $CuSO_4$, 0.10 M H_2SO_4 | Cu (+) for different distances between the edge of the electrode and the side wall of the cell indicated in the diagram. The inter-electrode distance: (**a**) 20 mm and (**b**) 150 mm (Reprinted from Ref. [2] with permission from the Serbian Chemical Society and Ref. [3] with kind permission from Springer)

the control becomes ohmic because the magnitude of the concentration overvoltage is negligible in a comparison with the ohmic voltage drop in the electrolyte. As the inter-electrode distance increases, the electrolyte resistance in the cell also increases, giving rise to an increase in the contribution of the ohmic control of the system which in turn causes a larger spreading out of the current lines between the edges of the electrodes and the side walls of the cell. Therefore, the difference between the $i - U$ curves for different distances between the edge of the electrode and the side wall of the cell becomes larger, meaning a worse current density distribution. At very large current densities, approaching to the limiting current density, the contribution of the concentration overvoltage to the total cell voltage becomes significant and the system is under mixed the diffusion–ohmic control. Finally, in the region of limiting current densities, the system is under the full diffusion control. The polarization curves for different distances between the edge of the electrode and the side wall of the cell are very similar to each other indicating a minimal spreading of the current lines out of the homogeneous electric field.

The Effect of Composition of the Solution and Deposition Conditions

Effect of Supporting Electrolyte Concentration

Polarization curves of the system (−) Cu | $CuSO_4$, H_2SO_4, H_2O | Cu (+) for the different inter-electrode and the electrode edge–the cell side wall distances, as well as for the different concentration of H_2SO_4 as the supporting electrolyte, are shown in Fig. 3.8a, b (0.50 M H_2SO_4) and Fig. 3.8c (0.010 M H_2SO_4). When the H_2SO_4 concentration is the high (Fig. 3.8a, b), all polarization curves have an "S" shape, which is a characteristic of dominant electrochemically controlled electrolysis. The spacing among the curve for $L_h = 0$ and the curves for $L_h > 0$ is small and slightly increases with inter-electrode distance. In addition, the spacing between all the curves at $L_h > 0$ is almost the same and practically negligible, indicating a very small penetration depth of the current lines. At high H_2SO_4 concentration, the conductivity of the electrolyte is also high, and consequently the ohmic potential drop is small. Hence, the overpotentials at the electrodes comprise the major part of the total cell voltage and electrolysis is practically under full electrochemical control at all l values. In this way, the increased importance of the activation overpotential makes the current density distribution more uniform [5].

The electrolyte resistivity increases with the decrease of the H_2SO_4 concentration (Fig. 3.8c). In this way, the ohmic potential drop becomes more significant. The polarization curves lose their "S" shape and become more straight lines. The limiting current density plateau is not so well pronounced, and it becomes shorter as the l values increase. Finally, the plateau of the limiting diffusion current density disappears at very large electrode edge–side wall distances. This is illustrated for $l = 50$ mm.

The spacing between the polarization curves increases with both l and L values indicating on an increase in the penetration depth of the current lines into the

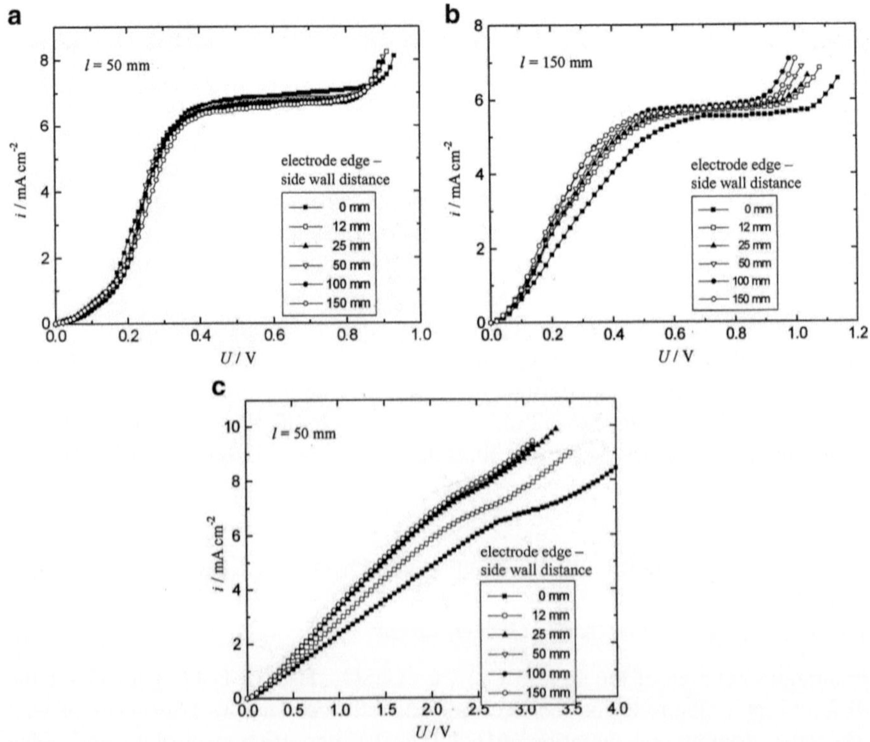

Fig. 3.8 Current density–cell voltage dependencies of the system (−) Cu | 0.10 M $CuSO_4$, 0.50 M H_2SO_4 | Cu (+) with an inter-electrode distance of (**a**) 50 mm, (**b**) 150 mm, and (**c**) current density–cell voltage dependencies of the system (−) Cu | 0.10 M $CuSO_4$, 0.010 M H_2SO_4 | Cu (+) with an inter-electrode distance of 50 mm. The distances between the edge of the electrode and the side wall of the cell are indicated in the diagrams (Reprinted from Ref. [5] with permission from the Serbian Chemical Society)

solution between the edges of the electrodes and the side walls of the cell. Thus, an increase in the ohmic resistance of the system leads to a decrease in the uniformity of the current density distribution.

Effect of Reaction Ion Concentration

The polarization curves of the system (−) Cu | $CuSO_4$, H_2SO_4, H_2O | Cu (+) obtained for a concentration of $CuSO_4$ as the source of the reacting Cu^{2+} ions of 0.030 and 0.30 M $CuSO_4$ for the different inter-electrode and the electrode edge–the cell side wall distances are shown in Fig. 3.9a (0.030 M $CuSO_4$) and Fig. 3.9b (0.30 M $CuSO_4$). The shape of the polarization curves with the appearance of the plateau of the limiting diffusion current density (Fig. 3.9a) clearly indicates that a decrease in Cu^{2+} ion concentration has the same effect as that of an increase in concentration of the supporting electrolyte. The limiting current density plateau is

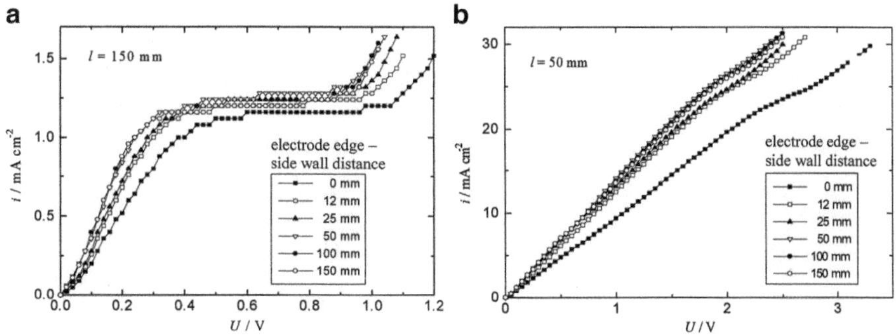

Fig. 3.9 Current density–cell voltage dependencies of the system: (**a**) (−) Cu | 0.030 M $CuSO_4$, 0.10 M H_2SO_4 | Cu (+); the inter-electrode distance: 150 mm and (**b**) (−) Cu | 0.30 M $CuSO_4$, 0.10 M H_2SO_4 | Cu (+); the inter-electrode distance: 50 mm. The distances between the edge of the electrode and the side wall of the cell are indicated in the diagrams (Reprinted from Ref. [5] with permission from the Serbian Chemical Society)

rather wide and well pronounced. The spacing between the systems is under dominant electrochemical control, and consequently, the penetration depth of the current lines into the solution between the edges of the electrode and the side walls of the cell is small. Hence, the uniformity of the current distribution is high.

When the concentration of the reacting Cu^{2+} ions is increased, the ohmic control of the process becomes dominant (Fig. 3.9b). The polarization curves, therefore, lose the "S" shape and become straight lines. The limiting current density plateau also gradually disappears as the inter-electrode distance increases and the spacing between the polarization curves increases. All these indicate that the penetration depth of the current lines is large and the uniformity of the current density distribution is bad.

Comparing the effects of concentrations of the reacting ions and the supporting electrolyte on the shape of the polarization curves, it can conclude that an increase in the concentration of the supporting electrolyte relative to the concentration of the reacting ion reduces the importance of the ohmic potential drop and enhances the importance of the overpotential as the electrochemical factor that makes the current density distribution more uniform.

Effects of Temperature and Stirring of the Electrolyte

The effect of electrolysis conditions (stirring of the electrolyte and temperature) is illustrated in Fig. 3.10. It can be seen that a stirring of the electrolyte leads to total ohmic control because of the strong increase in the limiting diffusion current density which in turn leads to a decrease in concentration overpotential.

Temperature does not have any effect on the shape of the polarization curves (Fig. 3.10b, c). The shape of the current density–cell voltage dependence does not change with temperature because an increase in temperature leads to both an increase in current density and a decrease in the resistivity of the electrolyte, making the $\rho \cdot l \cdot i$ product practically constant.

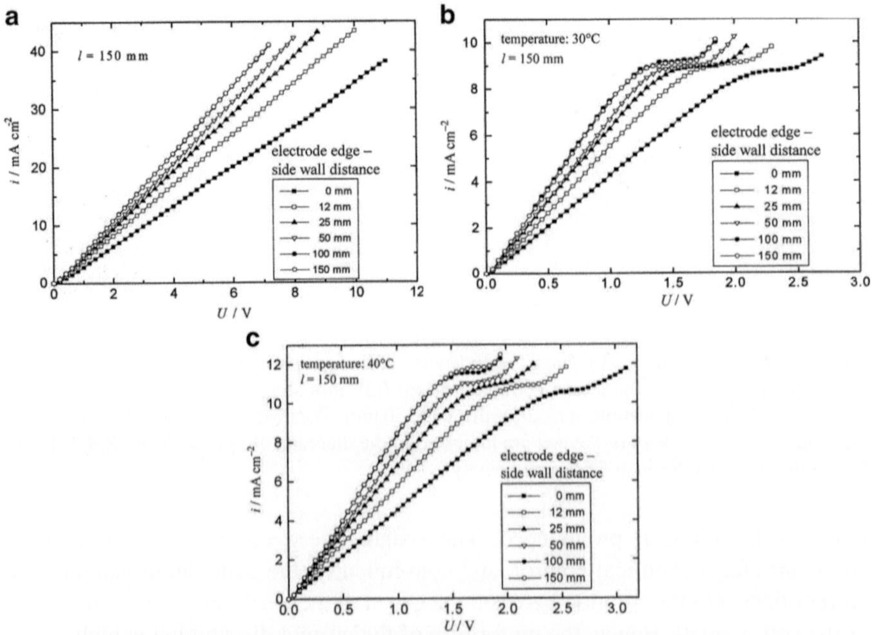

Fig. 3.10 Current density–cell voltage dependencies of the system (−) Cu | 0.10 M $CuSO_4$, 0.10 M H_2SO_4 | Cu (+): (**a**) the effect of stirring of the electrolyte and the effect of temperature of (**b**) 30 °C and (**c**) 40 °C. The inter-electrode distance: 150 mm. The different distances between the edge of the electrode and the side wall of the cell are indicated in the diagrams (Reprinted from Ref. [5] with permission from the Serbian Chemical Society)

Finally, it can be concluded from all presented diagrams that the greater the ohmic control, the deeper the penetration of the current lines between the electrode edges, and the cell side walls is observed.

3.4.5 *The Critical Current Density for Dendritic Growth Initiation at the Edges*

Substitution of η_i from Eq. (2.47) into Eq. (1.30) further rearranging produces

$$i_i = \frac{1}{2} i_L \tag{3.38}$$

where i_i is the critical current density for the dendritic growth initiation.

On the other hand, the edge current density i_e is given by the Eq. (3.19), where i' is given by the Eq. (3.18) and l' is given by the Eq. (3.30).

If $L_h > L'$, the edge current density could be obtained by combining Eqs. (3.18), (3.19), and (3.30) as

$$i_e = i \frac{\eta_a + \eta_e + \left(2 - \frac{\sqrt{2}}{2}\right)\rho l i}{\eta_a + \eta_e + \rho l i} \tag{3.39}$$

Assuming that maximum edge current density is given by Eq. (3.38), the substitution of i_i in Eq. (3.39) instead of i_e and further rearranging produce

$$i_{max} = \frac{1}{2} \frac{\eta_{a,max} + \eta_{e,max} + \rho l i_{max}}{\eta_{a,max} + \eta_{e,max} + \left(2 - \frac{\sqrt{2}}{2}\right)\rho l i_{max}} i_L \tag{3.40}$$

from which the maximum current density, i_{max}, in the homogeneous field at which dendrites at the edges do not grow can be calculated. It follows from Eq. (3.40) that for

$$\eta_{a,max} + \eta_{c,max} >> \rho l\, i_{max}, \quad i_{max} \tilde{} \, 0.5 i_L \tag{3.41}$$

and if

$$\eta_{a,max} + \eta_{c,max} << \rho l\, i_{max}, \quad i_{max} \tilde{} \, 0.4 i_L \tag{3.42}$$

being in both cases larger than the current density corresponding to the end of the Tafel linearity, which is the optimum current density for the deposition of compact metal (see section "The effect of the deposition current density on the electrode surface coarseness"). Hence, if deposition current density corresponds to the end of the Tafel linearity, dendrites will not grow at the edges of the electrode. It should be noted that in metal electrorefining, the working current density can be determined relative to the initial concentration of depositing ions, because it remains constant or increases during refining process. In electrowinning processes, the working current density must be determined relative to the final concentration of depositing ions, because it is lower than the initial one. The same reasoning is valid in the case of $L_h < L'$, meaning, in general, that if the current density in the cell is lower than $0.4 i_L$, the dendrites and probably carrot-like protrusion on the electrode edges cannot grow.

3.4.6 *Cells with Low Anode Polarization*

Cells with a small cathode and a large anode are often used in electroplating technology. In this case, a homogeneous distribution of the deposit over the entire cathode is required.

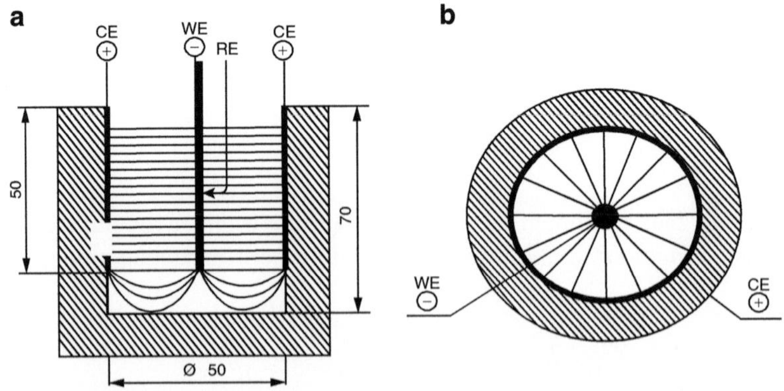

Fig. 3.11 Current distribution in cylindrical electrode geometry: (**a**) front cross section and (**b**) top cross section (Reprinted from Ref. [6] with permission from Elsevier and Ref. [3] with kind permission from Springer)

3.4.6.1 The Dependence of the Current Density at the Tip of a Stationary Wire Electrode on the Current Density in the Middle of the Electrode

It can be seen from Fig. 3.11 that the dissipation of current lines from the tip of a stationary wire electrode is more pronounced than in the case of the edges of plane parallel electrodes. This is because the dissipation in the former case occurs through the space, while in the latter case, it takes place in one plane, normal to the electrodes, to which two symmetrically positioned points belong. Hence, it can be taken that the overall resistance between the tip of the cathode and anode will be equal to an infinitely large number of resistances, as in the case of the edges of two plane parallel electrodes connected in parallel, being equal to zero.

The cell voltage, U, for the part of the system where the current density is homogeneously distributed is given by

$$U = E + \eta_{\mathrm{a}} + \eta_{\mathrm{c}} + RI \tag{3.43}$$

where R is the ohmic resistance of the electrolyte and I current in the cell, and for the tip of wire electrode:

$$U = E + \eta_{\mathrm{a,e}} + \eta_{\mathrm{c,t}} \tag{3.44}$$

where $\eta_{\mathrm{c,t}}$ and $\eta_{\mathrm{a,e}}$ are the overpotential at the tip of wire electrode and at the edge of the cylindrical anode, respectively, or after elimination of $U - E$ from Eqs. (3.43) and (3.44):

$$\eta_{a,e} + \eta_{c,t} = \eta_a + \eta_c + RI \tag{3.45}$$

In this way, the ohmic potential drop in a homogeneous field transforms into the electrochemical overpotential for points at the tip of a wire electrode or in some similar position. This means that a tip current density is larger than a current density in a homogeneous field. Finally, if the anode surface area is much larger than that of the cathode, *i.e.*, if

$$\eta_c >> \eta_a \tag{3.46}$$

Equation (3.45) can be rewritten in the form:

$$\eta_{c,t} = \eta_c + RI \tag{3.47}$$

The ability of an electrode to distribute uniformly current density on a whole cathode can be easily estimated by comparing the cathodic polarization curve with the cathodic current density–cell voltage dependence. The lower is the difference between them, the better distribution of the current density should be expected.

On the other hand, Eq. (3.47) can be rewritten in the form:

$$i_{tip} = \frac{i\ i_L \exp\left(\frac{2.3RI}{b_c}\right)}{i_L - i + i\exp\left(\frac{2.3RI}{b_c}\right)} \tag{3.48}$$

assuming that in mixed controlled deposition overpotential is given by Eq. (1.31), or

$$i_{tip} = i\exp\left(\frac{2.3RI}{b_c}\right) \tag{3.49}$$

assuming that in activation controlled deposition η is given by Eq. (1.29).

It can be taken to the first approximation that the same relation is also valid for the edge of a small, square stationary vertical cathode placed in the middle of a large cylindrical anode because the current line distribution is similar to the one from tip of a wire electrode.

It follows that for $R \rightarrow 0$, Eq. (3.48) can be rewritten in the form:

$$i_{tip} = i \tag{3.50}$$

This means that if the resistance of the electrolyte is lower, it is lower the difference between the current densities in the middle and at the tip of the electrode. The increase in b_c also leads to a more uniform current distribution. This can happen in the presence of strongly adsorbed species or during deposition from some complex salt solution. It is also seen that current density distribution on a macroprofile becomes uniform if $i \rightarrow i_L$, but in this case rough and dendritic deposit appears.

It follows from Eq. (3.48) that the exchange current density does not affect the current distribution, but there is an exception in the case of electrodeposition processes characterized by very large values of the exchange current density. When the deposition occurs under ohmic control, then the deposit will be formed only at the edges or in a similar position, where ohmic resistance is low. Decreasing the exchange current density by complexing of the depositing ion leads to a more homogeneous distribution, as illustrated and discussed by the following consideration.

3.4.6.2 Experimental Evidence

The Effect of Ohmic Resistance

The effect of the ohmic resistance of a cell is illustrated by the analysis of Cu electrodeposition at the room temperature on a stationary Cu wire electrodes (length: 40 mm and diameter: 0.8 mm) placed in the middle of a cylindrical cell (length: 5 cm and diameter: 6 cm). The surface of the cell was covered by a high purity copper plate as anode. Electrodepositions were performed from electrolytes containing 0.10 M $CuSO_4$, as well as 0.10 M $CuSO_4$ in 0.50 M H_2SO_4. The plot of cell voltage and overpotential versus current density for the copper deposition are shown in Fig. 3.12a, b.

The relation between the current density at the electrode tip and in the homogeneous field can be estimated by considering the cathodic overpotential–current density and cell voltage–current density dependencies. According to Eq. (3.47), the tip overpotential is equal to the cell voltage for a wire electrode.

It could be seen that the tip overpotential (cell potential) is larger than the overpotential in the middle of the electrode during electrodeposition from 0.10 M

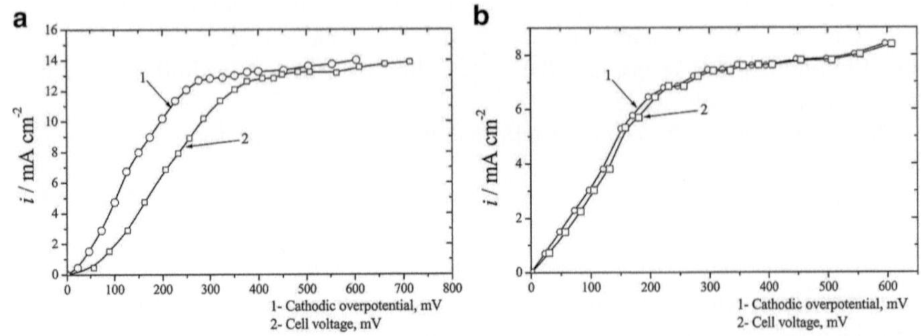

Fig. 3.12 Plots of overpotential (1) and cell voltage (2) versus current density for the Cu deposition from (**a**) 0.10 M $CuSO_4$ and (**b**) 0.10 M $CuSO_4$ in 0.50 M H_2SO_4 (Reprinted from Ref. [6] with permission from Elsevier and Ref. [3] with kind permission from Springer)

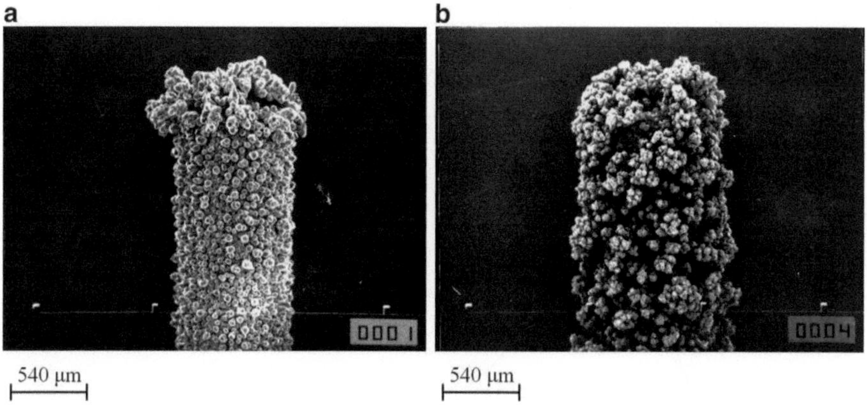

Fig. 3.13 Copper deposits obtained on a stationary Cu wire electrode from (**a**) 0.10 M $CuSO_4$ and (**b**) 0.10 M $CuSO_4$ in 0.50 M H_2SO_4. Quantity of electricity: 40 mA h cm^{-2}. Overpotential: 250 mV (Reprinted from Ref. [6] with permission from Elsevier and Ref. [3] with kind permission from Springer)

$CuSO_4$, while during electrodeposition from 0.10 M $CuSO_4$ in 0.50 M H_2SO_4, the overpotentials are practically the same. In the former case, there is a large difference in the morphology of the deposit at the tip and the rest of the electrode (Fig. 3.13a), while in the latter case, the quality of the deposit is the same over the whole surface (Fig. 3.13b).

This is in perfect agreement with Eq. (3.50).

Deposition in the Presence of Strongly Adsorbed Organic Additives (Effect of Increased Cathodic Tafel Slope)

The effect of strongly adsorbed organic additives is illustrated by the analysis of Cd electrodeposition process from 0.25 M $CdSO_4$ in 0.50 M H_2SO_4 to which 3.3 g dm^{-3} polyoxyethylene alkylphenol (9.5 mol ethylene oxide) was added [14]. Electrodeposition of Cd was performed onto a stationary vertical flat Cu electrode of surface area 1×1 cm^2 placed in the middle of a cylindrical cell of a diameter of 6 cm and a height of 5 cm. The cell surface was covered by the anode, which was made from a high purity Cd plate. The reference electrode was a high purity Cd wire.

The overpotential–current density and the cell voltage–current density plots for Cd deposition are presented in Fig. 3.14. The cathodic polarization curve obtained from potentiostatic polarization measurements has a similar shape to that found for an anode which can become passive; above a certain overpotential, increasing the cathode polarization leads to a decrease in the cathodic current density followed by a range of potential in which the overpotential has little effect on the current density. Current oscillations were observed at the beginning of this plateau in some cases (see also section "Organic compounds").

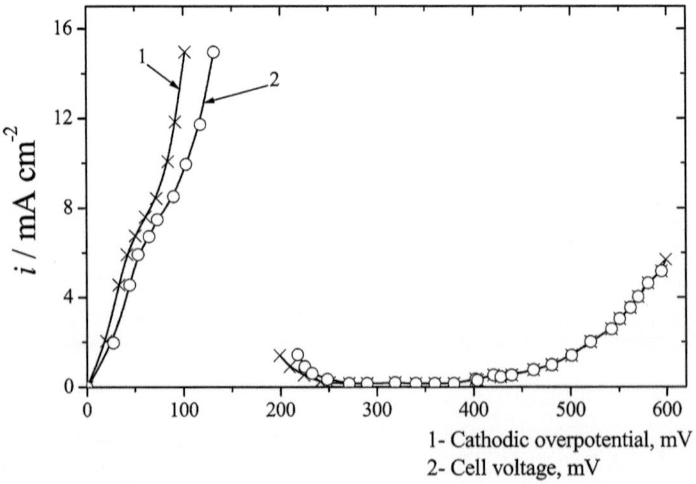

Fig. 3.14 Plots of overpotential (*1*) and cell voltage (2) versus current density for Cd electrodeposition from the sulfate electrolyte with the additive (Reprinted from Ref. [3] with kind permission from Springer and Ref. [14] with permission from Elsevier)

It was shown in section "The effect of the deposition current density on the electrode surface coarseness" that the optimum plating overpotential is determined by the upper limit of validity of the Tafel equation for the deposition process. In this case, as can be seen from Fig. 2.23, the optimum deposition overpotentials for Cd deposition are about 40 and 530 mV in the absence and in the presence of adsorption of additive, respectively.

Figure 3.14 shows that there is a large difference between the deposition overpotential and the cell voltage (tip overpotential) at low overpotentials which becomes negligible at high overpotentials, indicating a uniform current density distribution due to the additive adsorption, as illustrated in Fig. 3.15 [14].

Deposition from a Complex Salt Solution (Effect of Exchange Current Density)

Application of complex salt electrolytes in metal electrodeposition processes was examined by comparison of silver electrodeposition processes from the simple (nitrate) and complex (ammonium) electrolytes [15]. Silver was deposited from 0.10 M $AgNO_3$ in 0.20 M HNO_3 (the simple electrolyte) and 0.10 M $AgNO_3$ in 0.50 M $(NH_4)_2SO_4$ to which was added ammonium hydroxide to dissolve the precipitate of Ag sulfate (the complex electrolyte). The conductivities of both electrolytes were almost the same [3]. Silver was deposited onto a stationary vertical Pt cathode (1×1) cm^2 placed in the middle of a cylindrical cell (diameter: 6 cm and height: 5 cm). The surface of the cell was covered by anode of a high purity Ag plate. Polarization curves were recorded at the Pt wire electrodes at which Ag from the ammonium complex electrolyte was previously electrodeposited.

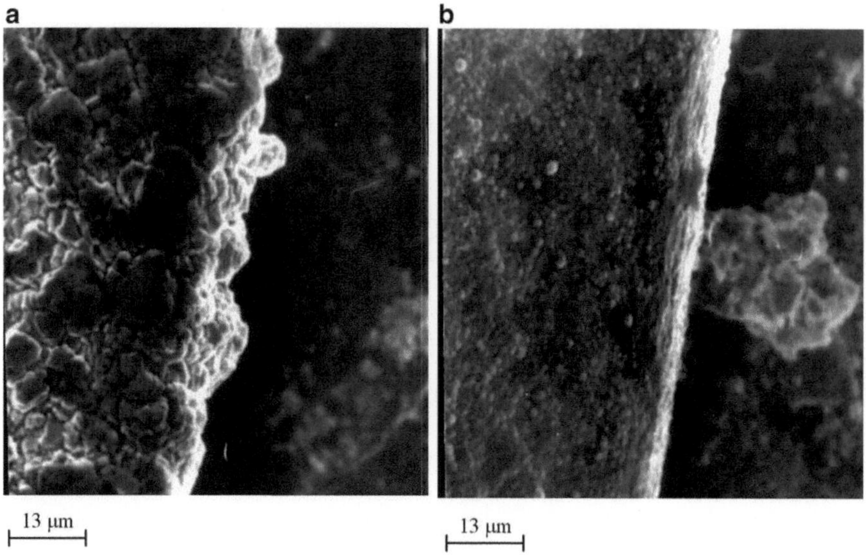

Fig. 3.15 Cd deposits obtained from a sulfate solution with the additive, at the edge of a Cu electrode. The thickness of deposit was 3 μm. Deposition overpotential: (**a**) 40 mV and (**b**) 530 mV (Reprinted from Ref. [3] with kind permission from Springer and Ref. [14] with permission from Elsevier)

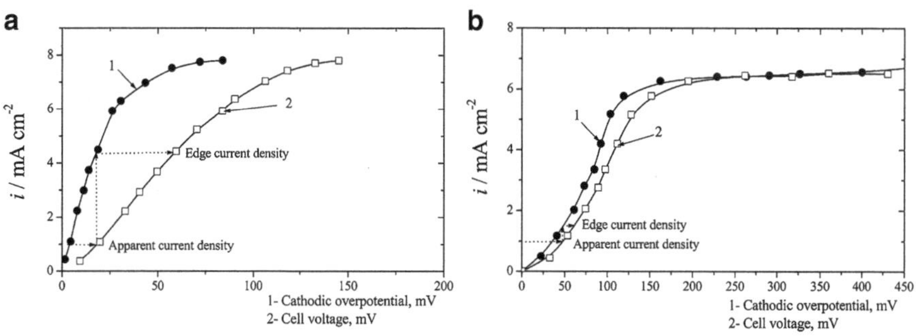

Fig. 3.16 Overpotential–apparent current density (1) and cell voltage–apparent current density (2) dependencies for deposition from (**a**) the nitrate and (**b**) the ammonium complex electrolytes (Reprinted from Ref. [3] with kind permission from Springer and Ref. [15] with permission from Elsevier)

The overpotential–apparent current density and the cell voltage (edge overpotential)–apparent current density plots for the Ag deposition from both the nitrate and the ammonium electrolytes are presented in Fig. 3.16a, b, respectively. For Ag deposition from the nitrate electrolyte, $i_0 >> i_L$ is valid, and at $i << i_L$, the nucleation in the middle of the electrode does not occur because the overpotential is very low. Hence, deposition from the nitrate electrolyte is only expected at the

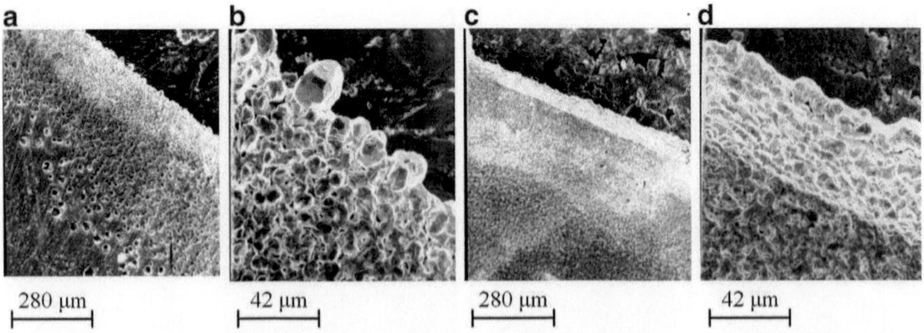

Fig. 3.17 Silver deposits obtained at 1 mA cm^{-2} (deposition time: 40 min): (**a**, **b**) the nitrate electrolyte and (**c**, **d**) the ammonium complex electrolyte (Reprinted from Ref. [3] with kind permission from Springer and Ref. [15] with permission from Elsevier)

edges where the overpotential is considerably larger, as predicted by Fig. 3.16a. This is illustrated in Fig. 3.17a, b.

For the ammonium electrolyte, there is a region where deposition is under the activation control because $i_0 < i < i_L$ (for this electrolyte, $i_0 = 0.25$ mA cm^{-2} and a slope of 60 mV dec^{-1}). Hence, nucleation occurs over the entire surface electrode. For deposition from the ammonium electrolyte (Fig. 3.16b), a more homogeneous distribution of the deposit is obtained, as is illustrated in Fig. 3.17c, d.

In this way, the influences of the deposition process parameters and the ohmic resistivity on the current density distribution are considered. It is obvious that the decrease of the ohmic resistance and the increase of the cathodic Tafel slope improve the current density distribution, as well as the decrease of i_0 from $i_0 >> i_L$ to $i_0 < i_L$.

3.4.7 *Corner Weakness Phenomena in Electroforming*

"Corner weakness" appears in deposits or electroforms at screened cathode parts, *i.e.*, corners. The deposit is thinner in the corners of the cathode, and at these areas, in extreme cases, there is no deposition at all along the line of the corner bisector [16].

The consequence of this effect is the emergence of a fracture under negligible load along the line of the corner bisection, instead of fracture at much higher loads across the narrowest cross section of an electroform normal to the line of pull.

A theoretical analysis of this phenomenon was performed using the following assumptions [7]:

- The potential difference between each of two points on the anode and cathode is equal to the cell voltage.
- The current lines are normal to the electrode surface.

- Along each current line, corresponding ohmic resistance exists and the current lines are independent of and insulated from each other.
- Current lines in the vicinity of a protrusion divide into components which are normal to the electrode surface.
- The Kirchhoff laws are valid for current line branching.

3.4.7.1 Ohmic-Controlled Deposition

According to the assumed model of current line division, it follows that there is no deposition along the line of bisection if the division of the current lines occurs along the line indicated by the dashed line in Fig. 3.18, as illustrated by Fig. 3.19. It can be seen that this configuration provides the same density of current lines at both the cathode and the anode.

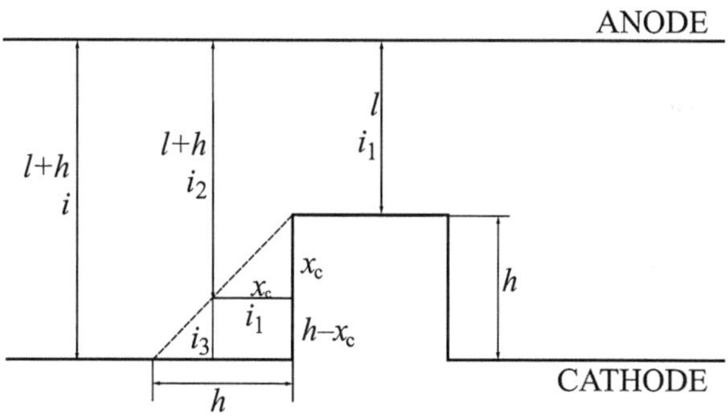

Fig. 3.18 Evaluation of a current distribution in the cell using the concept of current line division (Reprinted from Ref. [7] with permission from the Serbian Chemical Society and Ref. [3] with kind permission from Springer)

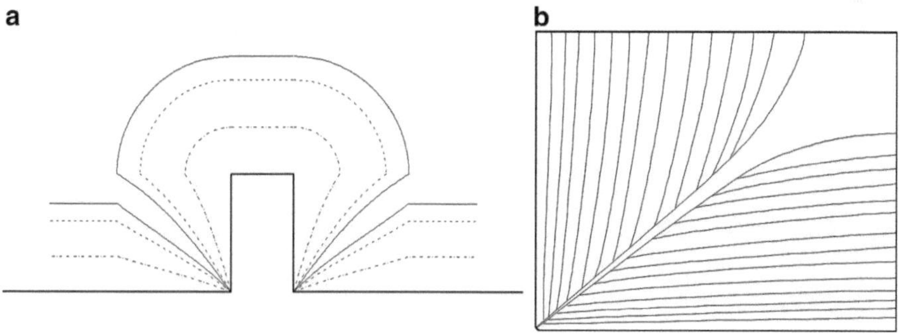

Fig. 3.19 (**a**) Simulation of a growth of the deposit from the model protrusion ($h = 5$ cm, $l = 15$ cm) calculated for a pure ohmic control employing Eqs. (3.57) and (3.58) and (**b**) schematic representation of microphotographs illustrating the "corner weakness" effect (Reprinted from Ref. [3] with permission from Springer and Ref. [7] with permission from the Serbian Chemical Society and adapted from [16])

The ohmic potential drops along the current lines i and i_1 are given by Eqs. (3.51) and (3.52), respectively, assuming $E = 0$:

$$U = \rho\,(l + h)i \tag{3.51}$$

$$U = \rho\, l\, i_1 \tag{3.52}$$

where U is the cell voltage and ρ is the resistivity of the solution. The ohmic resistance along the current line i_2 is somewhat different. It consists of the resistance between the anode and the dividing point (DP) and two resistances, proportional to x_c and $h - x_c$ connected in parallel between the DP and the cathode. Hence, the ohmic potential drop along current line i_2 can be written as

$$U = \rho\, i_2 \left[l + x_c + \frac{x_c(h - x_c)}{h} \right] \tag{3.53}$$

Elimination of U from Eqs. (3.51) and (3.52) and further rearrangement gives

$$i\,(l + h) = i_2 \frac{hl + 2hx_c - x_c^2}{h} \tag{3.54}$$

The current densities i_3 and i_4 are given by

$$i_3 = \frac{\rho\,(l + h)\,i - \rho\,(l + x_c)\,i_2}{\rho\,(h - x_c)} \tag{3.55}$$

and

$$i_4 = \frac{\rho\,(l + h)\,i - \rho\,(l + x_c)\,i_2}{\rho\, x_c} \tag{3.56}$$

Substitution of i_2 from Eq. (3.54) into Eqs. (3.55) and (3.56) and further rearrangement with use of Eq. (3.51) gives

$$i_3 = i\,\frac{(l + h)\,x_c}{hl + 2hx_c - x_c^2} \tag{3.57}$$

and

$$i_4 = i\,\frac{(l + h)(h - x_c)}{hl + 2hx_c - x_c^2} \tag{3.58}$$

which enable the calculation of the deposit profiles at the cathode.

The proposed model implies that there is no a current component in the direction of the corner vertex and that the appearance of a crack along the corner bisector is to be expected.

A compact deposit cannot be obtained directly but rather by the buildup of the deposit in the x and y direction. An overlap of the x and y oriented deposits should occur when the current density virtually does not depend on the distance from the very corner. However, if the current density decreases upon approaching to the corner vertex, the deposits would not overlap and a flaw would be created.

Equations (3.57) and (3.58) may be utilized for the calculation of the current density distribution at the beginning of deposition. The results of this calculation are shown in Fig. 3.19a, as well as for different deposition times.

It is obvious that there should be no overlap of the deposit upon prolonged deposition. Moreover, it should be noted that the profiles were calculated assuming a constant current density, which is not the case in a real system where the space in the vicinity of the corner vertex is increasingly screened as the deposit grows. This implies that the real distribution of the metal deposit in the corners is worse than that calculated and shown in Fig. 3.19a.

A number of microphotographs of deposit cross sections illustrating the "corner weakness" effect can be found in the literature. They are schematically exemplified by Fig. 3.19b. It can be seen that the calculated deposit profile (Fig. 3.19a), with a crack appearing along the corner bisector, looks very similar to that typically obtained in a plating practice (Fig. 3.19b).

3.4.7.2 Activation–Diffusion-Controlled Deposition

In this case, the current line between the DP and the cathode splits into two equal parts. Hence, for $i_{3,4} = i_3 = i_4$, one can write

$$U = \frac{b_a}{2.3} \ln \frac{2 i_{3,4}}{i_0} + \frac{b_c}{2.3} \ln \frac{i_{3,4}}{i_0} \frac{i_L}{i_L - i_{3,4}} \tag{3.59}$$

and simultaneously

$$U = \frac{b_a}{2.3} \ln \frac{i_1}{i_0} + \frac{b_c}{2.3} \ln \frac{i_1}{i_0} \frac{i_L}{i_L - i_1} \tag{3.60}$$

Elimination of U from Eqs. (3.59) and (3.60) permits the correlation between $i_{3,4}$ and i_1 in the form of Eq. (3.61), which cannot be solved explicitly if $b_a \neq b_c$:

$$b_a \ln \frac{2 i_{3,4}}{i_1} = b_c \ln \frac{i_1}{i_{3,4}} \frac{i_L - i_{3,4}}{i_L - i_1} \tag{3.61}$$

If $i_L >> i_{3,4}$ and $i_L >> i_1$, Eq. (3.61) can be rewritten in the form:

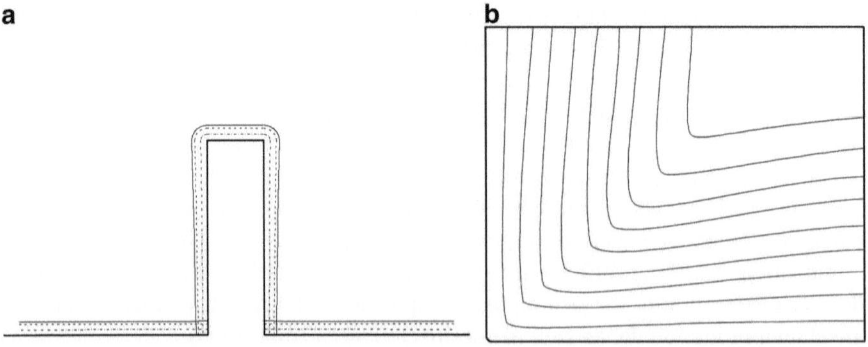

Fig. 3.20 (**a**) Simulation of the growth obtained at a cell voltage of 100 mV and (**b**) schematic representation of microphotographs of cross sections of deposits illustrating the effect of surface active agents on the elimination of "corner weakness" (Reprinted from Ref. [7] with permission from the Serbian Chemical Society and Ref. [3] with permission from Springer and adapted from [16])

$$b_a \ln \frac{2i_{3,4}}{i_1} = b_c \ln \frac{i_1}{i_{3,4}} \tag{3.62}$$

or

$$i_{3,4} = i_1 2^{-\frac{b_a}{b_a + b_c}} \tag{3.63}$$

which holds for a pure activation control.

Evenly distributed deposits, without "corner weakness," may be obtained only by deposition under complete activation control at high Tafel slopes (Fig. 3.20a). In practice, this is usually achieved by employing appropriate surface active additives, as seen in Fig. 3.20b.

In this way, not only the "corner weakness" effect is fully explained but also a new method of current density distribution evaluation in electrochemical cells is promoted.

References

1. Popov KI, Živković PM, Nikolić ND (2011) A mathematical model of the current density distribution in electrochemical cells. J Serb Chem Soc 76:805–822
2. Popov KI, Zečević SK, Pešić SM (1995) The current distribution in an electrochemical cell. Part I: the current voltage relationship for a cell with parallel plate electrodes. J Serb Chem Soc 60:307–316
3. Popov KI, Djokić SS, Grgur BN (2002) Fundamental aspects of electrometallurgy. Kluwer Academic/Plenum Publishers, New York, pp 101–143, Ch 4

4. Popov KI, Maksimović MD, Totovski DĆ, Nakić VN (1983) Some aspects of current density distribution in electrolytic cells I: dendritic growth of cadmium at the cathode edge in galvanostatic electrodeposition. Surf Technol 19:173–180
5. Popov KI, Zečević SK, Pešić SM (1996) The current distribution in an electrochemical cell. Part II: qualitative considerations of the basis of polarization curve shape. J Serb Chem Soc 61:583–590
6. Popov KI, Pavlović MG, Stojilković ER, Stevanović ZŽ (1997) The current density distribution on stationary wire electrodes during copper and lead electrodeposition. Hydrometallurgy 46:321–336
7. Popov KI, Stevanović RM (2000) A new line division concept for the determination of the current distribution in electrochemical cell. Part I. Theoretical background of the "corner weakness" effect in electroforming. J Serb Chem Soc 65:905–914
8. Ibl N (1983) Current distribution. In: Yeager EE, Bockris JO'M, Conway BE, Sarangapani S (eds) Comprehensive treatise of electrochemistry, vol 6. Plenum Press, New York
9. Ibl N (1975) Current distribution in electrolysis. Oberflähe-Surface 16:23–32 (in French)
10. Wagner C (1951) Theoretical analysis of the current density distribution in electrochemical cells. J Electrochem Soc 98:116–128
11. Newman J (1973) Electrochemical systems. Prentice Hall, Engelwood Clifts
12. Marathe V, Newman J (1969) Current distribution on a rotating disc electrode. J Electrochem Soc 116:1704–1719
13. Popov KI, Pešić SM, Kostić TM (1999) The current distribution in an electrochemical cell. Part V: the determination of the depth of the current line penetration between the edges of the electrodes and the side walls at the cell. J Serb Chem Soc 64:795–800
14. Popov KI, Rodaljević ZP, Krstajić NV, Novaković SD (1985) Fundamental aspects of plating technology. V. The effect of strongly adsorbed species on the morphology of metal deposits. Surf Technol 25:217–222
15. Popov KI, Krstajić NV, Popov SR (1984) Fundamental aspects of plating technology. III. The effect of electrodeposition from complex salt solutions on metal distribution over macroprofiles. Surf Technol 22:245–250
16. Spiro P (1968) Electroforming. Robert Draper Ltd, Teddington

Chapter 4
Electrodeposition at a Periodically Changing Rate

4.1 Introduction

The application of a periodically changing current in metal electrodeposition practice leads to improvements in the quality of electrodeposits. Three types of current variation have been found useful: reversing current (RC), pulsating current (PC), and sinusoidal, alternating current superimposed on a direct current (AC) [1–12]. The schematic presentation of the different current regimes of electrolysis is shown in Fig. 4.1. Also, the beneficial effects of pulsating overpotential (PO) have also been discussed [3]. Even though this kind of electrodeposition at a periodically changing rate (EPCR) is important from a theoretical point of view and offers a variety of experimental possibilities, it is as yet not frequently used in metal electrodeposition practice.

4.1.1 Reversing Current

Reversing current (RC) is represented schematically in Fig. 4.1a. It is characterized by the cathodic current density, i_c, and the anodic current density, i_a, as well as by the duration of flow of the current in the cathodic and the anodic direction, t_c and t_a, respectively. Naturally,

$$t_c + t_a = T_p \tag{4.1}$$

where T_p is the full period of the RC wave.

K.I. Popov et al., *Morphology of Electrochemically and Chemically Deposited Metals*, DOI 10.1007/978-3-319-26073-0_4

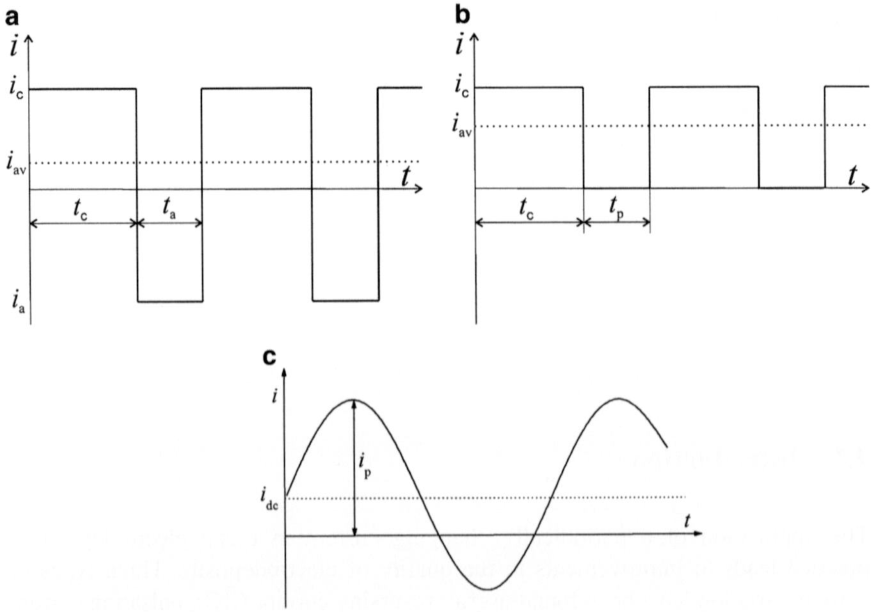

Fig. 4.1 The current waveforms in EPCR: (**a**) reversing current cycle, (**b**) pulsating current cycle, and (**c**) the sinusoidal alternating current superimposed on the direct current (Reprinted from Refs. [7, 12] with kind permission from Springer)

The average current density is then given by:

$$i_{av} = \frac{i_c t_c - i_a t_a}{t_c + t_a} \tag{4.2}$$

and for $i_c = i_a = i_A$

$$i_{av} = i_A \frac{1 - r_{RC}}{1 + r_{RC}} \tag{4.3}$$

where

$$r_{RC} = \frac{t_a}{t_c} \tag{4.4}$$

RC is used in the second and millisecond range [7, 12], and cathodic current density is taken as positive.

4.1.2 Pulsating Current

Pulsating current (PC) consists of a periodic repetition of square pulses. It is similar in shape to RC except for the absence of the anodic component, as is shown in Fig. 4.1b. The PC is characterized by the amplitude of the cathodic current density, i_c, the cathodic deposition time, t_c (on period), and the time interval t_p, in which the system relaxes (off period).

The full period, T_p, is given by:

$$t_c + t_p = T_p \tag{4.5}$$

and the average current density by:

$$i_{av} = \frac{i_c t_c}{t_c + t_p} \tag{4.6}$$

or

$$i_{av} = \frac{i_c}{1 + p} \tag{4.7}$$

where

$$p = \frac{t_p}{t_c} \tag{4.8}$$

It should be noted that rectified sinusoidal AC, especially half-rectified sinusoidal AC, often termed pulsating current in the literature, shows similar effects to those of the PC [7].

4.1.3 Alternating Current Superimposed on Direct Current

Sinusoidal AC superimposed on a direct cathodic current (DC) is represented in Fig. 4.1c. It is characterized by i_{dc}, i_p, and the frequency, which is usually 50 or 60 Hz. The resultant is termed an asymmetric sinusoidal current. The average current is equal to i_{dc}.

At a given DC value, three different types of current can be obtained, which can be denoted as follows: $i_p < i_{dc}$ "rippling current"; $i_p = i_{dc}$, "pulsating current"; and $i_p > i_{dc}$, "current with an anodic component." The last type is mainly used in plating practice.

4.1.4 Pulsating Overpotential

Pulsating overpotential consists of a periodic repetition of overpotential pulses of different shapes. Square-wave PO is defined in the same way as PC except that the overpotential pulsates between the amplitude value η_A and zero instead of current density. Non-rectangular pulsating overpotential is defined by the amplitude of the overpotential, η_A, frequency, and overpotential waveform [7].

4.1.5 Reversing Overpotential

Reversing overpotential is defined in the same way as RC except that overpotential pulsates between the cathodic amplitude value and anodic amplitude value.

Also, there are numerous other different current and overpotential waveforms used in EPCR [13, 14], but the most important have been mentioned above.

4.2 Surface Concentration of Depositing Ions in the Periodic Conditions

4.2.1 Electrodeposition with Periodically Changing Range in the Millisecond Range

Electrodeposition with a periodically changing rate can be described in terms of time-and distance-dependent concentrations:

$$\frac{\partial C}{\partial t} = D\frac{\partial^2 C}{\partial x^2} \tag{4.9}$$

$$C(0,x) = C_0 \tag{4.10}$$

$$C(t,\delta) = C_0 \tag{4.11}$$

$$\frac{\partial C}{\partial x}\|_{x=0} = \frac{i(t)}{nFD} \tag{4.12}$$

where n, F, D, C_0, and δ have already mentioned meanings, while x is the coordinate in a horizontal direction and C is concentration in a time t.

Equations (4.9), (4.10), (4.11), and (4.12) are solved for different $i(t)$ shapes and the solutions applied to different types of problems [7].

The current density $i(t)$ is the periodic function of a time, which for periodic reverse currents is given by [15]:

$$i(t) = \begin{cases} i_c, & \text{for } mT_p < t \le [m + 1/(r_{RC} + 1)]T_p \\ -i_a, & \text{for} [m + 1/(r_{RC} + 1)]T_p < t \le (m + 1)T_p \end{cases} \quad m = 0, 1, 2\ldots \tag{4.13}$$

for pulsating currents by [16]:

$$i(t) = \begin{cases} i_c, & \text{for } mT_p < t \le [m + 1/(p + 1)]\,T_p \\ 0, & \text{for } [m + 1/(p + 1)]\,T_p < t \le (m + 1)T_p \end{cases} \quad m = 0, 1, 2\ldots \tag{4.14}$$

and for AC superimposed on DC by [17]:

$$i(t) = i_{dc} + i_p \sin(\omega t) \tag{4.15}$$

In the case of pulsating overpotential, $i(t)$ is given by [3]:

$$i(t) = i_0 \left[\frac{C(0,t)}{C_0} \exp\left(2.3 \frac{\eta(t)}{b_c} \right) - \exp\left(-2.3 \frac{\eta(t)}{b_a} \right) \right] \tag{4.16}$$

The surface concentration under periodic conditions can be evaluated as follows. For $i(t)$ given by Eq. (4.13), the solution of Eqs. (4.9), (4.10), (4.11), and (4.12) for $x = 0$, $t = [m + 1/(r_{RC} + 1)]\,T_p$, and $m \to \infty$, *i.e.*, at the end of the cathodic pulses, under the periodic conditions is given by [15]:

$$\begin{aligned} C_c = C_0 - \frac{8\delta}{\pi^2 nFD} \sum_{k=0}^{\infty} \frac{1}{(2k+1)^2} & \\ \times \left[i_c \frac{1 - \exp[-\lambda_k T_p/(r_{RC} + 1)]}{1 - \exp(-\lambda_k T_p)} \right. & \\ \left. - i_a \frac{\exp[-\lambda_k T_p/(r_{RC} + 1)] - \exp(-\lambda_k T_p)}{1 - \exp(-\lambda_k T_p)} \right] & \end{aligned} \tag{4.17}$$

where $\lambda_k = \frac{(2k+1)^2 \pi^2 D}{4\delta^2}$ and $k = 0, 1, 2, \ldots..$

The surface concentration, C_a, at the end of the anodic pulses under the same conditions, *i.e.*, for $x = 0$, $t = (m + 1)\,T_p$, and $m \to \infty$ is given by:

$$C_a = C_0 - \frac{8\delta}{\pi^2 nFD} \sum_{k=0}^{\infty} \frac{1}{(2k+1)^2} \times \left[i_c \frac{\exp\left[-\lambda_k r_{RC} T_p/(r_{RC}+1)\right] - \exp\left(-\lambda_k T_p\right)}{1 - \exp\left(-\lambda_k T_p\right)} - i_a \frac{1 - \exp\left(-\lambda_k r_{RC} T_p/(r_{RC}+1)\right)}{1 - \exp\left(-\lambda_k T_p\right)} \right] \tag{4.18}$$

For a sufficiently long period T_p, ($T_p \gg t_0$), where $t_0 = \delta^2/(\pi^2 \cdot D)$, the system behaves as under DC conditions. For $T_p \to \infty$, and taking into account Eq. (1.14), Eqs. (4.17) and (4.18) become:

$$\lim_{T_p \to \infty} C_c = C_0 - \frac{\delta i_c}{nFD} = C_0 \left(1 - \frac{i_c}{i_L}\right) \tag{4.19}$$

and

$$\lim_{T_p \to \infty} C_a = C_0 + \frac{\delta i_a}{nFD} \tag{4.20}$$

For a sufficiently small value of T_p, ($T_p \ll t_0$):

$$\lim_{T_p \to 0} C_c = \lim_{T_p \to 0} C_a = C_s = C_0 - \frac{\delta}{nFD} \frac{i_c - r_{RC}\, i_a}{r_{RC} + 1} \tag{4.21}$$

For $i(t)$ given by Eq. (4.14), solution of Eqs. (4.9), (4.10), (4.11), and (4.12) for $x = 0, t = [m + 1/(p+1)]T_p$, and $m \to \infty$, *i.e.*, at the end of the cathodic pulses under periodic conditions, is given by [16]:

$$C_{on} = C_0 - \frac{8\delta i_c}{\pi^2 nFD} \sum_{k=0}^{\infty} \frac{1}{(2k+1)^2} \frac{1 - \exp\left(-\lambda_k T_p/(p+1)\right)}{1 - \exp\left(-\lambda_k T_p\right)} \tag{4.22}$$

The surface concentration at the end of pauses, C_{off}, under the same conditions $[x = 0, (m+1)T_p, m \to \infty]$ is given by:

$$C_{off} = C_0 - \frac{8\delta i_c}{\pi^2 nFD} \sum_{k=0}^{\infty} \frac{1}{(2k+1)^2} \frac{\exp\left(-\lambda_k p T_p/(p+1)\right) - \exp\left(-\lambda_k T_p\right)}{1 - \exp\left(-\lambda_k T_p\right)} \tag{4.23}$$

As in the previous case for $T_p \gg t_0$, the system behaves as under DC conditions where

$$\lim_{T_p \to \infty} C_{on} = C_0 - \frac{\delta i_c}{nFD} \tag{4.24}$$

and

$$\lim_{T_p \to \infty} C_{\text{off}} = C_0 \tag{4.25}$$

For $T_p \ll t_0$, it follows from Eqs. (4.22) and (4.23) that

$$\lim_{T_p \to 0} C_{\text{on}} = \lim_{T_p \to 0} C_{\text{off}} = C_s = C_0 - \frac{\delta i_c}{nFD\,(p+1)} \tag{4.26}$$

It is obvious from Eqs. (4.2), (4.4), and (4.7) as well as from Eqs. (4.21) and (4.26) that in both cases

$$C_s = C_0 - \frac{\delta i_{\text{av}}}{nFD} = C_0\left(1 - \frac{i_{\text{av}}}{i_L}\right) \tag{4.27}$$

taking into account also Eq. (1.14).

For $i(t)$ given by Eq. (4.15), the surface concentration under periodic conditions is approximately given by [17]:

$$C_s = C_0\left(1 - \frac{i_{\text{dc}}}{i_L}\right) - \frac{i_p}{nF\,(D\omega)^{1/2}} \sin\left(\omega t - \frac{\pi}{4}\right) \tag{4.28}$$

Hence, at sufficiently small value of T_p and for not extremely high values of i_p, C_s, in AC will also be given by Eq. (4.27), implying that at sufficiently high frequencies, surface concentration is determined by the average current density regardless of the shape of the current wave.

In the case of a rectangular pulsating overpotential, $\eta(t)$ as a function of a time is given by [18]:

$$\eta(t) = \left\{ \begin{array}{ll} \eta_A, & \text{for } mT_p < t \le [m + 1/(p+1)]\ T_p \\ 0, & \text{for } [m + 1/(p+1)]\ T_p < t \le (m+1)\ T_p \end{array} \right\} \tag{4.29}$$

where η_A is the overpotential amplitude, and $m = 0,1,2,\ldots\ldots$

Assuming that the surface concentration is determined by the average current density, i_{av}, Eq. (4.16) can be rewritten in the form

$$i = i_0\left(1 - \frac{i_{\text{av}}}{i_L}\right) \exp\left(2.3\frac{\eta(t)}{b_c}\right) - i_0 \exp\left(-2.3\frac{\eta(t)}{b_a}\right) \tag{4.30}$$

For a sufficiently high value of η_A, Eq. (4.30) reduces during the on periods to:

$$i_{\text{on}} = i_0\left(1 - \frac{i_{\text{av}}}{i_L}\right) \exp\left(2.3\frac{\eta_A}{b_c}\right) \tag{4.31}$$

and during the off periods to:

$$i_{\text{off}} = -i_0 \frac{i_{\text{av}}}{i_{\text{L}}} \tag{4.32}$$

The average current density in the PO deposition can easily be determined by:

$$i_{\text{av}} = \frac{i_0}{p+1} \left[\left(1 - \frac{i_{\text{av}}}{i_{\text{L}}} \right) \exp\left(2.3 \frac{\eta_{\text{A}}}{b_{\text{c}}} \right) - p \frac{i_{\text{av}}}{i_{\text{L}}} \right] \tag{4.33}$$

The overpotential amplitude is then given by:

$$\eta_{\text{A}} = \eta_{\text{dc}} + \frac{b_{\text{c}}}{2.3} \ln\left(p + 1 + p \frac{i_0}{i_{\text{L}}} \right) \tag{4.34}$$

and

$$\eta_{\text{av}} = \frac{\eta_{\text{dc}}}{p+1} + \frac{b_{\text{c}}}{2.3(p+1)} \ln\left(p + 1 + p \frac{i_0}{i_{\text{L}}} \right) \tag{4.35}$$

where

$$\eta_{\text{dc}} = \frac{b_{\text{c}}}{2.3} \ln\left(\frac{i_{\text{av}}}{i_0} \right) - \frac{b_{\text{c}}}{2.3} \ln\left(1 - \frac{i_{\text{av}}}{i_{\text{L}}} \right) \tag{4.36}$$

and

$$\eta_{\text{av}} = \frac{\eta_{\text{A}}}{p+1} \tag{4.37}$$

Polarization curves for the average values for the copper deposition have been successfully calculated from the stationary polarization curve using Eq. (4.35) for $i_0 \ll i_{\text{L}}$ [18]. This is good evidence that in PO deposition, the average current density also determines the surface concentration of the depositing ion.

The overpotential amplitude is larger than in the DC regime for one and the same average current density. Simultaneously, the diffusion overpotential remains constant, depending on the average current density only. Hence, the part of activation control in the overall amplitude overpotential increases with increasing pause to pulse ratio.

The situation is similar in pulsating or reversing current electrodeposition.

4.2.2 Capacitance Effects

From the above discussion, it can be concluded that the useful range of frequencies is limited by mass-transfer effects at low frequencies. At high frequencies, the useful range is limited by the effect of the capacitance of the electrical double layer [16]. This is shown here for the PC deposition. The time dependencies of the overpotential during the current pulses are shown in Fig. 4.2.

Mass-transfer limitations cause an increase in the overpotential at deposition times longer than the transition time as shown in Fig. 4.2a; the system enters full diffusion control at low frequencies if $i_c > i_L$. This is followed by an increase in the average overpotential [10, 16]. At high frequencies, the PC is used both for double-layer charging and discharging and for the deposition process, as illustrated by Fig. 4.3. The capacitance current during periodic charging and discharging of the double layer, at frequencies at which the effect of the double layer cannot be neglected, produces a smearing effect on the Faradic current wave, as illustrated by Fig. 4.2b.

Hence, as the frequency increases, the faradic current wave flattens, approaching to a DC shape, and gives the same quality of deposit as DC even though the overall

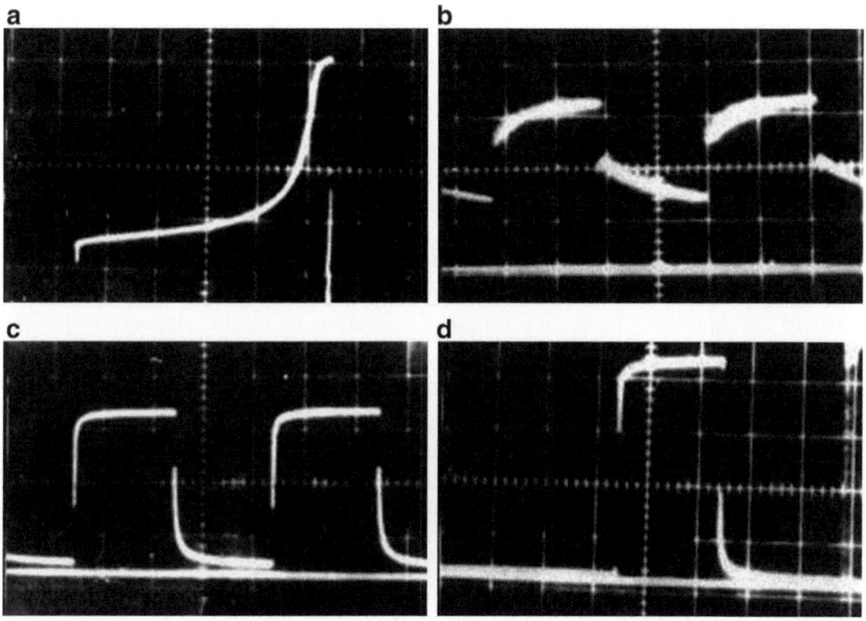

Fig. 4.2 The time dependence of the overpotential during current pulses in PC copper deposition from 0.50 M $CuSO_4$ in 1.0 M H_2SO_4: (**a**) $t_c = 10$ s, $p = 1$, $i_c = 1.2\, i_L$; x-axis, 2 s/div.; y-axis, 0.2 V/div., (**b**) $t_c = 10^{-4}$ s, $p = 1$, $i_c = 1.2\, i_L$; x-axis, 5×10^{-5} s/div.; y-axis, 0.1 V/div., (**c**) $t_c = 10^{-2}$ s, $p = 1$, $i_c = 1.2\, i_L$; x-axis, 5×10^{-3} s/div.; y-axis, 0.1 V/div., and (**d**) $t_c = 10^{-2}$ s, $p = 9$, $i_c = 6\, i_L$; x-axis, 5×10^{-3} s/div.; y-axis, 0.2 V/div (Reprinted from Ref. [12] with kind permission from Springer and Ref. [16] with permission from Elsevier)

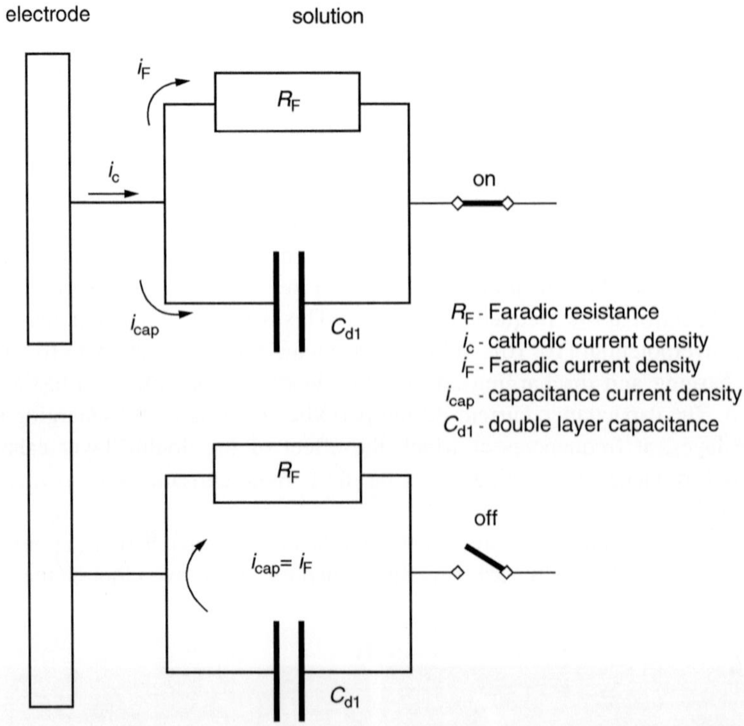

Fig. 4.3 Schematic representation of the effect of the double-layer capacitance on the faradic current during pulsating current electrodeposition (Reprinted from Ref. [12] with kind permission from Springer)

current appears to be pulsating one. This is also followed by an increase in the average overpotential. Hence, the minimum average overpotential is a good indicator of the optimum frequency range of pulsation in the PC deposition [3, 10, 16]. This range depends on the average current density and p, but, in general, the frequency lies in the range between 10 and 100 Hz, as illustrated in Fig. 4.2c, d.

In the PO deposition, the effect of the double-layer capacitance becomes less pronounced at higher frequencies compared to the other cases [10]. Also, at very high frequencies, the shape of the PO wave changes; for example, a square-wave PO becomes similar to a triangular one [10, 18].

4.2.3 Reversing Current in the Second Range

For T_p close to t_0, the behavior of the system under RC conditions has to be analyzed using Eq. (4.18) [7]. In this case, the concentration distribution inside the diffusion layer at the end of the anodic pulse is close to that given by Eq. (4.10). It follows from Eq. (4.18) that this will occur at:

$$C_{\mathrm{a}} = C_0 \tag{4.38}$$

or

$$\sum_{k=0}^{x} \frac{1}{(2k+1)^2}\left[i_{\mathrm{c}} \frac{\exp\left(\frac{-\lambda_{\mathrm{k}} r_{\mathrm{RC}} T_{\mathrm{p}}}{r_{\mathrm{RC}}+1}\right) - \exp(-\lambda_{\mathrm{k}} T_{\mathrm{p}})}{1-\exp(-\lambda_{\mathrm{k}} T_{\mathrm{p}})} - i_{\mathrm{a}} \frac{1-\exp\left(\frac{-\lambda_{\mathrm{k}} r_{\mathrm{RC}} T_{\mathrm{p}}}{r_{\mathrm{RC}}+1}\right)}{1-\exp(-\lambda_{\mathrm{k}} T_{\mathrm{p}})}\right] = 0 \tag{4.39}$$

It is known [15] that for $r_{\mathrm{rc}} T_{\mathrm{p}}/(r_{\mathrm{rc}}+1) \geq 1.5\, t_0$, the series in Eq. (4.39) can be approximated using only the first term ($k=0$). Hence, for $i_{\mathrm{c}} = i_{\mathrm{a}}$,

$$\exp\left(-\frac{r_{\mathrm{RC}}}{r_{\mathrm{RC}}+1}\cdot\frac{T_{\mathrm{p}}}{4t_0}\right) - \exp\left(-\frac{T_{\mathrm{p}}}{4t_0}\right) = 1 - \exp\left(-\frac{r_{\mathrm{RC}}}{r_{\mathrm{RC}}+1}\cdot\frac{T_{\mathrm{p}}}{4t_0}\right) \tag{4.40}$$

or

$$r_{\mathrm{RC}} = \frac{\frac{4t_0}{T_{\mathrm{p}}}\ln\frac{2}{1+\exp\left(-T_{\mathrm{p}}/4t_0\right)}}{1-\frac{4t_0}{T_{\mathrm{p}}}\ln\frac{2}{1+\exp\left(-T_{\mathrm{p}}/4t_0\right)}} \tag{4.41}$$

It is easy to show that for $T_{\mathrm{p}} = 3\,t_0$, $r_{\mathrm{RC}} = 0.7$ and for $T_{\mathrm{p}} = 16\,t_0$, $r_{\mathrm{RC}} = 0.2$ by assuming that Eq. (4.40) is valid for $T_{\mathrm{p}} > 3\,t_0$ and that for $T_{\mathrm{p}} > 16\,t_0$, the system behaves as under the DC conditions. The optimum ratio $t_{\mathrm{c}}/t_{\mathrm{a}}$ is given by:

$$1.5 \leq t_{\mathrm{c}}/t_{\mathrm{a}} \leq 5 \tag{4.42}$$

for periods T_p such that

$$3\ \mathrm{s} \leq T_{\mathrm{p}} \leq 16\ \mathrm{s} \tag{4.43}$$

if $t_0 = 1$ s for $\delta = 10^{-2}$ cm and $D = 10^{-5}$ cm^2 s^{-1}. A good agreement between the shape and the frequency of the RC calculated in this way and the literature data is obtained, because in practically all cases is according to Bakhvalov [2]:

$$1\ \mathrm{s} \leq T_{\mathrm{p}} \leq 30\ \mathrm{s}. \tag{4.44}$$

On the other hand, the solution of Eqs. (4.9), (4.10), (4.11), and (4.12) for

$$i(t) = i_{\mathrm{c}} \tag{4.45}$$

is given by [19]:

$$C_s = C_0 - \frac{i_c\delta}{nFD}\left\{1 - \frac{8}{\pi^2}\sum_{k=0}^{\infty}\frac{1}{(2k+1)^2}\exp\left[-\frac{(2k+1)^2 t}{4t_0}\right]\right\} \tag{4.46}$$

It follows from Eq. (4.46) that the surface concentration of depositing ions for $t \geq t_0$ can be given by:

$$C_s = C_0 - \frac{i_c\delta}{nFD}\left[1 - \frac{8}{\pi^2}\exp\left(-\frac{t}{4t_0}\right)\right] \tag{4.47}$$

The maximum amplitude of the current density variation, $i_{A,max,}$ corresponding to $C_s = 0$ after a deposition time t_c, is given by:

$$i_{A,max} = \frac{i_L}{1 - (8/\pi^2)\exp(-t_c/4t_0)} \tag{4.48}$$

It is obvious that using Eqs. (4.1) and (4.4) and $r_{RC} = f(T_p)$ given by Eq. (4.41), Eq. (4.48) can be rewritten in the form:

$$i_{A,max} = \frac{i_L}{1 - (16/\pi^2)\frac{\exp\left(-T_p/4t_0\right)}{1+\exp\left(-T_p/4t_0\right)}} \tag{4.49}$$

In this way, the complete RC wave can be estimated or precisely calculated without approximations using a computer.

4.3 Prevention of the Formation of Spongy Deposits and the Effect on Dendritic Particles

4.3.1 *Basic Facts*

EPCR is used in the charging of silver-zinc storage batteries, to prevent, or to delay, the formation of spongy and dendritic deposits of zinc [20, 21]. It is impossible to obtain smooth deposits of zinc from alkaline zincate solutions during prolonged deposition at a constant rate due to formation of spongy deposits at lower and dendritic deposits at higher overpotentials [21, 22].

It is well known that the reversible potential of a surface with radius of curvature r_{cur} would depart from that of a planar surface by the quantity [23]:

$$\Delta E_r = \frac{2\gamma V}{F r_{cur}} \tag{4.50}$$

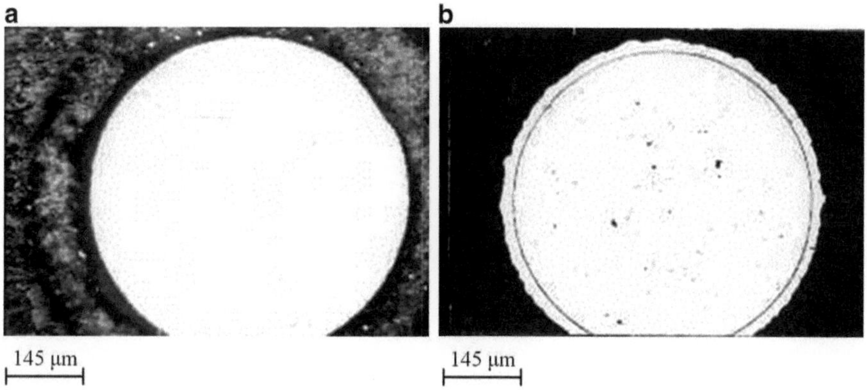

Fig. 4.4 Cross section photomicrographs of Zn deposits plated out from a 50 g dm^{-3} ZnO in 10 M KOH solution onto a copper wire: (**a**) constant overpotential, $\eta = 22$ mV, initial $i_{av} = 4.0$ mA cm^{-2}; deposition time: 40 s, and (**b**) PO, $p = 1$, $\eta_A = 40$ mV, $\nu = 100$ Hz, initial $i_{av} = 4.0$ mA cm^{-2}; deposition time: 4 h (Reprinted from Refs. [12, 21] with kind permission from Springer)

where γ is the interfacial energy between metal and solution. The filaments which form spongy deposits have extremely small tip radii. This makes the equilibrium potential of the spongy deposit to be 7–10 mV more cathodic than that of zinc foil [24, 25].

Spongy deposit formation can, however, be completely prevented by the PO deposition [21], as illustrated in Fig. 4.4.

Obviously, more negative filaments dissolve faster during the off period than the flat surface, resulting in a smooth deposit. This is also valid for deposition using current or overpotential waveforms that are characterized by some anodic current flow [3, 7]. This means that the dissolution of a protrusion with tip radius r_{cur} is faster relative to the flat surface or relative to a protrusion with a sufficiently large value of r_{cur}. It is obvious that spongy filaments can be completely dissolved (Fig. 4.4), while dendrites with low tip radii can be either partially or completely dissolved during the pause (Fig. 4.5). This means that both branching of dendrites and the formation of agglomerates can be prevented in the square-wave pulsating overpotential deposition. In this way, even powder particle like that in Fig. 4.6 can be formed.

The monocrystal surfaces of silver powder particles can be explained by the assumption that during the off period in the PO, the adatoms in nonstable positions will dissolve faster than atoms in a stable position in the lattice. The similar effect on the morphology of powder particles can be seen in the RC deposition [29, 30], which leads to the strong effect on the apparent density of copper powders [30].

Fig. 4.5 Pb deposits obtained from 0.10 M $Pb(CH_3COO)_2 + 1.5$ M $NaCH_3COO + 0.15$ M CH_3COOH on Cu substrate at a constant overpotential of: (**a**) 75 mV for 50 s, (**b**) the same as in (**a**) but for 350 s at 0 mV, (**c**) the same as in (**a**) but for 100 s at -10 mV (anodic), and (**d**) by PO with $\eta_A = 75$ mV, pause-to-pulse ratio: 3; pulse duration, 0.5 s (Reprinted from Ref. [12] with kind permission from Springer and Ref. [26] with permission from Elsevier)

4.3.2 Quantitative Treatment

4.3.2.1 One Electron Transfer Process

The one electron transfer processes were quantitatively discussed for the regime of the square-wave pulsating overpotential [31, 32]. The square-wave pulsating overpotential is described by Eqs. (4.9), (4.10), and (4.11), as well as by Eq. (4.51) [18, 31]:

$$\frac{\partial C(0,t)}{\partial x} = \frac{i_0}{nFD}\left[\frac{C(0,t)}{C_0}\exp\left(\frac{2.3\eta}{b_c}\right) - \exp\left(-\frac{2.3\eta}{b_a}\right)\right] \tag{4.51}$$

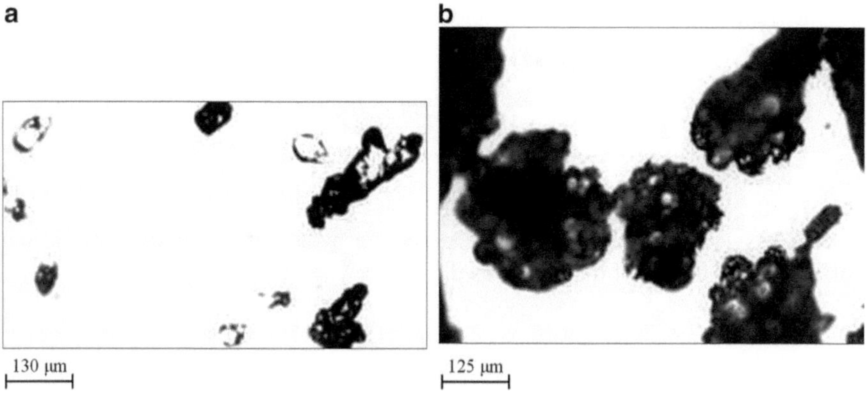

Fig. 4.6 Powder particles obtained by a square-wave PO regime: (**a**) silver. Pulse-to-pause ratio of 1:5. Pulse duration: 50 ms. $\eta_A = 160$ mV. Deposition was carried out from an electrolyte containing 10 g dm^{-3} $AgNO_3$ in 100 g dm^{-3} $NaNO_3$ onto a graphite electrode and (**b**) copper. Pulse-to-pause ratio of 1:5. Pulse duration: 5 ms. $\eta_A = 600$ mV. Deposition was carried out from 0.10 M $CuSO_4$ in 0.50 M H_2SO_4 onto a Pt electrode painted with shellac (Reprinted from Refs. [12, 27, 28] with kind permission from Springer)

In the case of a rectangular pulsating overpotential, $\eta(t)$ as a function of a time is given by Eq. (4.29).

Assuming that at sufficiently high frequencies the surface concentration in the pulsating overpotential deposition does not vary with time, it is easy to show that response of the current density, i, to the input overpotential:

$$i = \begin{cases} i_0\left(1 - \frac{i_{av}}{i_L}\right)\exp\left(\frac{2.3\eta_A}{b_c}\right) - i_0\exp\left(-\frac{2.3\eta_A}{b_a}\right) & \\ \quad \text{for } mT_p < t \le \left(m + \frac{1}{p+1}\right)T_p & \text{(a)} \\ -i_{av}\frac{i_0}{i_L} \text{ for} \left(m + \frac{1}{p+1}\right)T_p < t \le (m+1)T_p & \text{(b)} \end{cases} \tag{4.52}$$

Equations (4.9), (4.10), (4.11), (4.12), (4.29), (4.51), and (4.52) are valid for the flat electrode surfaces or protrusions with sufficiently large tip radii where the surface energy term, γ, can be neglected [23]. If it cannot be neglected, then the surface energy term affects the reaction rate [3], and for one electron transfer process, it is valid Eq. (4.53):

$$\frac{\partial C}{\partial x} = \frac{i_0}{nFD}\left[\left(1 - \frac{i_{av}}{i_L}\right)\exp\left(-\frac{2\beta\gamma V}{RTr_{cur}}\right)\exp\left(\frac{2.3\eta}{b_c}\right) - \exp\left(\frac{2(1-\beta)\gamma V}{RTr_{cur}}\right)\exp\left(-\frac{2.3\eta}{b_a}\right)\right] \tag{4.53}$$

where β, T, V, and R have already mentioned meanings.

The right-hand side of Eq. (4.51) should be transformed by taking Eq. (4.53) into account. The output current during pauses ($\eta = 0$) becomes:

$$i = -i_0 \exp\left(\frac{2(1-\beta)\gamma V}{RT r_{\mathrm{cur}}}\right) \tag{4.54}$$

if $r_{\mathrm{cur}} \to 0$.

It is easy to show that the difference between the current density on the flat surface and at the tip of the dendrites during the "off" period is given by:

$$\Delta i = i_0 - i_0 \exp\left(\frac{2(1-\beta)\gamma V}{RT r_{\mathrm{cur}}}\right) \tag{4.55}$$

and, if $i_{\mathrm{av}} \approx i_{\mathrm{L}}$, which leads to

$$h_{\mathrm{p}} = h_{\mathrm{p},0} + \frac{V i_0}{F}\left[1 - \exp\left(\frac{2(1-\beta)\gamma V}{RT r_{\mathrm{cur}}}\right)\right] t \tag{4.56}$$

because of

$$\frac{\mathrm{d}h_{\mathrm{p}}}{\mathrm{d}t} = \frac{V \Delta i}{F} \tag{4.57}$$

where h_{p} is the height of protrusion, and $h_{\mathrm{p},0}$ is the initial height of protrusion and taking Eq. (4.55) into account:

$$\frac{\mathrm{d}h_{\mathrm{p}}}{\mathrm{d}t} = \frac{V i_0}{F}\left[1 - \exp\left(\frac{2(1-\beta)\gamma V}{RT r_{\mathrm{cur}}}\right)\right] \tag{4.58}$$

Equation (4.56) represents the change of the height of the protrusion with tip radius r_{cur} relative to the flat surface or relative to the protrusion with a sufficiently large r_{cur}. In square-wave PO electrodeposition [7, 21], the filaments on the growing grains formed in the spongy electrodeposition can be completely dissolved during the pause duration leading to the formation of the compact deposit. In powder electrodeposition by the same regime, the dissolution of branches on the dendrite stalk is also expected.

Hence, the larger the "off" period, the less dendritic particles are obtained. On the other hand, the current density during the "on" period on the tip of dendrites growing inside the diffusion layer is given by:

$$i = i_0 \frac{h_{\mathrm{p}}}{\delta} \exp\left(\frac{2.3\eta_{\mathrm{A}}}{b_{\mathrm{c}}}\right) \tag{4.59}$$

which is a somewhat modified Eq. (4.52) [33]. For the same "on" period, the particles will be more dendritic with increasing overpotential amplitude. In the millisecond range, the ratio between the overpotential corresponding to the bulk diffusion control and the activation overpotential can be reduced to the value corresponding to electrodeposition at the lower overpotentials in the constant overpotential regime. Hence, the deposits obtained in the PO regimes (at the same η_A and with different p used) are more similar to those obtained in the constant overpotential regime ($p = 0$) at the lower overpotentials than those obtained in the constant regime at the overpotential corresponding to η_A in the PO regimes. The degree of diffusion control decreases with increasing p, even at the limiting diffusion current density, and it can become sufficient to produce the quality of deposits corresponding to a mixed, activation, or a surface energy control. This will be discussed in more detail in the case of lead electrodeposition.

4.3.2.2 Two Electron Transfer Process

Equations (4.9), (4.10), (4.11), (4.12), (4.29), (4.51), and (4.52) are also valid for the two electron transfer processes. These equations are valid for flat electrode surfaces or protrusions with sufficiently large tip radii, where the surface energy term can be neglected [23]. If it cannot be neglected, the effect of the surface energy term on the reaction rate [3] for two electron reaction steps is described by Eq. (4.60) [32]:

$$\frac{\partial C(0,t)}{\partial x} = \frac{i_0}{nFD}\left[\left(1 - \frac{i_{av}}{i_L}\right)\exp\left(\frac{2.3\eta}{b_c}\right) - \exp\left(\frac{2\gamma V}{RTr_{cur}}\right)\exp\left(-\frac{2.3\eta}{b_a}\right)\right] \tag{4.60}$$

The output current density, i, during pauses ($\eta = 0$) at the tip of the dendrite is presented by Eq. (4.61):

$$i = i_0\left(1 - \frac{i_{av}}{i_L}\right) - i_0\exp\left(\frac{2\gamma V}{RTr_{cur}}\right) \tag{4.61}$$

The corresponding output current density on the flat surface is given by Eq. (4.52b). The difference between the current density at the tip of the dendrite and on the flat surface during the "off" period is given by Eq. (4.62):

$$\Delta i = i_0 - i_0\exp\left(\frac{2\gamma V}{RTr_{cur}}\right) \tag{4.62}$$

if $i_{av} \approx i_L$, that is satisfied in most cases of dendrite growth.

Then, the change of height of surface protrusions with tip radius r_{cur} relative to the flat surface is given by Eq. (4.63) [21]:

$$\frac{dh_p}{dt} = \frac{V\Delta i}{nF} \tag{4.63}$$

and finally, for the two electron reaction steps:

$$h_{\mathrm{p}} = h_{\mathrm{p},0} + \frac{Vi_0}{2F}\left[1 - \exp\left(\frac{2\gamma V}{RTr_{\mathrm{cur}}}\right)\right]t \tag{4.64}$$

where

$$\left(m + \frac{1}{p+1}\right)T_{\mathrm{p}} < t \leq (m+1)T_{\mathrm{p}} \tag{4.65}$$

Equation (4.64) represents the height of the dendrite with tip radius r_{cur} as a function of time, relative to the flat surface or relative to the protrusion with a sufficiently large r_{cur}. It is obvious that dendrites with very low tip radii can be completely dissolved during the pause. This means that the branching of dendrites can be prevented in square-wave pulsating overpotential deposition. Obviously, the larger p, the greater the degree of dissolution, as followed from Eq. (4.65).

4.4 Compact Deposits

4.4.1 Surface Film

The first stage of metal film formation is nucleation on a foreign substrate. The nucleation rate in DC regime, J, is given by Eq. (2.86) and depends strongly on the deposition overpotential. The nucleation overpotential is larger than the stationary one in galvanostatic deposition, and the stationary values of the overpotential can be used in discussions of the effect of EPCR in galvanostatic as well as in potentiostatic deposition processes. It is obvious that in all types of EPCR, the overpotential amplitude, η_{A}, is larger than in constant-current or constant overpotential deposition, η, for the same average current density [34].

Therefore,

$$\eta_{\mathrm{A}} = \eta + \Delta\eta \tag{4.66}$$

where $\Delta\eta > 0$ and

$$J_{\mathrm{EPCR}} = K_1 \exp\left(-\frac{K_2}{(\eta + \Delta\eta)^2}\right) \tag{4.67}$$

Hence, for the same current density:

$$J_{\mathrm{EPCR}} > J \tag{4.68}$$

as illustrated in Fig. 4.7.

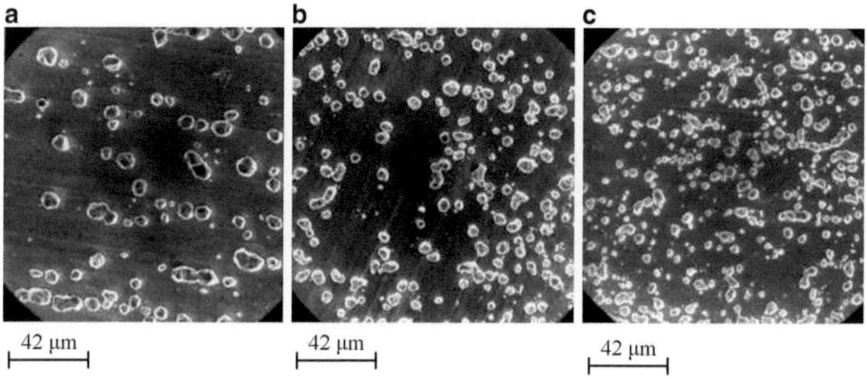

Fig. 4.7 SEM photomicrographs of cadmium deposits obtained from 1.0 M $CdSO_4$ in 0.50 M H_2SO_4 on a plane Cu cathode by the DC and the PC depositions at the different pause-to-pulse ratios. $i_{av} = 10$ mA cm^{-2}, $t = 120$ s: (**a**) DC, (**b**) PC; $p = 4$; $t_c = 10$ ms, and (**c**) PC; $p = 9$; $t_c = 10$ ms (Reprinted from Ref. [12] with kind permission from Springer and Ref. [34] with permission from Elsevier)

On the other hand, the increased amplitude on the current density leads to an increase in the ohmic potential drop during the pulses in EPCR relative to the constant regimes and Eq. (2.81) can be rewritten in the form:

$$r_{\text{EPCR}} = f\frac{U_{\Omega(\text{EPCR})}}{U_{\Omega(\text{EPCR})} - \eta_{\text{crit}}} r_{\text{N}} \tag{4.69}$$

It follows from Eqs. (2.81) and (4.69) that

$$r_{\text{EPCR}} < r_{\text{sz}} \tag{4.70}$$

because with increasing p

$$U_{\Omega(\text{EPCR})} > U_{\Omega} \tag{4.71}$$

Hence, the increasing nucleation density is also due to the decreasing zero nucleation zone radii. This effect leads to an increased coverage of the foreign substrate by the same quantity of deposited metal and to a decreased porosity, a surface resistance and an increased density of a deposit. Also, it can be expected that increase in compactness of a deposit is associated with a decrease in internal stresses and increased ductility and hardness of metal deposits [7].

4.4.2 Electrode Surface Coarsening

The general equation of the polarization curve for the flat part of an electrode is given by Eq. (1.13) and by Eq. (2.23) for the tip of a protrusion ($\delta \gg h_p$), around which the lateral diffusion flux can be neglected.

Using Eq. (4.27) in the form:

$$C_0 - C_s = C_0 \frac{i}{i_L} \tag{4.72}$$

it is easy to show that

$$C_0 - C_s = C_0 \frac{i_0 f_c}{i_L + i_0 f_c} \tag{4.73}$$

and

$$C_0 - C_{s,t} = C_0 \frac{i_0 f_c}{\frac{i_L \delta}{\delta - h_p} + i_0 f_c} \tag{4.74}$$

if $f_c \gg f_a$, where C_s and $C_{s,t}$ are the surface concentration of depositing ions on the flat electrode surface and on the tip of a protrusion, respectively.

The rate of growth of the tip of a protrusion relative to the flat surface is given by [35]:

$$\frac{dh_p}{dt} = VD\left(\frac{C_0 - C_{s,t}}{\delta - h_p} - \frac{C_0 - C_s}{\delta}\right) \tag{4.75}$$

or

$$\frac{dh_p}{dt} = \frac{i^2}{i_L} \frac{V}{\delta n F} h_p \tag{4.76}$$

after substitution of $C_0 - C_{s,t}$ and $C_0 - C_s$ from Eqs. (4.72) and (4.74) into Eq. (4.75) and further rearranging, assuming $\delta \gg h_p$.

It was shown earlier that at sufficiently high frequencies, the average current density in electrodeposition at a periodically changing rate produces the same concentration distribution inside the diffusion layer as a constant current density of the same intensity. Hence, Eq. (4.76) is valid for all cases of electrodeposition at a constant and periodically changing rate at sufficiently high frequencies.

However, an increase in surface coarseness in a deposition using a rectangular pulsating overpotential or pulsating current is only possible during the pulses of current or overpotential [35] and the integral form of Eq. (4.76) can be written as:

$$h_{\mathrm{p}} = h_{\mathrm{p},0}\exp\left(\frac{i}{i_{\mathrm{L}}}\frac{V}{\delta nF}\frac{Q}{p+1}\right) \tag{4.77}$$

and, if $i = i_{\mathrm{av}}$, and Q is given by Eq. (2.27).

Equation (4.77) is valid for pulsating current, pulsating overpotential, and reversing current in the millisecond range under the assumption that the entire surface dissolves uniformly during the pauses. The deposits obtained by constant and pulsating overpotential in the mixed control under the other conditions that are the same are shown in Fig. 4.8a, b. The deposit obtained by pulsating overpotential is considerably less rough.

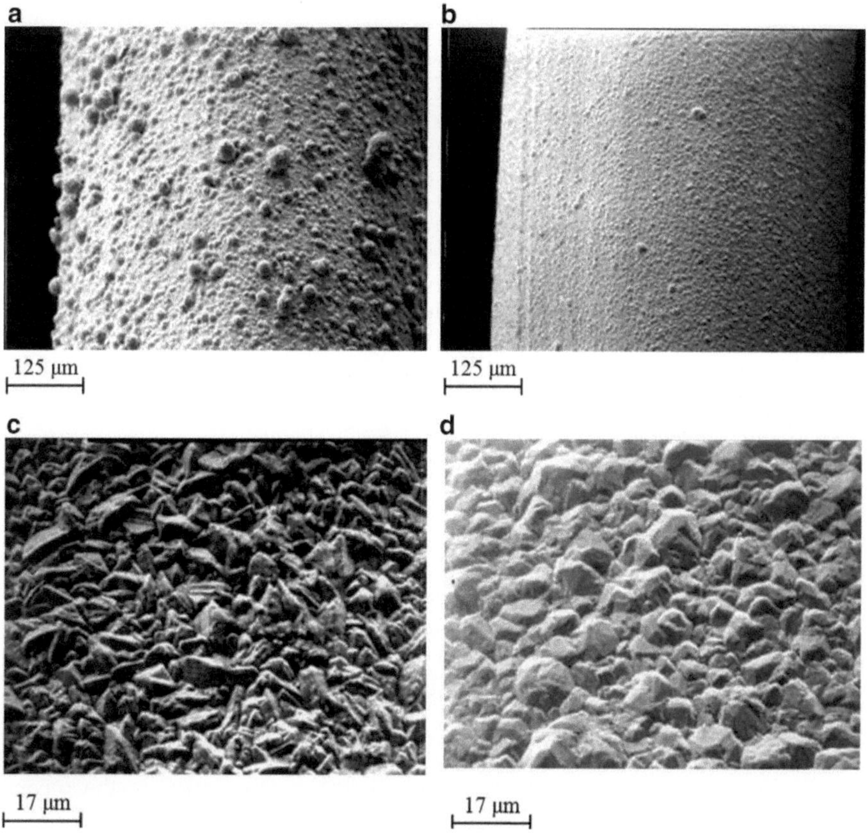

Fig. 4.8 Copper deposits obtained from 0.10 M $CuSO_4$ in 0.50 M H_2SO_4 on a Cu wire electrode with a quantity of electricity of 20 mA h cm^{-2} in: (**a**) the constant potentiostatic regime; an overpotential: 210 mV; initial current density: 6.5 mA cm^{-2}, as well as by the different pulsating overpotential (PO) regimes: (**b**) initial average current density: 6.5 mA cm^{-2}; $\eta_A = 322$ mV; pulse-to-pause ratio: 3, (**c**) initial average current density: 2.9 mA cm^{-2}; $\eta_A = 210$ mV; pause-to-pulse ratio: 3, and (**d**) initial average current density: 6.5 mA cm^{-2}; $\eta_A = 322$ mV; pause-to-pulse ratio: 3 (Reprinted from Ref. [12] with kind permission from Springer and Ref. [35] with permission from Elsevier)

The copper deposits obtained under activation and mixed controls as those shown in Fig. 2.4 are shown in Fig. 4.8c, d. A considerable decrease in the grain size of deposit obtained at the low current densities (in the activation controlled region; Figs. 2.4ac and 4.8) due to the increase of the amplitude of the overpotential relative to the corresponding value in constant overpotential deposition can be seen. There is no qualitative change, however, in the structure of the deposit.

A qualitative change in the structure of the deposit appears in the mixed-controlled region (Figs. 2.4c and 4.8d). It can be seen that the protrusions caused by mass transport limitations are strongly reduced relative to the deposits shown in Fig. 2.4c, but the grain size is enlarged. It is obvious that the grains obtained by the PO regime with current densities belonging to the region of the mixed control (Fig. 4.8d) are almost regular in comparison with those deposited under the activation control (Fig. 4.8c). This is due to the increase of both the degree of activation control during the overpotential pulses and the grain size as a result of the selective dissolution during the "off" periods (in relation to those shown in Fig. 2.4b, c). The smaller nuclei formed during the overpotential pulse will be completely or partially dissolved during the overpotential pause, and the current density on the partially dissolved nuclei will be considerably lower during the next overpotential pulse than the one on larger ones because of their more negative reversible potentials. In this way, the growth of larger grains will be favored.

Then, the deposit shown in Fig. 2.4c changes and becomes like the one shown in Fig. 4.8d, which is formed by use of the PO deposition at the same average current density with the same quantity of deposited metal. It can be also seen from Fig. 4.8c, d that a good deposit can be obtained by the PO deposition over wide range of current densities. This means that in EPCR deposition, current density can be considerably increased relative to the DC regime. Nevertheless, it seems that the PO and PC regimes are not suitable for the prolonged depositions, due to the formation of large grains.

On the other hand, it is known that the orientation of nuclei strongly depends on the depositing overpotential and that the electrode reaction parameters can be different for different crystal planes. Therefore, it is not surprising that the effect of structure on EPCR has been reported for many cases. In some cases, deposits which behave as monocrystals [7] and deposits with improved crystal perfection can be obtained.

It is obvious that the same reasoning is valid for the RC in the millisecond range and the PC. Some different situations appear in the case of RC in the second range [36].

The surface concentration changes during the cathodic pulse in RC deposition in the second range according to Eq. (4.46).

$$\frac{C_\mathrm{s}}{C_0} = 1 - \frac{i_\mathrm{c}}{i_\mathrm{L}} f(t) \tag{4.78}$$

where

$$f(t) = 1 - \frac{8}{\pi^2}\sum_{k=1}^{\infty}\frac{1}{(2k+1)^2}\exp\left[-\frac{(2k+1)^2 t}{4t_0}\right] \quad \text{and} \quad t_0 = \frac{\delta^2}{\pi^2}D \tag{4.79}$$

In this case

$$i_c = i_0\frac{C_s}{C_0}f_c \tag{4.80}$$

is also valid and substitution of $\frac{C_s}{C_0}$ from Eq. (4.78) in Eq. (4.80) gives

$$i_{tip} = i_0 f_c\left[1 - \frac{i_{tip}}{i_L}\frac{\delta - h_p}{\delta}f(t)\right] \tag{4.81}$$

for the tip of a protrusion and

$$i_c = i_0 f_c\left[1 - \frac{i_c}{i_L}f(t)\right] \tag{4.82}$$

for a position on a flat surface. If the whole surface is iso-potential, elimination of f_c from Eqs. (4.81) and (4.82) after rearranging produces

$$i_{tip} = \frac{i_c}{1 - \frac{i_c}{i_L}\frac{h_p}{\delta}f(t)} \tag{4.83}$$

The difference in the current densities between the tip of a protrusion and the flat portion of the electrode surface is then given by:

$$\Delta i = i_{tip} - i_c = \frac{i_c^2}{i_L}\frac{h_p}{\delta}f(t) \tag{4.84}$$

for $\delta \gg h_p$.

Now, according to Eqs. (4.63) and (4.84) it follows:

$$\frac{dh_p}{dt} = \frac{i_c^2}{i_L}\frac{h_p}{\delta}\frac{V}{nF}f(t) \tag{4.85}$$

or in the integral form:

$$h_p = h_{p,0} + \frac{i_c^2}{i_L{}^L}\frac{V}{\delta nF}F(t) \tag{4.86}$$

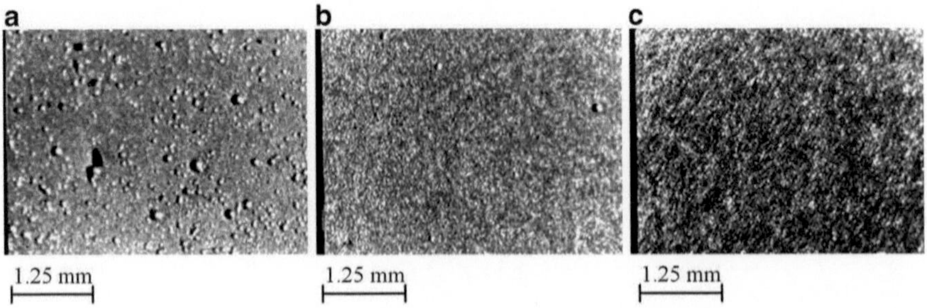

Fig. 4.9 Photomicrographs of copper deposited on a flat Pt electrode from 0.50 M $CuSO_4$ in 0.50 M H_2SO_4; $i_{av} = 0.9\ i_L$; $t = 60$ min: (**a**) DC; (**b**) RC, $r_{RC} = 1/7$, $T_p = 16$ s, and (**c**) PC, $p = 1$, $\nu = 10$ Hz (Reprinted from Ref. [12] with kind permission from Springer and Ref. [37] with permission from Elsevier)

where

$$F(t) = \int_0^{t_c} f(t)\,\mathrm{d}t \tag{4.87}$$

for the first phase.

Assuming that the surface will dissolve uniformly during the anodic period (it is because $f(0) = 0$ and $f(\infty) = 1$), the increase in the surface coarseness in the RC regime will be lower than in the DC regime until the condition

$$t_c << \infty \tag{4.88}$$

is satisfied. The above reasoning is valid if a polycrystalline deposit is obtained and the same derivation as for the case of electrodeposition at a constant overpotential can be used.

It can be seen from Fig. 4.9 that the structure of the deposit obtained by the RC in the second range is more similar to the one obtained in the DC than in the PC regime. However, the surface coarseness of this deposit is considerably lower than in the DC regime, and it is close to the surface coarseness obtained in the PC deposition. This is because in the RC deposition, there is a considerable concentration polarization, producing polycrystalline deposit.

4.5 Current Density and Morphology Distribution on a Macroprofile

The current density distribution on a macroprofile in EPCR has been treated in Ref. [7]. It seems that the current density distribution is improved by application of current waves with an anodic flow, but without this flow, it is worse than in the DC deposition. This behavior can be successfully explained [38].

Assuming that Eq. (3.49) is also valid for the flat electrode in a cell with a low anode polarization, the current densities in the middle, i, and at the edge, i_e of a flat electrode in a cell with a low anode polarization can be related by:

$$i_e = i\exp\left(\frac{2.3RI}{b_c}\right) \tag{4.89}$$

Equation (4.89) is valid if the electrodeposition process is an under activation control in the Tafel region, the limiting diffusion current density is the same in both the middle and at the edge of the electrode, and I is the cell current corresponding to i. During the current pulses, the amplitude values of current densities and current should be substituted in Eq. (4.89) producing

$$i_{e,A} = i_c\exp\left(\frac{2.3RI_A}{b_c}\right) \tag{4.90}$$

where $i_{e,\,A}$ and i_c are the amplitude values of current densities at the edge and in the middle of electrode and I_A is the amplitude of the current in the cell. On the other hand, the amplitude in pulsating current deposition and the average current density are related by Eq. (4.7), so Eq. (4.90) can be rewritten in the form:

$$i_{e,\,av} = i_{av}\exp\left(\frac{2.3RI(p+1)}{b_c}\right) \tag{4.91}$$

assuming that $i = i_{av}$, if

$$\exp\left(\frac{2.3RI(p+1)}{b_c}\right) > \exp\left(\frac{2.3RI}{b_c}\right) \tag{4.92}$$

then $i_{av,e} > i_{av}$, meaning worse current density distribution in the PC than in the DC conditions.

The effect of reversing current on the current distribution at the macroprofile level can easily be discussed for the case of activation-controlled deposition if the Tafel slopes of the anodic and cathodic processes are different, as they are for copper deposition and dissolution from sulfate solutions. With the assumption that the current density in the RC deposition is sufficiently high so that the effect of the opposing processes can be neglected, the limiting diffusion current density is the same over all electrode surface. The difference between the current density at the edge and the one in the middle of the electrode in a cathodic deposition is:

$$\Delta i_c = i_c\left[\exp\left(\frac{2.3RI_c}{b_c}\right) - 1\right] \tag{4.93}$$

and for an anodic case

$$\Delta i_a = i_a \left[\exp\left(\frac{2.3 \cdot 3RI_a}{b_c} \right) - 1 \right] \tag{4.94}$$

since $b_a = b_c/3$ for copper deposition, where I_c and I_a are the cell currents corresponding to i_c and i_a. It is obvious that for:

$$\Delta i_c t_c = \Delta i_a t_a \tag{4.95}$$

which occurs when

$$\frac{t_c}{t_a} = \frac{\Delta i_a}{\Delta i_c} = \frac{\exp\left(\frac{2.3 \cdot 3RI}{b_c}\right) - 1}{\exp\left(\frac{2.3RI}{b_c}\right) - 1} \tag{4.96}$$

(if $i = i_a = i$ and $I_c = I_a = I$), deposits of equal thickness can be obtained at the edge and in the middle of the electrode. In this way, a completely uniform average current density distribution on the macroprofile level can be obtained in RC deposition. The diffusion limitations of the cathodic processes will improve the distribution in the RC, but this approach is sufficient to explain the essence of the effect, as illustrated in Fig. 4.10 [38].

The best current distribution is expected in the case of PO if the whole electrode surface can be taken as an iso-potential. Under the assumption that the limiting diffusion current density does not vary over the electrode surface area, the same current density can be expected over all points of the electrode. A good approximation of the PO deposition can be the RC deposition by the current wave optimized relative to the current density distribution on both the microprofile and macroprofile [39]. Hence, it seems that the RC should be the optimum regime of EPCR. Besides, the crack-free chromium deposits with an improved current density

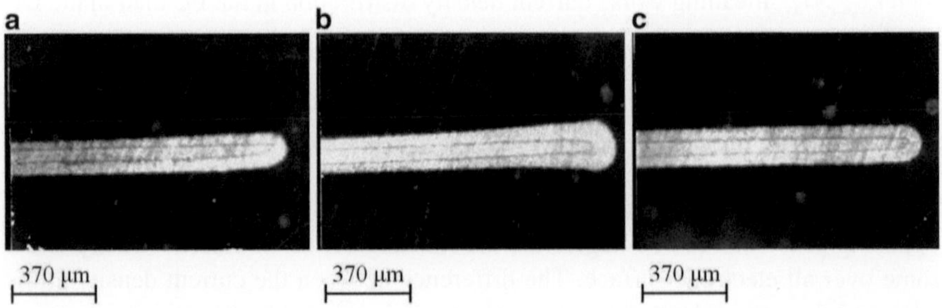

Fig. 4.10 Crosssection of copper deposits obtained from 1.0 M $CuSO_4$ in 1.0 M H_2SO_4 at the edge of plane Cu electrodes, previously plated with bright nickel. $i_{av} = 10$ mA cm^{-2}, $t = 4$ h: (**a**) DC, (**b**) PC, $p = 1$, 50 Hz, and (**c**) RC, $r_{RC} = 1/7$, $T_p = 8$ s (Reprinted from Ref. [12] with kind permission from Springer and Ref. [38] with permission from Elsevier)

distribution on both the microprofile and macroprofile and with practically no reduced hardness were obtained by the RC deposition [40, 41]. This means that the formation of unstable chromium hydride can also be prevented by the RC, but this phenomenon has been not treated semiquantitatively so far.

References

1. Bibikov N (1961) Electrodeposition of metals by AC. Mashgiz, Moscow/Leningrad (in Russian)
2. Bakhvalov G (1966) New technology of metal electrodeposition. Metallurgiya, Moscow (in Russian)
3. Despić AR, Popov KI (1972) Transport controlled deposition and dissolution of metals. In: Conway BE, Bockris JO'M (eds) Modern aspects of electrochemistry, vol 7. Plenum, New York, pp 199–313
4. Ibl N, Puippe JA, Angerer H (1978) Electrocrystallization in pulse electrolysis. Surf Technol 6:287–330
5. Polukarov YM, Grinina VV (1985) Electrodeposition of metals by periodically changing rate and single pulses. In: Polukarov YM (ed) Itogi nauki i tekhniki, Seriya Elektrokhimiya, vol 22. Acad Sci Moscow, Moscow (in Russian)
6. Puippe JC, Leaman F (eds) (1986) Theory and practice of pulse plating. American Electroplate and Surface Finishing Society, Orlando
7. Popov KI, Maksimović MD (1989) Theory of the effect of electrodeposition at a periodically changing rate on the morphology of metal deposits. In: Conway BE, Bockris JO'M, White RE (eds) Modern aspects of electrochemistry, vol 19. Plenum Press, New York, pp 193–250
8. Pesco A, Cheh H (1989) Theory and application of periodic electrolysis. In: Conway BE, Bockris JO'M, White RE (eds) Modern aspects of electrochemistry, vol 19. Plenum Press, New York, pp 251–293
9. Popov KI, Pavlović MG (1993) Electrodeposition of metal powders with controlled grain size and morphology. In: White RE, Bockris JO'M, Conway BE (eds) Modern aspects of electrochemistry, vol 24. Plenum Press, New York, pp 299–391
10. Maksimović MD (1995) The influence of the charge and discharge of the electrical double layer in electrodeposition by periodically changing rate. J Serb Chem Soc 60:449–458
11. Maksimović MD, Popov KI (1999) Mass transfer during electrodeposition of metals at a periodically changing rate. J Serb Chem Soc 64:317–340
12. Popov KI, Djokić SS, Grgur BN (2002) Fundamental aspects of electrometallurgy. Kluwer Academic/Plenum Publishers, New York, pp 101–143, Ch 4
13. Viswanathan K, Cheh HY (1979) Mass transfer aspect of electrolysis by periodic currents. J Electrochem Soc 126:398–401
14. Chin DT (1983) Mass transfer and current-potential relation in pulse electrolysis. J Electrochem Soc 130:1657–1667
15. Popov KI, Maksimović MD, Simić MS (1982) The effect of periodic reverse current on the surface rougneses of metal deposits and maximal deposition rate. Surf Technol 16:209–218
16. Popov KI, Maksimović MD, Ocokoljić BM, Lazarević BJ (1980) Fundamental aspects of pulsating current metal electrodeposition I: the effect of the pulsating current on the surface roughness and the porosity of metal deposits. Surf Technol 11:99–109
17. Maksimović MD, Totovski DĆ, Ivić AP (1983) The effect of AC superimposed on DC in the electrodeposition of metals. Surf Technol 18:233–241
18. Popov KI, Maksimović MD, Zečević SK, Stojić MR (1986) Surface roughening and dendritic growth in pulsating overpotential copper electrodeposition. Surf Coat Technol 27:117–129

19. Popov KI, Maksimović MD (1991) Maximum deposition rate in metal electrodeposition by reversing current in the second range. J Serb Chem Soc 56:25–31
20. Romanov VV (1963) The Zinc spongy electrodeposits formation during deposition from zincate alkaline solution. Zh Prikl Khim 36:1057–1063 (in Russian)
21. Popov KI, Keča DN, Anđelić MD (1978) Electrodeposition of zinc on copper from alkaline zinkate solutions. J Appl Electrochem 8:19–23
22. Diggle JW, Despić AR, Bockris JO'M (1969) The mechanism of the dendritic electrocrystallization of zinc. J Electrochem Soc 116:1503–1514
23. Barton JL, Bockris JO'M (1962) The electrolytic growth of dendrites from ionic solutions. Proc Roy Soc London A 268:485–505
24. Bek RY, Kudryavtsev NT (1961) The effect of a periodically changing rate on the zinc electrodeposition from zincate alkaline solutions. Zh Prikl Khim 34:2020–2027 (in Russian)
25. Arouete S, Blurton KF, Oswin HG (1969) Controlled current deposition of zinc from alkaline solution. J Electrochem Soc 116:166–169
26. Popov KI, Stojilković ER, Radmilović V, Pavlović MG (1997) Morphology of lead dendrites electrodeposited by square-wave pulsating overpotential. Powder Technol 93:55–61
27. Pavlović MG, Maksimović MD, Popov KI, Kršul MB (1978) The effect of pulsating overpotential on the morphology of electrodeposited silver powder particles. J Appl Electrochem 8:61–65
28. Popov KI, Maksimović MD, Pavlović MG, Ostojić GR (1977) Formation of powdered copper deposits by square-wave pulsating overpotential. J Appl Electrochem 7:331–337
29. Pavlović MG, Pavlović LjJ, Stojilković ER, Radmilović V, Popov KI (2001) The effect of particle structure on apparent density of electrolytic copper powder. J Serb Chem Soc 66:923–933
30. Popov KI, Pavlović LjJ, Stojilković ER, Radmilović V, Pavlović MG (2002) The effect of reversing current deposition on apparent density of electrolytic copper powder. J Serb Chem Soc 67:61–67
31. Popov KI, Pavlović MG, Stojilković ER, Radmilović V (1996) Silver powder electrodeposition by constant and pulsating overpotential. J Serb Chem Soc 61:47–55
32. Popov KI, Pavlović MG, Remović GŽ (1991) The effect of the pause-to-pulse ratio on the morphology of metal powder particles electrodeposited by square-wave pulsating overpotential. J Appl Electrochem 21:743–745
33. Popov KI, Pavlović MG, Jovičević JN (1989) Morphology of tin powder particles obtained in electrodeposition on copper cathode by constant and square-wave pulsating overpotential from Sn(II) alkaline solution. Hydrometallurgy 23:127–137
34. Popov KI, Maksimović MD, Stevanović RM, Krstajić NV (1984) Fundamental aspect of pulsating current metal electrodeposition. VIII: the effect pulse to pause ratio on microthrowing power of metal deposits. Surf Technol 22:155–158
35. Popov KI, Pavlović MG, Pavlović LjJ, Čekerevac MI, Remović GŽ (1988) Electrode surface coarsening in pulsating overpotential copper electrodeposition. Surf Coat Technol 34:355–363
36. Popov KI (1992) The effect of periodically changing rate in electrodeposition on the morphology of metal deposits. In: Dekanski A, Pavlović MG (eds) Application of pulsating regimes in electroplating. ICTM–Department of Electrochemistry, Belgrade, pp 49–73 (in Serbian)
37. Popov KI, Maksimović MD, Totovski DĆ (1982) Fundamental aspects of pulsating current metal electrodeposition. VI: the comparison of electrode surface roughening in pulsating current and periodic reverse current electrodeposition of metals. Surf Technol 17:125–129
38. Popov KI, Totovski DĆ, Maksimović MD (1983) Fundamental aspects of pulsating current metal electrodeposition. VII: the comparison of current density distribution in pulsating current and periodic reverse current electrodeposition of metals. Surf Technol 19:181–185

39. Popov KI, Kostić TM, Stojilković ER, Nikolić ND, Pavlović MG (1998) The determination of the optimum current wave in reversing current metal electrodeposition. J Serb Chem Soc 63:537–544
40. Petrović BM, Kostić TM (2000) The properties of chromium electrodeposited with programed currents. Part I: direct current. J Serb Chem Soc 65:55–63
41. Petrović BM, Kostić TM (2000) The properties of chromium electrodeposited with programed currents. Part II: reversing current. J Serb Chem Soc 65:65–72

Chapter 5
Electrodeposition of Metals with Hydrogen Evolution

5.1 Introduction

Hydrogen generated during electrodeposition processes can achieve a significant influence on morphology of electrodeposited metal. This effect is especially important during electrodeposition of metals characterized by low (so-called *the intermediate metals*, like Cu) and very low (so-called *the inert metals*, like Ni, Co, Fe, Pt, Cr) overpotentials for hydrogen discharge [1]. In the case of Cu, hydrogen evolution commences at some overpotential belonging to the plateau of the limiting diffusion current density. The increase of the overpotential intensifies this reaction (see Fig. 1.10a). For copper solution containing 0.10 M $CuSO_4$ in 0.50 M H_2SO_4, the plateau of the limiting diffusion current density corresponds to the range of overpotentials between 300 and 750 mV, and hydrogen evolution as the second reaction commences at an overpotential of 590 mV [2]. The quantity of evolved hydrogen is determined by the current efficiency for hydrogen evolution reaction, $\eta_{I,av}(H_2)$. As presented in Table 5.1, the current efficiency of hydrogen evolution increases with an increase in overpotential.

Figure 5.1 shows morphologies of Cu deposits obtained at overpotentials inside the plateau of the limiting diffusion current density (Fig. 5.1a – 550 mV and Fig. 5.1b – 650 mV) and outside this plateau (Fig. 5.1c, d – 850 mV and Fig. 5.1e, f – 1000 mV).

The cauliflower-like agglomerates of copper grains were formed at an overpotential of 550 mV, where there was no hydrogen evolution (Fig. 5.1a). The very branchy 3D (three dimensional) dendrites were formed at an overpotential of 650 mV where hydrogen evolution was very small and corresponds to $\eta_{I,av}(H_2)$ of about 2.0 % (Fig. 5.1b) [2]. Aside from very branchy dendrites, small cauliflower-like forms and individual holes formed from detached hydrogen bubbles (part in circles) were also obtained at this overpotential. The quantity of evolved hydrogen at 650 mV ($\eta_{I,av}(H_2) = 2.0$ %) was not sufficient to achieve any effect on the Cu electrodeposition process. The strong nonuniform electrode surface obtained at

K.I. Popov et al., *Morphology of Electrochemically and Chemically Deposited Metals*, DOI 10.1007/978-3-319-26073-0_5

Table 5.1 The values of the average current efficiency of hydrogen evolution, $\eta_{I,av}(H_2)$, at overpotentials, η, within the range from 550 to 1000 mV

η/mV	550	650	850	1000
$\eta_{I,av}(H_2)$/%	0	2.0	30.7	46.8

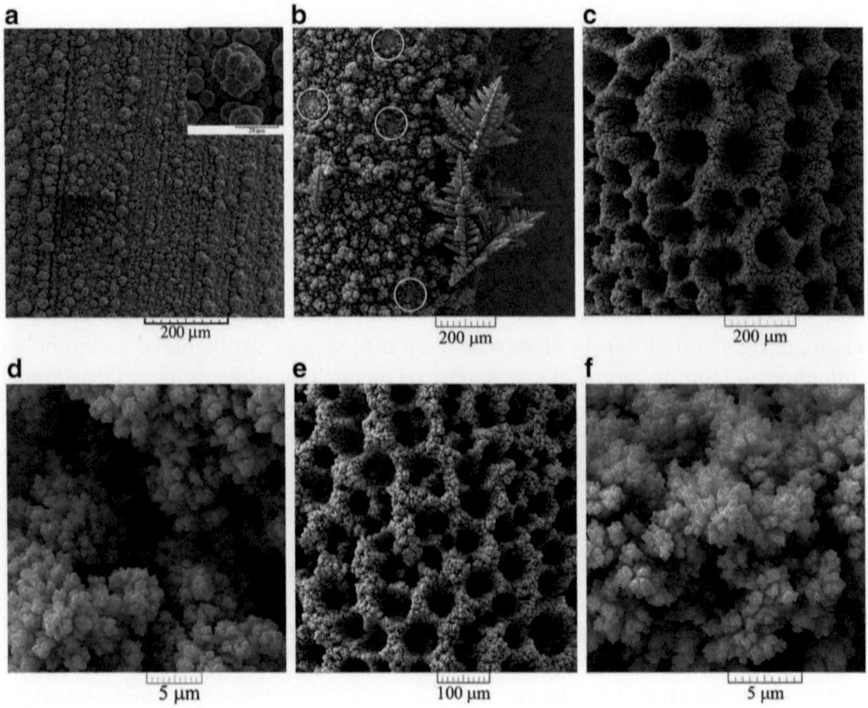

Fig. 5.1 Morphologies of Cu deposits obtained by electrodeposition at overpotentials of (**a**) 550 mV, (**b**) 650 mV, (**c, d**) 850 mV, and (**e, f**) 1000 mV (Reprinted from Ref. [2] with permission from Society of Chemists and Technologists of Macedonia)

650 mV (very branchy dendrites, small cauliflower-like agglomerates of Cu grains, and rare holes with an origin of detached hydrogen bubbles) clearly confirms that the diffusion layer of the macroelectrode is not disturbed by this quantity of generated hydrogen. A completely different situation was observed during the electrodeposition of Cu at overpotentials outside the plateau of the limiting diffusion current density in the zone of the fast increase in current density with increasing overpotential (Fig. 5.1c–f). In both cases, holes formed of detached hydrogen bubbles (Fig. 5.1c, e) surrounded by cauliflower-like agglomerates of copper grains (Fig. 5.1d, f) were obtained. It is necessary to note that the number of holes formed at an overpotential of 1000 mV (Fig. 5.1e) was larger than the number of holes formed at an overpotential of 850 mV (Fig. 5.1c). It is understandably due to more vigorous hydrogen evolution at 1000 mV ($\eta_{I,av}(H_2) = 46.8$ %) than at

850 mV ($\eta_{I,av}(H_2) = 30.7$ %) [2]. This type of deposits is denoted as the 3D (three dimensional) foam [3] or the honeycomb-like structures [4, 5]. The uniform distribution of morphological forms clearly points out the strong effect of evolved hydrogen on formation of this deposit type. The concept of "effective overpotential" is proposed to explain formation of this structure type [4, 5].

5.2 Mechanism of Formation of the Honeycomb-Like Structure: The Concept of "Effective Overpotential"

The hydrogen evolution influences the hydrodynamic conditions inside electrochemical cell [6–8]. The increase in hydrogen evolution rate leads to the decrease of the diffusion layer thickness and hence to the increase of limiting diffusion current density of electrode processes. It was shown [6] that if the rate of gas evolution at the electrode is larger than 100 cm^3/cm^2 min (>5 A cm^{-2}), the diffusion layer becomes only a few micrometers thick. A coverage of an electrode surface with gas bubbles can be about 30 % [6]. If the thickness of the diffusion layer in conditions of natural convection is ~$5 \cdot 10^{-2}$ cm and in strongly stirred electrolyte ~$5 \cdot 10^{-3}$ cm [9], it is clear that gas evolution is the most effective way of the decrease of mass transport limitations for electrochemical processes in mixed activation–diffusion control.

For electrochemical process in mixed activation–diffusion control, the overpotential η and the current density i are related by Eq. (1.31) [10]:

The first term in Eq. (1.31) corresponds to the activation part of deposition overpotential, and the second one is due to the mass transfer limitations. If one and the same process occurs under two different hydrodynamic conditions, characterized by two different values of the limiting diffusion current densities $i_{L,1}$ and $i_{L,2}$, the Eq. (1.31) can be rewritten in the forms:

$$\eta_1 = \frac{b_c}{2.3} \ln \frac{i_1}{i_0} + \frac{b_c}{2.3} \ln \frac{1}{1 - \frac{i_1}{i_{L,1}}} \tag{5.1}$$

and:

$$\eta_2 = \frac{b_c}{2.3} \ln \frac{i_2}{i_0} + \frac{b_c}{2.3} \ln \frac{1}{1 - \frac{i_2}{i_{L,2}}} \tag{5.2}$$

where η_1 and η_2 and i_1 and i_2 are the corresponding values of overpotentials and current densities. The same degree of diffusion control is obtained if:

$$\frac{i_1}{i_{L,1}} = \frac{i_2}{i_{L,2}} \tag{5.3}$$

or:

$$i_2 = i_1 \frac{i_{L,2}}{i_{L,1}} \tag{5.4}$$

and substitution of i_2 from the Eq. (5.4) in the Eq. (5.3) and further rearranging gives:

$$\eta_2 = \frac{b_c}{2.3} \ln \frac{i_1}{i_0} + \frac{b_c}{2.3} \ln \frac{1}{1 - \frac{i_1}{i_{L,1}}} + \frac{b_c}{2.3} \ln \frac{i_{L,2}}{i_{L,1}} \tag{5.5}$$

and if the Eq. (5.1) is taken into account:

$$\eta_2 = \eta_1 + \frac{b_c}{2.3} \ln \frac{i_{L,2}}{i_{L,1}} \tag{5.6}$$

Hence, if

$$i_{L,2} > i_{L,1} \tag{5.7}$$

in order to obtain the same degree of diffusion control in two hydrodynamic conditions, the Eq. (5.6) must be satisfied, meaning that:

$$\eta_2 > \eta_1 \tag{5.8}$$

In the absence of strong hydrogen evolution, the diffusion layer is due to the natural convection and does not depend on the overpotential of electrodeposition. The vigorous hydrogen evolution changes the hydrodynamic conditions and decreases the degree of diffusion control. Hence, the Eq. (5.6) should be rewritten in the form:

$$\eta_1 = \eta_2 - \frac{b_c}{2.3} \ln \frac{i_{L,2}}{i_{L,1}} \tag{5.9}$$

where η_1 becomes the effective overpotential, $\eta_1 = \eta_{eff}$, related to conditions of natural convection at which there is the same degree of diffusion control as at overpotential η_2 with the hydrogen evolution. Hence, the dendritic growth can be delayed or completely avoided due to the strong hydrogen evolution, as can be seen from Fig. 5.1d, f.

The concept of "effective overpotential" can be summarized as follows. When hydrogen evolution is vigorous enough, the electrodeposition process occurs at an overpotential which is effectively lower than specified, and this overpotential is denoted as "effective" in a deposition process. The vigorous hydrogen evolution causes a strong stirring of the electrolyte in the near-electrode layer, leading to a decrease in the thickness of the diffusion layer, an increase in the limiting diffusion current density, and a decrease in the degree of diffusion control of the electrodeposition process. From the morphological point of view, this means that the morphologies of metal deposits become similar to those obtained at some lower overpotentials at which hydrogen evolution does not occur or is very slow. The formation of cauliflower-like agglomerates of copper grains instead of branchy dendrites clearly indicates the existence of a smaller degree of diffusion control at overpotentials of 850 and 1000 mV than at 650 mV.

5.2.1 The Concept of "Effective Overpotential" Applied for Metal Electrodeposition Under an Imposed Magnetic Field

The concept of "effective overpotential" is applicable for all cases where there is the change of the hydrodynamic conditions in the near-electrode layer, which can be induced by the agitation of electrolyte not only by evolving hydrogen but also by application of ultrasonic [11] and magnetic [12, 13] fields or simply by vigorous stirring of an electrolyte [14].

Of course, the influence of magnetic field appears to be restricted to the diffusion-limited regions. Unlike the very branchy Cu dendrites (the full diffusion control), cauliflower-like structures (a dominant diffusion in mixed control of electrodeposition process) are formed with a magnetic field of 39.8 kA m^{-1} applied to be parallel to the electrode surface (Fig. 5.2) [13]. Actually, application of parallely oriented magnetic fields (the largest magnetohydrodynamic (MHD) effect) of 39.8 kA m^{-1} leads to shifting of the formation of characteristic morphological forms toward lower cathodic potentials for about 100–200 mV. During electrolysis under parallel fields, the Lorentz force induces convective flow of the electrolyte close to electrode surface. A magnetically stimulated convection leads

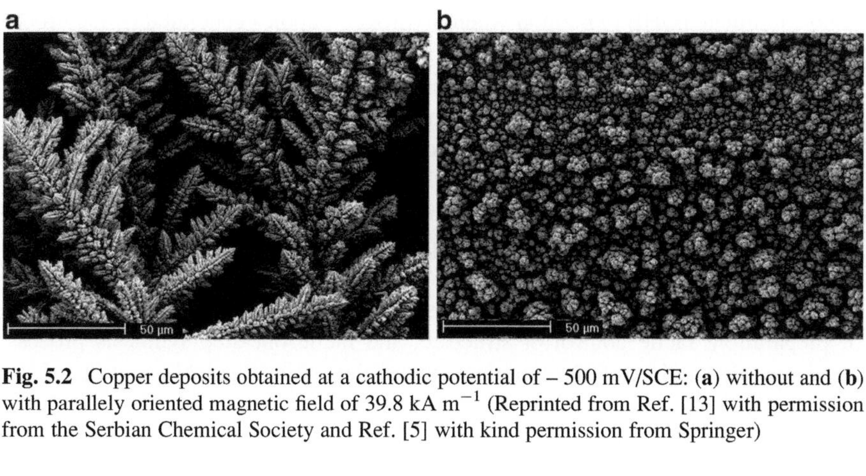

Fig. 5.2 Copper deposits obtained at a cathodic potential of – 500 mV/SCE: (**a**) without and (**b**) with parallely oriented magnetic field of 39.8 kA m^{-1} (Reprinted from Ref. [13] with permission from the Serbian Chemical Society and Ref. [5] with kind permission from Springer)

to a decrease of the diffusion layer thickness thus increasing the diffusion-limited current density. As a rule, it was adopted that the limiting diffusion current density depends on magnetic field, as $i_L \propto {B_{mf}}^{1/3}$, where B_{mf} is the magnetic field [15]. The increase of the limiting current density caused by the effect of applied magnetic fields with a parallel orientation leads to a decrease of the degree of diffusion control of the deposition process, and then, the electrodeposition process occurs at some overpotential which is effectively lower. This overpotential at which a metal electrodeposition occurs when the change of hydrodynamic conditions is caused by the effect of imposed magnetic fields (*i.e.*, by the magnetohydrodynamic effect) also represents "effective overpotential" of electrodeposition process.

Similar effects can be observed during electrodeposition in an ultrasonic field [11]. Copper deposits obtained in an ultrasonic field were compact and more ordered structure than copper deposits obtained without an effect of ultrasonic fields.

5.3 The Honeycomb-Like Structures: Basic Facts, Phenomenology, and Factors Affecting Their Formation (Cu as a Model System)

5.3.1 Basic Facts

The 3D foam or honeycomb-like structure is the type of irregular metal deposits which is of high both academic and technological significance. Due to open porous structure with the extremely high surface area, 3D foam or the honeycomb-like structures are very suitable to be used as electrodes in many electrochemical devices, such as fuel cells, batteries, and sensors [3, 16, 17], as well as in catalytic purpose [18]. The procedure of formation of these electrodes by electrodeposition techniques is known as gas bubble dynamic template method or dynamic hydrogen bubble template (DHBT) method where hydrogen bubbles generated by electrodeposition at high overpotentials or current densities act as dynamic template around which metal electrodeposition occurs [16, 19]. Hydrogen evolution responsible for the formation of the 3D foam or honeycomb-like electrodes is enough vigorous to cause such stirring of the solution which leads to the decrease of the cathode diffusion layer thickness and to the increase of the limiting diffusion current density and hence to the change of the hydrodynamic conditions in the near-electrode layer. Compared with hard templates, such as porous polycarbonate membranes, anodic alumina membranes, colloidal crystals, and echinoid skeletal structures, the hydrogen bubble dynamic template method possesses several advantages: low cost, ease of preparation, facile control of structure, and facile one-step synthesis process, including preparation of the template, metal deposition, and elimination of the template [20].

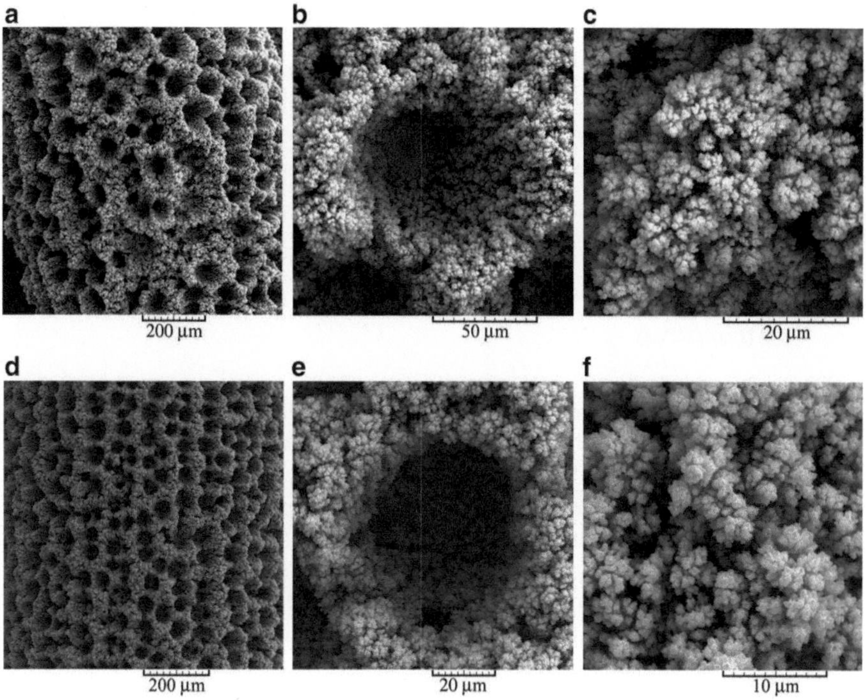

Fig. 5.3 (**a**, **d**) The honeycomb-like structures and the typical elements constructing this type of a structure: (**b**, **e**) holes, (**c**, **f**) cauliflower-like agglomerates of Cu grains formed around holes. (**a–c**) the potentiostatic regime of electrolysis. $\eta = 1000$ mV. (**d–f**) The galvanostatic regime of electrolysis. $i = 0.44$ A cm^{-2}. Solution: 0.15 M $CuSO_4$ in 0.50 M H_2SO_4 (Reprinted from Refs. [21, 22] with permission from Elsevier and Ref. [23] with kind permission from Springer)

The main characteristics of this deposit type are micro-sized holes or pores formed due to attached hydrogen bubbles and cauliflower-like agglomerates of metal grains or dendrites formed around them [3–5]. In the potentiostatic regime of electrolysis, the honeycomb-like structures are formed at overpotentials outside the plateau of the limiting diffusion current density in the zone of the fast increase in the current density with increasing the overpotential after the inflection point, while in the galvanostatic regime of electrolysis are formed at current densities larger than the limiting diffusion current one. There is no any difference between the honeycomb-like structures obtained in the potentiostatic and galvanostatic regimes of electrolysis (Fig. 5.3) [21–23].

The number, size, and distribution of holes in 3D foam or the honeycomb-like structures can be easily controlled by the choice of appropriate electrolysis parameters [20, 24, 25]. Also, the size of grains constructing dendrites or cauliflower-like agglomerates formed around holes, as well as the size of channels between small metal grains agglomerates, is controlled by conditions of electrolysis. The cauliflower-like agglomerates and dendrites among holes define the wall width in this structure type. The specific surface area of the 3D foam or the honeycomb-like structures is determined by the number and size of holes, as well as by wall width among them [26]. The critical quantity of evolved hydrogen leading to the change of hydrodynamic conditions in the near-electrode layer and formation of the honeycomb-like structure corresponds to the average current efficiency of hydrogen evolution of 10.0 %, and this value is valid for copper solutions containing concentrations of Cu^{2+} ions of 0.15 M $CuSO_4$ and less (in 0.50 M H_2SO_4) [27].

5.3.2 Phenomenology of Formation of the Honeycomb-Like Structures

In the initial stage of the electrodeposition process, both nuclei of deposited metal and "nuclei" of hydrogen bubbles are formed at the active sites of the electrode surface (Fig. 5.4a) [28]. The hydrogen bubbles isolate the substrate and then the current lines are concentrated around them making rings consisted of agglomerates of grains of deposited metal. The current lines are also concentrated at the metal nuclei formed in the initial stage between the hydrogen bubbles forming copper grain agglomerates of them. In the growth process, due to current density distribution effect, both hydrogen evolution and copper nucleation primarily occur at top of

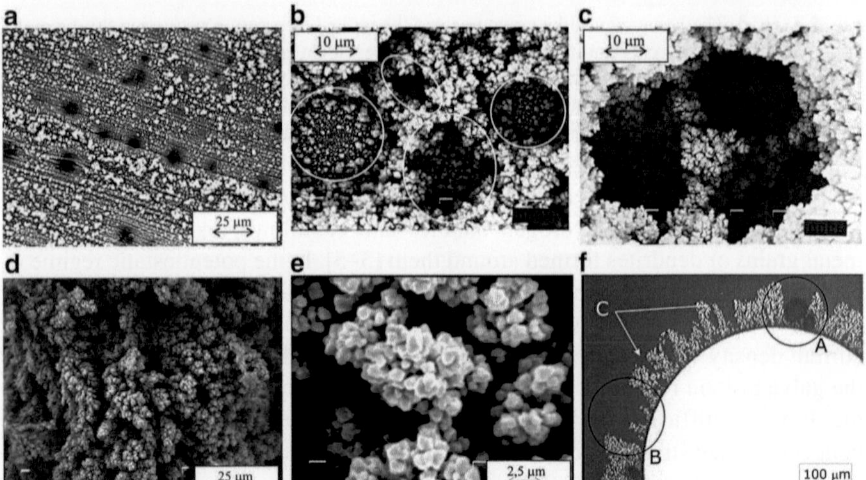

Fig. 5.4 (**a–c**) Cu deposits obtained from 0.15 M $CuSO_4$ in 0.50 M H_2SO_4 at 1000 mV. The time of electrolysis: (**a**) 10 s, (**b**) 30 s, and (**c**) 120 s, (**d, e**) Cu deposits obtained from 0.075 M $CuSO_4$ in 0.50 M H_2SO_4 at 1000 mV. The quantity of electricity: 20 mA h cm^{-2} and (**f**) crosssection of Cu deposit obtained from 0.15 M $CuSO_4$ in 0.50 M H_2SO_4, at 1000 mV with a quantity of the electricity of 10 mA h cm^{-2} (Reprinted from Refs. [5, 23, 28] with kind permission from Springer, Ref. [24] with permission from Elsevier and Ref. [29] with permission from the Serbian Chemical Society)

these agglomerates. Some of new, freshly formed hydrogen bubbles will coalesce with hydrogen bubbles formed in the initial stage of electrodeposition process, leading to their growth with electrolysis time. When the critical size of these hydrogen bubbles to detach from electrode surface is reached, they will detach from electrode surface forming holes of regular shapes at electrode surface. This "regular" type of holes is shown in Fig. 5.4b in circle. Simultaneously, holes of irregular shape are formed at electrode surface of agglomerate copper grains formed between hydrogen bubbles [28]. These "irregular" holes are situated between regular holes, and they are shown in Fig. 5.4b in ellipse. For longer electrodeposition time, coalescence of closely formed hydrogen bubbles occurs, leading to the formation of large, so-called coalesced holes (Fig. 5.4c) [28].

Meanwhile, some of new freshly formed hydrogen bubbles will not coalesce with previously formed hydrogen bubbles because they are situated between freshly formed copper nucleus, and these hydrogen bubbles have not enough place to develop in large hydrogen bubbles. These hydrogen bubbles will detach very fast from electrode surface forming channel structure through the interior of the deposit [24]. The typical channel structure formed by simultaneous hydrogen evolution and copper nucleation is shown in Fig. 5.4d. The "top view" of the part of the honeycomb-like structure shown in Fig. 5.4d is given in Fig. 5.4e and from it can be seen cauliflower-like agglomerates of copper grains surrounded by irregular channels which the origin is of evolved hydrogen.

All elements constructing the honeycomb-like structure can be also seen from Fig. 5.4f showing a cross section of this type of deposit [29]. The "regular holes" formed by both attached hydrogen bubbles (part in circle denoted with *A* in Fig. 5.4f) and coalescence of neighboring hydrogen bubbles (part in circle denoted with *B* in Fig. 5.4f),"irregular holes" formed due to the effect of current distribution at the growing surface (parts denoted by arrow labeled *C* in Fig. 5.4f), as well as channel structures formed through the interior of the deposit, can be easily observed by crosssection analysis of this deposit.

5.3.3 Factors Affecting the Size and Distribution of Holes in the Honeycomb-Like Structures

Electrodeposition technique is a suitable way to obtain open porous structures because it is very easy to control number, size, and distribution of holes by the choice of appropriate electrodeposition conditions [5, 23, 24].

Factors affecting number, size, and distribution of holes are (a) regime of electrolysis, (b) preparation of working electrode, (c) concentration of the depositing (Cu^{2+}) ions, (d) concentration of the supporting electrolyte (H_2SO_4), (e) temperature of electrolysis, and (f) time of electrolysis.

5.3.3.1 Regime of Electrolysis

The increase of overpotential or current density intensifies hydrogen evolution reaction leading to the increase in number of holes formed of detached hydrogen bubbles [5, 23, 30]. For example, for copper solution containing 0.15 M $CuSO_4$ in 0.50 M H_2SO_4, the average current efficiency of hydrogen evolution, $\eta_{I,av}(H_2)$ was about three times larger at an overpotential of 1000 mV ($\eta_{I,av}(H_2) = 30.0$ %) than at 800 mV ($\eta_{I,av}(H_2) = 10.8$ %) [4]. It is manifested by the formation of the honeycomb-like structures with the considerably larger number of holes formed of detached hydrogen bubbles at 1000 mV than at 800 mV.

5.3.3.2 Preparation of Working Electrode

The number, size, and distribution of holes in the honeycomb-like electrodes strongly depend on type of working electrode used for copper electrodeposition [28, 30]. The strong difference in the initial stage of their formation, as well as in the formed honeycomb-like structures depends on the type of used working electrode. The number of hydrogen bubbles formed at the electrode with large number of active centers, where irregularities at electrode surface represent active centers for formation of both hydrogen bubbles and agglomerates of copper grains, is several times higher than the number of holes formed at the electrode with "killed" active centers, where active centers were removed by the formation of uniform thin copper film by electrodeposition at some lower overpotential [28].

5.3.3.3 Concentration of Cu^{2+} Ions

The increase of Cu^{2+} concentration causes a sharp decrease in the quantity of evolved hydrogen and, hence, the decrease of the average current efficiencies for hydrogen evolution reaction [24, 27]. This is in a good agreement with the prediction of the ionic equilibrium of the species in the $CuSO_4$-H_2SO_4-H_2O system [31–33]. Electrodeposition processes from copper solutions containing the concentration of Cu^{2+} ions above 0.15 M $CuSO_4$ (in 0.50 M H_2SO_4) lead to the formation of new type of holes, denoted as dish-like hole [5, 23, 24]. The typical dish-like holes obtained by electrodeposition from 0.60 M $CuSO_4$ in 0.50 M H_2SO_4 at an overpotential of 1000 mV are shown in Fig. 5.5a. The appearance of very branchy dendrites between dish-like holes (Fig. 5.5b), as well as at shoulders of holes with longer electrodeposition times (Fig. 5.5c), clearly points out that the diffusion layer of the macroelectrode is not disturbed during copper electrodeposition from this solution and that the quantity of evolved hydrogen was not enough to cause stirring of solution in the near-electrode layer.

Fig. 5.5 Copper deposits electrodeposited from 0.60 M $CuSO_4$ in 0.50 M H_2SO_4 at an overpotential of 1000 mV. The quantity of electricity: (**a**, **b**) 2.5 mA h cm^{-2} and (**c**) 20 mA h cm^{-2} (Reprinted from Ref. [24] with permission from Elsevier and Refs. [5, 23] with kind permission from Springer)

5.3.3.4 Concentration of Sulfuric Acid

In accordance with the ionic equilibrium of the species in the $CuSO_4$-H_2SO_4-H_2O system [31–33], the increasing H_2SO_4 concentration leads to an intensification of hydrogen evolution reaction. For solutions containing 0.15 M $CuSO_4$, the honeycomb-like structures are formed at an overpotential of 1000 mV with concentrations larger than 0.25 M H_2SO_4 [25]. The number of holes increased, while the hole size decreased with the increasing H_2SO_4 concentration. Also, orientation of holes was changed from random oriented holes to holes which were lined up in parallel row. At lower concentrations of H_2SO_4 (0.15 M $CuSO_4$ in 0.125 M H_2SO_4), aside from holes and cauliflower-like agglomerates of Cu grains among them, degenerate dendrites, a channel structure around dendritic and cauliflower-like particles and holes with the shoulders of degenerate dendrites were electrodeposited at 1000 mV [25].

5.3.3.5 Temperature of Electrolysis

The increase of temperature of electrolysis leads to the intensification of hydrogen evolution reaction [5, 23, 34]. Meanwhile, despite intensification of hydrogen evolution with the increasing temperature, the decrease of number of holes formed per mm^2 surface area of electrodes and the increase of their average diameter were observed during copper electrodeposition at an overpotential of 800 mV (Fig. 5.6). This unexpected trend in the development of morphology of electrodeposited copper can be explained by the effect of temperature on some properties of electroplating solution, such as viscosity and surface tension [34]. The increase of temperature causes the decrease of both the viscosity [35] and the surface tension of this solution [36]. The decrease of the surface tension of the solution lowers the break-off diameter of hydrogen bubble from the electrode surface [36], while the decreased viscosity of the solution probably facilitates a transport of the detached hydrogen bubbles through the interior of the deposit forming the channel structure through it. Anyway, increasing temperature leads to redistribution of evolved

Fig. 5.6 Copper deposits electrodeposited at an overpotential of 800 mV from 0.15 M $CuSO_4$ in 0.50 M H_2SO_4 at a temperature of (**a**) 14.0, (**b**) 35.0, (**c**, **d**) 50.0 ± 0.5 °C. The quantity of electricity: 10 mA h cm^{-2} (Reprinted from Ref. [34] with permission from the Serbian Chemical Society and Refs. [5, 23] with kind permission from Springer)

hydrogen from those creating honeycomb-like structure (Fig. 5.6a, b) to structure with dish-like holes (Fig. 5.6c) and by dominant presence of agglomerates of copper grains surrounded by irregular channels of evolved hydrogen (that is a channel structure) (Fig. 5.6d).

5.3.3.6 Time of Electrolysis

The size of holes increases with electrolysis time due to the growth of the hydrogen bubbles with time, as well as due to a coalescence of neighboring hydrogen bubbles. Simultaneously, the number of holes decreases with the electrolysis time not only due to coalescence of hydrogen bubbles, but as well, due to the current density distribution effect. Namely, in the growth process, due to the current density distribution effect, some hydrogen bubbles can remain captive in the interior of deposit making the honeycomb-like structure very porous [23, 29].

5.4 Effect of Additives on Micro- and Nanostructural Characteristics of the 3D Foam or the Honeycomb-Like Electrodes

The increasing overpotential, the decreasing concentration of Cu^{2+} ions, and the increasing H_2SO_4 concentration intensify hydrogen evolution reaction. Intensification of hydrogen evolution leads to an increase in the number of formed holes, as well as to a decrease of hole size. Meanwhile, the ratio of the coalesced holes to the overall number of formed holes increases with the intensification of hydrogen evolution. The process of a coalescence of closely formed hydrogen bubbles is undesired process during formation of the honeycomb-like electrodes because this process causes both the decrease of the overall number of the formed holes and the increase of the hole size and, hence, leads to the decrease of the specific surface area of electrodes.

In order to increase the specific surface area and enhance the effectiveness/activity of the porous electrodes, it is necessary to reduce the size of the pores, as well as the branches in the foam or agglomerates of copper grains in the honeycomb-like structures [26]. One of the ways to improve micro- and nanostructural characteristics of open porous electrodes is the addition of additives to the electroplating solution [26]. The decrease of diameter of holes, as well as the increase of their number in 3D foam copper structures, can be realized by the addition of acetic acid to the copper sulfate solution [26]. Also, the addition of chloride ions dramatically reduces the size of the copper branches in the walls of holes. The reduction in pore size is a result of lowering hydrophobic force of the generated hydrogen gas by adding bubble stabilizer (*e.g.*, acetic acid) that suppresses the coalescence of bubbles, while the decrease in branch size in the foam wall is a consequence of the catalytic effect of chloride ions on the copper deposition reaction.

Mechanical strength of the foam structure can be improved by the addition of $(NH_4)^+$, Cl^-, polyethylene glycol, and 3-mercapto-1-propane sulfonic acid to the deposition bath [37]. The foam structure obtained by a combination of these additives was a highly porous with better mechanical strength than the one obtained without additives, owing to higher compactness of crystallites.

Unlike of corncob-like forms obtained from the additive free bath, the porous walls in the 3D foam structures prepared from the $(NH_4)_2SO_4$ containing Cu bath were constructed from needle-like nanodendrites [38]. The mechanical strength and stiffness of the copper foam prepared with the addition of $(NH_4)_2SO_4$ was significantly improved due to the self-supporting effects of the tightly interlocked needle-like nanodendrites. In contrast, the Cu foam prepared from the solution with $(NH_4)_2SO_4$ and benzotriazole (BTA) showed high ductility but low mechanical strength due to the formation of grape-like copper deposits. Both the copper foams exhibited higher mechanical properties than the one with corncob-like deposits formed in the additive-free solution.

5.5 Structural Characteristics of the 3D Foam or the Honeycomb-Like Structures of the Other Metals

Metals from all three groups according to Winand classification of metals [1] can be obtained in the honeycomb-like or the 3D (three dimensional) foam forms under the appropriate electrodeposition conditions.

5.5.1 The Honeycomb-Like or 3D Foam Structures of the Normal Metals

The group of normal metals (Pb, Sn, Tl, Cd, Hg, Zn, Ag (simple electrolytes)) is characterized by low melting points, T_m, high exchange current densities, i_0 ($i_0 > 1$ A dm^{-2}), and high overpotentials for hydrogen discharge [1]. The overall macroscopic features of the 3D foam structures of Sn are similar to those of Cu [3]. The foam walls of Sn were composed of dendritic particles, with a clear difference in the shape of branches constructing the dendrites of Sn and Cu. The branches of the Sn dendrites are longer, relatively dense, and straighter than those constructing Cu dendrites. The branches of Cu dendrites are corncob-like form composed of small agglomerates of Cu grains. The individual grains in the dendritic tin branches are not formed because no hydrogen was generated on Sn branches due to the high hydrogen overvoltage on Sn [3]. The Cu deposition rate and the size of surface pore are much smaller than those for Sn under similar conditions of formation. For example, to achieve similar thickness of the foam structure, the time needed for deposition of Cu is about four times of that for deposition of Sn.

Another metal from the group of "normal metals" obtained in the 3D foam is Pb [39]. The perchloric acid showed as suitable source of H_3O^+ ions, with the strong effect of its concentration on macroscopic characteristics of the honeycomb-like structures (Fig. 5.7a). With an increased amount of perchloric acid in the electrolyte, the hydrogen evolution rate and, thus, the deposition current grow. In the dependence of concentration of H_3O^+ ions, the two types of surface morphology of Pb are formed around holes: the sub-micrometer wires of the diameters of about 200–300 nm on the lower and particles or wires covered by particles at the higher concentrations of $HClO_4$ (Fig. 5.7b).

There is an optimal concentration of $HClO_4$ which allows formation of the honeycomb-like structures with the maximal surface area [39]. The decrease of number of holes and the increase of their diameter observed after the optimal concentration of this acid are due to coalescence of neighboring hydrogen bubbles. The effect of the different parameters of electrolysis was comparable with those observed in the case of Cu. The mean diameter of the pores increased as the Pb^{2+} concentration increased. As the current density increased, the porous Pb films became uniform while, at the same time, the pore size, wall thickness, and wire diameter became smaller. The pore size of the deposited porous film was closely related to the size of the hydrogen bubbles that departed from the electrode. A high current density facilitates the departure of the H_2 bubbles with a smaller bubble size.

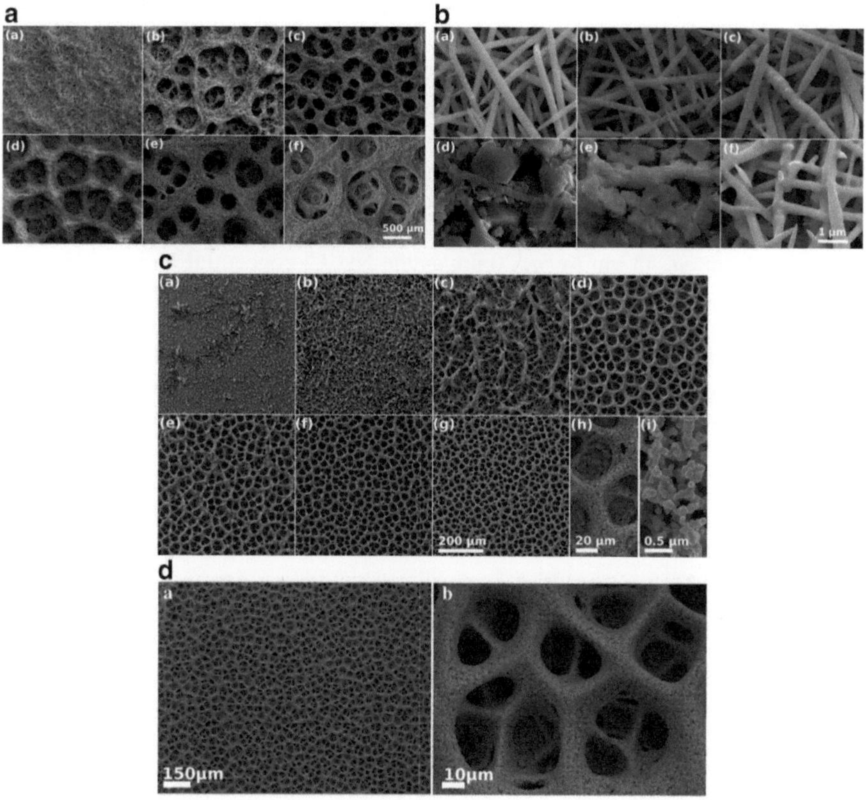

Fig. 5.7 The honeycomb-like and elements constructing this structure type: (**a**) Pb; Solution: 0.010 M $Pb(ClO_4)_2 \cdot 3H_2O$ with $HClO_4$ concentrations of: (*a*) 0.030, (*b*) 0.30, (*c*) 0.60, (*d*) 0.90, (*e*) 1.2, and (*f*) 1.8 M. The voltage: – 2 V. Time: 2 min, (**b**) the magnified details of the Pb porous films presented in (**a**), (**c**) SEM images of Ag structures deposited from electrolytes 0.010 M Ag_2SO_4 in 1.5 M KSCN containing (*a*) 0 M, (*b*) 0.10 M, (*c*) 0.25 M, (*d*) 0.5 M, (*e*) 0.75 M, (*f*) 1.0 M, and (*g*) 2.0 M NH_4Cl. (*h*) and (*i*) show the magnified views taken from the wall part of the Ag foam deposited from the electrolyte with 0.50 M NH_4Cl, and (**d**) SEM images of a typical as-deposited multimodal Au foam ($E = -$ 4 V, $t = 20$ s, 0.10 M $HAuCl_4 + 2.0$ M NH_4Cl) (Reprinted from Refs. [39–41] with permission from Elsevier)

5.5.2 The Honeycomb-Like or 3D Foam Structures of the Intermediate Metals

Copper is the typical representative of the intermediate metals, and mechanism of formation of the honeycomb-like structures is only established for this metal by far. Aside from Cu, Au and complex electrolytes of Ag also belong to this group of metals. This group of metals is characterized by moderate T_m, medium i_0 (i_0 in the interval from 10^{-2} to 1 A dm^{-2}), and lower hydrogen overpotentials [1].

Silver makes complexes with many ligands which cause the transfer of this metal to the group of the intermediate metals [42]. One of the electrolyte-enabling formations of Ag in the honeycomb-like form is the thiocyanate one to which NH_4Cl was added [40]. By adding NH_4^+ ions to the electrolyte, the quantity of H_3O^+ is increased, and there is a critical concentration of NH_4Cl above which Ag is deposited in the honeycomb-like form (Fig. 5.7c). The decrease in the diameter and wall thickness of the surface pores, as well as the increase of the number of formed holes, was observed with increasing concentration of NH_4Cl. The pore size decreased from *ca.* 45 μm to *ca.* 20 μm as the NH_4Cl concentration was increased from 0.50 to 2.0 M during formation of the honeycomb-like structures in the galvanostatic regime of electrolysis. The pore size in this structure type can be decreased to 10 μm by application of the potentiostatic regime of electrolysis. The structure of deposits among holes was constructed from dendrites or coral-like structures with the smallest branches being in the nanoscale range. Generally, in the dependence of the regime and parameters of electrolysis, the morphology of Ag deposits formed around holes can be categorized into three groups: dendrites, a framework of identical particles, and agglomerates of inhomogeneous particles with big Ag granules distributed on the foam surface [43]. The change of surface morphology of Ag from dendrites to small agglomerates of particles with simultaneous increase in the foam wall compactness was observed with intensification of hydrogen evolution reaction, *i.e.*, with increasing NH_4Cl concentration. The role of NH_4^+ ions as a crucial factor in formation of the honeycomb-like structures of Ag is ascribed to the increase of the rate of hydrogen evolution which results in an increase of the Ag deposition rate with increasing of the NH_4Cl concentration [43].

The morphology of the Au foam (the third representative of this group of metals) is similar to that of Ag and Cu foams (Fig. 5.7d). The concentration of NH_4Cl is also crucial for the production of the Au foam structures from a solution containing 0.10 M $HAuCl_4$ [41]. The foam walls are composed of nanometer scale Au particles, while pores were on both the microscale (tens of micrometers) and nanoscale (tens of nanometers). The pore sizes range from 10 to 50 μm in diameter, and the internal wall structure consisted of long dendrites with many branches and nodules where the backbone tapers to a fine point of around 30 nm which were a characteristic of the honeycomb-like structures of Au obtained from the 0.10 M $KAuBr_4$ in 1.5 M H_2SO_4 [44]. The change of the internal wall structure from highly branched fern-like dendrites to more elongated rods with smaller secondary nodules is observed with increasing the electrodeposition time. The increasing current density of electrodeposition has similar beneficial effect on macroscopic characteristics of the honeycomb-like structures.

5.5.3 *The Honeycomb-Like or 3D Foam Structures of the Inert Metals*

Due to a low overpotential for hydrogen discharge, electrodeposition of the inert metals (Fe, Co, Ni, Mn, Cr, Pt; high T_m, low i_0, and very low hydrogen overpotentials) occurs together with hydrogen evolution in the whole range of potentials and current densities. In this way, formation of metals from this group in the honeycomb-like form is facilitated.

The honeycomb-like structures of Pt consist of interconnecting pores with sizes from 5 to 10 μm, with fine structure consisting of nodules of 100–200 nm with sharp extrusions that exhibited excellent electrocatalytic activity for methanol oxidation [45].

The foam walls of the honeycomb-like electrodes of Pd which show a high activity toward the oxidation of C_2 alcohols were composed of a tiny, highly dispersed, relatively uniform network of Pd dendrites [46]. The width of the dendrite branches was on the 10 nm scale. The increase of concentration of Pd^{2+} ions shows the useful effect on formation of the honeycomb-like electrodes [46].

The NH_4Cl concentration also plays an important role in the creating of the honeycomb-like structures of Ni [47]. The role of NH_4Cl of Ni^{2+} electrolytes is to prevent formation of $Ni(OH)_2$ during deposition process. The size of macropores in the honeycomb-like electrodes which shows an excellent mechanical resistance was from 1 to 40 μm. The micropores (10–100 nm) and nanopores (<10 nm) were also components of the honeycomb-like electrodes of this metal. Formation of compact cauliflower-like forms and globules around holes or pores can be ascribed to a very vigorous hydrogen evolution which follows electrodeposition of Ni in the honeycomb-like form. Namely, the current for hydrogen evolution can reach 60–70 % of the total current [48]. This very vigorous hydrogen evolution lowers sharply a potential at which Ni electrodeposition occurs causing formation of a compact deposit around holes (the concept of "effective overpotential" or "effective potential").

5.6 Application of Periodically Changing Regimes of Electrolysis on Metal Electrodeposition in the Hydrogen Evolution Range

Application of periodically changing regimes of electrolysis on electrodeposition of metals in the hydrogen evolution range showed the strong effect on hydrogen evolution reaction and, hence, morphology of metal deposits [5, 23]. In the PO and PC regimes, this effect is usually attained by regulation of the pause-to-pulse ratios, as well as by the durations of deposition pulse and pause. In the RC regime, the quantities of evolved hydrogen with the strong consequences on morphology of metal deposits are primarily affected by regulation of the anodic to cathodic ratios, as well as by the durations of both cathodic and anodic pulses. For example, the pyramid-like precursors of dendrites (Fig. 5.8a, b) and dendrites (Fig. 5.8c, d) are obtained by the PO regimes with an overpotential amplitude of 1000 mV and the pause-to-pulse ratio 10, but with the different lengths of deposition pulse and pause ($t_c = 1$ ms and $t_p = 10$ ms for Fig. 5.8a, b, and $t_c = 10$ ms and $t_p = 100$ ms for Fig. 5.8c, d). As a reminder, in the constant potentiostatic regime (Fig. 5.3a–c), the honeycomb-like structures constructed from cauliflower-like agglomerates of Cu

Fig. 5.8 (**a, b**) The pyramid-like and (**c, d**) dendritic structures of Cu obtained by the pulsating overpotential (PO) regimes with the same pause-to-pulse ratio 10, but with the different durations of deposition pulse (t_c) and pause (t_p): (**a, b**) $t_c = 1$ ms and $t_p = 10$ ms, (**c, d**) $t_c = 10$ ms and $t_p = 100$ ms (Reprinted from Refs. [49, 50] with permission from Elsevier and Refs. [5, 23, 51] with kind permission from Springer)

grains formed around holes were obtained at an overpotential of 1000 mV which is equal to the applied overpotential amplitude in the PO regime.

The periodically changing regimes also strongly affect micro- and nano-structural characteristics of the honeycomb-like electrodes. The increase of the specific surface area of this electrode type manifested by the increase of number of holes formed of detached hydrogen bubbles and by the decrease of both hole size and wall width among holes can be attained by application of the appropriate square-waves PO, PC, and RC [21, 23, 52].

Figure 5.9 shows the honeycomb electrodes obtained at a constant overpotential of 1000 mV (Fig. 5.9a) and by the PO regime with the overpotential amplitude of 1000 mV, deposition pulse, t_c, of 10 ms, and pause duration, t_p, of 50 ms (Fig. 5.9b). Holes obtained by these regimes of electrolysis of detached hydrogen bubbles are shown in Fig. 5.9c (the constant potentiostatic regime) and 5.9d (the PO regime). The bottom of the hole obtained at the constant overpotential was relatively smooth (Fig. 5.9c), while the one obtained by the PO regime was constructed from the small

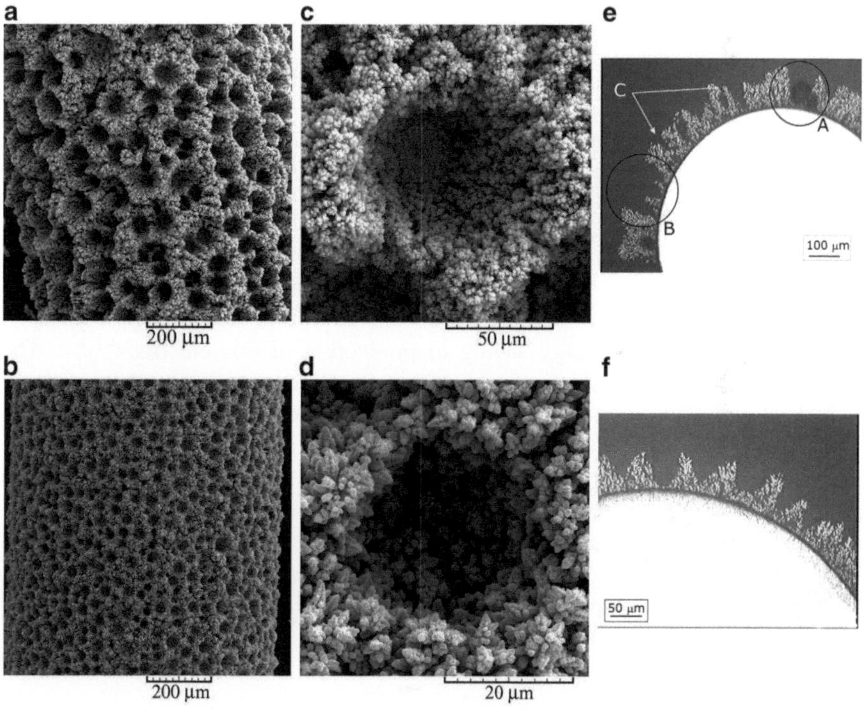

Fig. 5.9 The honeycomb-like electrodes, holes, and cross sections of the honeycomb-like electrodes obtained by (**a**, **c**, **e**) the constant potentiostatic regime at an 1000 mV and (**b**, **d**, **f**) the regime of pulsating overpotential (PO). Overpotential amplitude: 1000 mV. Deposition pulse: 10 ms. Pause duration: 50 ms. Solution: 0.15 M $CuSO_4$ in 0.50 M H_2SO_4 (Reprinted from Ref. [29] with permission from the Serbian Chemical Society, Ref. [21] with permission from Elsevier and Refs. [5, 23, 51] with kind permission from Springer)

agglomerates of Cu grains (Fig. 5.9d). A noticeable increase in the number of holes and a decrease in their size were attained by application of the PO regime. This improvement of structural characteristics by application of the square-wave PO can be primarily ascribed to suppressed coalescence of neighboring hydrogen bubbles and to the improvement of the current density distribution effect, as seen by comparison of cross sections of these deposits (Fig. 5.9e, f) [21, 29, 51]. Aside from holes obtained of detached hydrogen, crosssection analysis of the honeycomb-like deposit electrodeposited at the constant overpotential (Fig. 5.9e) showed the presence of both coalesced holes (part in circle denoted with B in Fig. 5.9e) and irregular holes (part in circle denoted with C in Fig. 5.9e). Unlike this interior structure, the ordered interior structure consisted only of regular holes was obtained by the PO regime (Fig. 5.9f).

The increase of compactness of the honeycomb-like electrodes obtained by the PO regime is due to the effect of a current density during "off" periods (*i.e.*, during duration of pause). Although this current density can be neglected in comparison with the current density during "on" periods (*i.e.*, during the duration of deposition pulse), it is clear that its effect on the formation of these deposits is very important [21, 23, 29, 51].

5.6.1 The Regime of Pulsating Overpotential in the Hydrogen Evolution Range: Optimization of Formation of the Honeycomb-Like Electrodes

Aside from the increase of the specific surface area of the honeycomb-like electrodes, the other benefits of application of the PO regime are the improvement of structural stability and energy saving in production of this electrode type. The structural stability or mechanical strength of metal deposits is closely associated with the quantity of produced hydrogen and by its effect on adhesion of the deposit with electrode surface [53]. The increase of mechanical strength is related with the decrease in the quantity of evolved hydrogen spent for formation of the honeycomb-like electrodes. It was necessary to find optimal square-wave PO enabling formation of the honeycomb-like electrodes with maximal number of holes formed of detached hydrogen bubbles and with minimal quantity of evolved hydrogen spent for its formation.

Figure 5.10 shows the honeycomb-like structures obtained by the PO regimes with an overpotential amplitude of 1000 mV, pause duration of 10 ms, and deposition pulses of 3 ms (Fig. 5.10a), 5 ms (Fig. 5.10b), and 20 ms (Fig. 5.10c). The quantities of evolved hydrogen spent for formation of these electrodes corresponded to the average current efficiencies of hydrogen evolution, $\eta_{I,av}(H_2)$ of 16.4, 22.4, and 28.1 %, respectively [49]. As a reminder, the honeycomb-like electrode was obtained at the constant overpotential equal to this overpotential amplitude with $\eta_{I,av}(H_2)$ of 30.0 % [4, 5, 23]. Analysis of the honeycomb-like

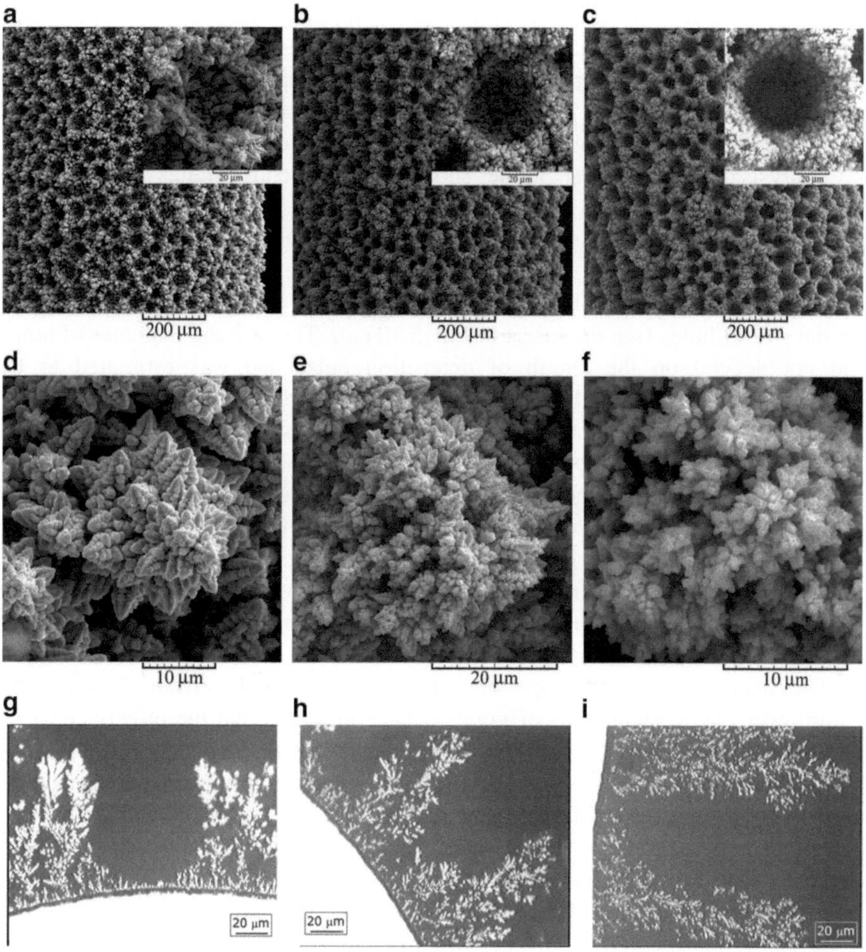

Fig. 5.10 Honeycomb-like copper structures obtained by the PO regimes with deposition pulses of (**a**) 3 ms, (**b**) 5 ms, and (**c**) 20 ms. Morphology of deposits among holes: (**d**) 3 ms, (**e**) 5 ms, and (**f**) 20 ms. Cross sections of the holes (**g**) 3 ms, (**h**) 5 ms, and (**i**) 20 ms. Overpotential amplitude: 1000 mV. Pause duration: 10 ms. Solution: 0.15 M $CuSO_4$ in 0.50 M H_2SO_4 (Reprinted from Ref. [23] with kind permission from Springer and Ref. [49] with permission from Elsevier)

structures showed that the number of holes formed by the attached hydrogen bubbles did not change considerably with the length of deposition pulses of 3 and 5 ms [23, 49], which can be ascribed to suppressed coalescence of closely formed hydrogen bubbles. The mild decrease of the number of the holes observed with a deposition pulse of 20 ms is due to the enhanced coalescence of neighboring hydrogen bubbles and the effects related to current density distribution at the growing electrode with deposition pulses larger than pause durations [23].

Morphology of deposits around holes strongly depended on the length of deposition pulse. Dendrites were formed with a deposition pulse of 3 ms (Fig. 5.10d).

Dendrites were also formed with a deposition pulse of 5 ms (Fig. 5.10e), but they were more branchy than those obtained with a deposition pulse of 3 ms. Very disperse cauliflower-like agglomerates of Cu grains were formed with a deposition pulse of 20 ms (Fig. 5.10f). The numerous micro- and nanopores are formed around small Cu agglomerates of grains inside large cauliflower-like agglomerates.

Analysis of cross sections of holes showed that the depth of holes did not change noticeably with deposition pulses of 3 and 5 ms (Fig. 5.10g, h, respectively). The depth of the hole obtained with a deposition pulse of 20 ms (Fig. 5.10i) was larger than those obtained with 3 and 5 ms, which can be ascribed to the enhanced dispersity of this deposit. Also, there was not any effect of deposition pulse duration on the size of holes (see the insert in Fig. 5.10a–c). The average diameter of holes did not depend on the length of deposition pulse and was estimated to be 65 ± 2.0 μm [23, 49]. With the decreasing pause-to-pulse ratios, the bottoms of holes changed from those constructed from dendrites and cauliflower-like agglomerates of Cu grains to smooth ones.

Dendrites formed around hydrogen bubbles are clearly seen from Fig. 5.10g ($t_c = 3$ ms). Analysis of cross section shown in Fig. 5.10h confirmed formation of more branchy dendrites with a deposition pulse of 5 ms than with 3 ms. The increase of a dispersity of the internal structure is observed with the further increase of deposition pulse. The numerous channels formed around relatively small copper particles and degenerate dendrites are easily observed from Fig. 5.10i ($t_c = 20$ ms).

The change of morphology of deposits around holes from dendrites to cauliflower-like agglomerates of grains clearly indicates that the effectiveness of a stirring of solution by evolved hydrogen increased with the decreasing pause-to-pulse ratios. On the basis of the same hole sizes and their unchanged number, as well as of the same depth of holes (Fig. 5.10), it follows that the overall quantity of hydrogen evolved during formation of the honeycomb-like electrodes can be divided in the two parts [54]. The one part is spent for a creating of macropores or holes, and this part determines the overall specific surface area of the honeycomb-like electrodes. This quantity of evolved hydrogen does not contribute to the stirring of electrolyte in the near-electrode layer, and it is estimated to contribute to $\eta_{I,av}(H_2)$ of 10.0 % [54]. The difference between the overall quantity of evolved hydrogen and quantity spent for creation of macropores causes the stirring of electrolyte and, hence, affects the hydrodynamic conditions in the near-electrode layer. Morphology of deposits around holes is just determined by this quantity of evolved hydrogen, which is also responsible for dispersity of deposits among holes, *i.e.*, formation of micro- and nanopores.

From the above consideration, it is clear that the maximal surface area of the honeycomb-like electrodes can be attained by the careful selection of parameters of the PO regimes. This is followed by the decrease in the quantity of evolved hydrogen and, hence, by the improvement of the structural stability of the honeycomb-like electrodes. Simultaneously, analysis of the specific energy consumption shows that energy saving is achieved in the production of the honeycomb-like electrodes by the shortening of deposition pulse and by the decrease in the quantity of evolved hydrogen [23, 49].

5.6.2 The Pulsating Current in the Hydrogen Evolution Range

In the hydrogen evolution range, Eqs. (4.6) and (4.7) should be corrected because the current efficiency for metal electrodeposition is less than 1. Then, the average current density in the conditions of hydrogen evolution or the effective average current density, $i_{av}*$, can be presented by Eq. (5.10) [23, 55]:

$$i_{av}* = \frac{\eta_{I,av}(M) i_c}{1 + p} \tag{5.10}$$

where $\eta_{I,av}(M)$ is the average current efficiency for metal electrodeposition.

Since:

$$\eta_{I,av}(M) + \eta_{I,av}(H_2) = 1 \tag{5.11}$$

Equation (5.10) can be rewritten as:

$$i_{av}* = \frac{[1 - \eta_{I,av}(H_2)] i_c}{1 + p} \tag{5.12}$$

and according to Eqs. (4.7) and (5.12):

$$\eta_{I,av}(H_2) = 1 - \frac{i_{av}*}{i_{av}} \tag{5.13}$$

In the hydrogen evolution range, if $i_{av}* > i_L$, the amplitude of overpotential is related to the hydrogen reduction, increasing with the current density of hydrogen evolution [56].

The overpotential amplitude adapted to metal electrodeposition in the hydrogen evolution range is given by Eq. (5.14):

$$\eta_{A,eff} = \frac{b_c}{2.3} \ln \frac{i_{av}*(p + 1)}{i_0} + \frac{b_c}{2.3} \ln \frac{1}{1 - \frac{i_{av}*}{i_L*}} \tag{5.14}$$

where $\eta_{A,eff}$ represents the effective overpotential amplitude and i_L* is the effective limiting diffusion current density which depends on the hydrodynamic conditions in the near-electrode layer caused by a parallel process of hydrogen evolution. Eq. (5.14) is valid if the condition $i_{av}*/i_L* < 1$ is fulfilled. The effective average current density, $i_{av}*$, can be calculated by the use of Eq. (5.12) if the value of the average current efficiency of hydrogen evolution is known. The determination of i_L*, when the change of hydrodynamic conditions in the near-electrode layer is caused by hydrogen evolution, is not possible. For that reason, morphological analysis of metal deposits obtained in the hydrogen evolution range is an excellent auxiliary tool for the estimation of the change of effectiveness of solution stirring by evolved hydrogen.

In PC regimes, intensification of hydrogen evolution reaction can be achieved by [23, 55]:

(a) The increase of the current density amplitude and keeping both the durations of the deposition pulse and pause constant
(b) The prolonging of a deposition pulse duration and keeping both the current density amplitude and a pause duration constant
(c) The shortening of a pause duration and keeping both the current density amplitude and a deposition pulse constant

Figure 5.11 shows Cu deposits obtained with the current density amplitudes of 0.20 A cm^{-2} (Fig. 5.11a) and 0.44 A cm^{-2} (Fig. 5.11b). In both cases, a deposition pulse of 1 ms and a pause duration of 10 ms were applied [22, 23, 55, 57]. Formation of these deposits was accompanied by the quantity of evolved hydrogen which corresponded to the average current efficiency of hydrogen evolution, $\eta_{I,av}(H_2)$, of 5.5 % with the applied current density amplitude, i_c, of 0.20 A cm^{-2} [57], and

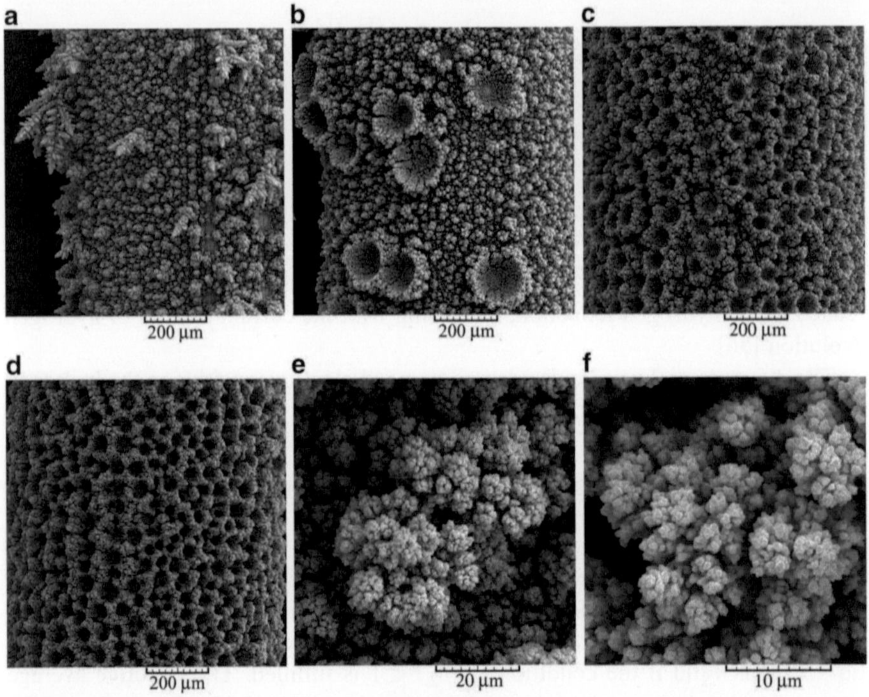

Fig. 5.11 Copper deposits obtained by the different PC regimes: (**a**) $i_c = 0.20$ A cm^{-2}, $t_c = 1$ ms, $t_p = 10$ ms, (**b**) $i_c = 0.44$ A cm^{-2}, $t_c = 1$ ms, $t_p = 10$ ms, (**c, e**) $i_c = 0.44$ A cm^{-2}, $t_c = 4$ ms, $t_p = 10$ ms, (**d, f**) $i_c = 0.44$ A cm^{-2}, $t_c = 10$ ms, $t_p = 10$ ms. Solution: 0.15 M $CuSO_4$ in 0.50 M H_2SO_4 (Reprinted from Refs. [22, 57] with permission from Elsevier and Refs. [23, 55] with kind permission from Springer)

13.7 % with i_c of 0.44 A cm^{-2} [22]. Very branchy dendrites, small cauliflower-like forms, and shallow holes formed from detached hydrogen bubbles are formed when the amplitude of current density of 0.20 A cm^{-2} was applied (Fig. 5.11a). On the other hand, dish-like holes and small cauliflower-like agglomerates of copper grains were formed with a current density amplitude of 0.44 A cm^{-2} (Fig. 5.11b).

The number of holes increased with a prolongation of duration of deposition pulses in order to the honeycomb-like structure was formed as the final shape with both current density amplitudes [22, 57]. Keeping i_c of 0.44 A cm^{-2} and pause of 10 ms, the mixture of dish-like holes and holes constructing the honeycomb-like structure was formed with a deposition pulse of 4 ms (Fig. 5.11c) and the honeycomb-like structure with a deposition pulse of 10 ms (Fig. 5.11d). The maximal number of holes is obtained with a pause-to-pulse ratio 1 and corresponds to suppressed coalescence of hydrogen bubbles. The size of non-coalesced holes does not depend on the length of deposition pulse. The decrease in number of holes with pause-to-pulse ratios smaller than 1 (deposition pulses larger than pause duration) is due to the appearance of coalesced holes. The decrease in number of holes with a pause-to-pulse ratio larger than 1 (deposition pulses shorter than pause duration) is due to the decrease in the quantity of evolved hydrogen and the appearance of larger dish-like holes.

In the dependence of duration of deposition pulses, the two groups of cauliflower-like agglomerates of Cu grains are formed [22, 23]. In the first group are situated independently formed agglomerates of Cu grains formed during the electrodeposition process with the shorter deposition pulses (Fig. 5.11e). They are larger and more compact than those formed among holes in the honeycomb-like structures (the second group; Fig. 5.11f).

The change in morphology of electrodeposited copper from very branchy dendrites and dish-like holes to the honeycomb-like structures is a consequence of the influence of parameters of PC regimes on the quantity of evolved hydrogen and effectiveness of solution stirring in the near-electrode layer. With an intensification of hydrogen evolution, the thickness of the cathodic diffusion layer decreases and the limiting diffusion current density increases, causing the decrease of the degree of diffusion control of electrodeposition process. The degree of diffusion control of electrodeposition process additionally decreases due to the smaller values of the effective average current densities, i_{av}*, in relation to those obtained in the absence of hydrogen evolution (Eqs. (4.7) and (5.10)). Then, the overpotential amplitude corresponding to metal electrodeposition will be smaller than the one specified by pulse rectifiers, and this value is denoted as effective overpotential amplitude and it is presented by Eq. (5.14) [23, 55]. Hence, the effective overpotential amplitude, $\eta_{A,\ eff}$, decreases with the intensification of hydrogen evolution reaction due to the decrease of the degree of diffusion control of electrodeposition process. So, the validity of the concept of "effective overpotential" can be expanded to include metal electrodeposition in the hydrogen evolution range by the regime of pulsating current (PC), and then, the concept applied for the PC regimes is denoted as "effective overpotential amplitude" one [23, 55].

All benefits attained by the application of the pulsating overpotential (PO) regime on optimization of formation of the honeycomb-like electrodes are valid for a technologically more attractive pulsating current (PC) regime. The following conveniences in the production of these electrodes can be attained since the appropriate square-wave PC parameters were selected: (a) energy saving, (b) the increase of the specific surface area of the electrodes, and (c) the improvement of the deposit structural stability or mechanical strength due to the decrease in the quantity of evolved hydrogen needed for their formation. In this moment, the largest problem for the commercial manufacturing of open porous structures by the PC regimes is high cost of a pulse rectifier which is much greater than a DC unit [58]. It is a highly regulated and sophisticated design that costs more to manufacture. At the first sight, energy savings attained by the application of the PC regimes and high cost of production of pulse rectifier are in contradiction, but the further development of electronic industry will probably decrease a cost of these rectifiers and will enable their larger application in electrochemical industry.

5.6.3 The Reversing Current in the Hydrogen Evolution Range

The modified Eq. (4.2) to include hydrogen evolution reaction is presented by Eqs. (5.15) and (5.16) [59]:

$$i_{av}{}^{*} = \frac{\eta_{I,c}(M)\, i_c - i_a r_{RC}}{1 + r_{RC}} \tag{5.15}$$

where $i_{av}{}^{*}$ is the effective average current density and $\eta_{I,c}(M)$ is the current efficiency for metal electrodeposition during cathodic pulses, and:

$$i_{av}{}^{*} = \frac{\left[1 - \eta_{I,c}(H_2)\right] i_c - i_a r_{RC}}{1 + r_{RC}} \tag{5.16}$$

where $\eta_{I,c}(H_2)$ is the current efficiency for hydrogen evolution reaction during cathodic pulses. The cathodic current density pulses are larger than the limiting diffusion current density, there is not any gas evolution during anodic pulses [59], and $i_{av}{}^{*}$ can be calculated using Eq. (5.16) if $\eta_{I,c}(H_2)$ is known.

For $i_{av}{}^{*} > i_L$, the overpotential amplitude is related to the hydrogen evolution [56]. The overpotential amplitude adjusted to metal electrodeposition in the hydrogen evolution range is represented by Eq. (5.17):

$$\eta_{A,eff} = \frac{b_c}{2.3} \ln \frac{i_{av}^*(r_{RC} + 1) + i_a r_{RC}}{i_0} + \frac{b_c}{2.3} \ln \frac{1}{1 - \frac{i_{av}^*}{i_L^*}} \tag{5.17}$$

where $\eta_{A,eff}$ is the effective overpotential amplitude and i_L^* is the effective limiting diffusion current density. Naturally, $i_{av}^* < i_L^*$.

The quantity of evolved hydrogen decreases with the increasing anodic current density values, causing the strong consequences on morphology of deposits [23, 59]. Figure 5.12 shows copper deposits with the anodic current densities of 40 mA cm^{-2} (Fig. 5.12a), 240 mA cm^{-2} (Fig. 5.12b), and 640 mA cm^{-2} (Fig. 5.12c). The cathodic current density of 440 mA cm^{-2}, a cathodic pulse of 10 ms, and an anodic pulse of 5 ms were applied in the all experiments. The honeycomb-like structures were formed with the cathodic current density larger than the anodic current density (Fig. 5.12a, b). In the galvanostatic regime at a current density of 440 mA cm^{-2} (Fig. 5.3d–f), as well as in the PC regime with the same values of deposition pulse and pause and the cathodic current density (the PC regime is the special case of the RC regime for $i_a = 0$ mA cm^{-2}), the honeycomb-like

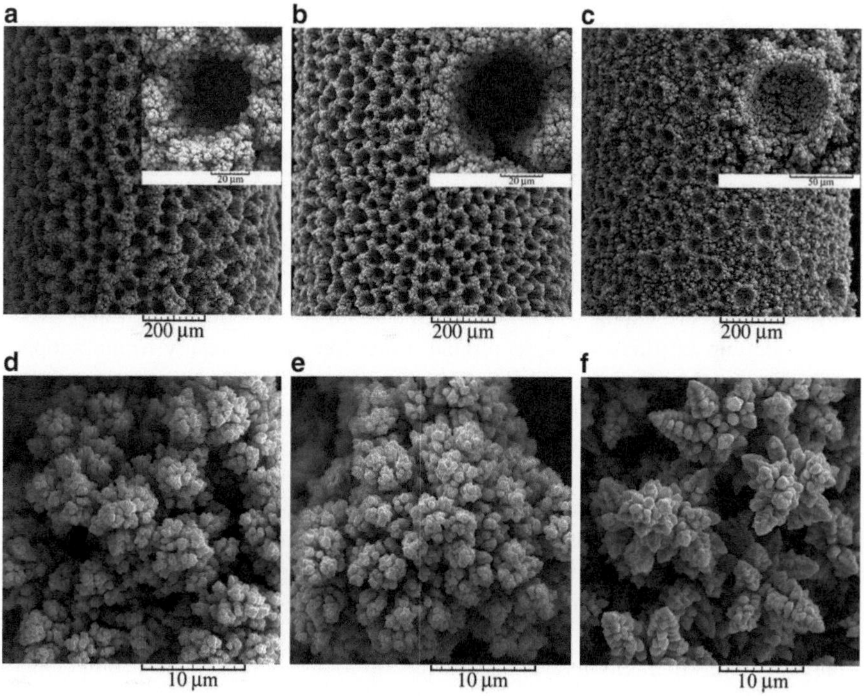

Fig. 5.12 Copper deposits obtained by the RC regimes with the anodic current densities of (**a, d**) $i_a = 40$ mA cm^{-2}, (**b, e**) $i_a = 240$ mA cm^{-2}, and (**c, f**) $i_a = 640$ mA cm^{-2}. The cathodic current density: 440 mA cm^{-2}. The cathodic pulse: 10 ms. The anodic pulse: 5 ms. Solution: 0.15 M $CuSO_4$ in 0.50 M H_2SO_4 (Reprinted from Ref. [23] with kind permission from Springer and Ref. [59] with permission from Elsevier)

structures were also formed. With the applied anodic current density larger than the cathodic current density, dish-like holes were predominately formed (Fig. 5.12c).

The maximum number and minimum size of holes are obtained in the honeycomb-like structure with $i_a = 240$ mA cm^{-2}. The smaller number of holes obtained with the anodic current densities smaller than 240 mA cm^{-2} (as well as in the PC regime) is due to the appearance of coalesced holes [23, 59]. The decrease in number of holes and the increase in their size after maximum are a consequence of the strong effect of the anodic current densities on hydrogen evolution reaction and the appearance of dish-like holes.

The effect of the anodic current density is also visible in the structure of deposits among holes. Cauliflower-like agglomerates of Cu grains were formed with i_a of 40 mA cm^{-2} (Fig. 5.12d) and 240 mA cm^{-2} (Fig. 5.12e) and dendrites with i_a of 640 mA cm^{-2} (Fig. 5.12f). However, there is a clear difference between cauliflower-like agglomerates obtained with 40 and 240 mA cm^{-2}. The cauliflower-like agglomerates of Cu grains obtained with i_a of 40 mA cm^{-2} are very disperse with deep irregular channels originated from detached hydrogen bubbles formed around small Cu grain agglomerates [5, 24]. With the increasing anodic current density, deep irregular channels around small Cu agglomerates are lost making the cauliflower-like agglomerates of Cu grains obtained with i_a of 240 mA cm^{-2} more compact than those obtained with smaller anodic current density. As a result, the wall width among holes decreased and holes approached to each other additionally increasing the specific surface area of the honeycomb-like electrodes.

The increase in the compactness of the honeycomb-like structures can be explained by the effect of the anodic current density on both hydrogen evolution reaction and copper electrodeposition rate [23, 59]. During cathodic pulses, both copper nuclei and hydrogen bubbles are primarily formed at the top of agglomerates of grains formed around initially formed hydrogen bubbles (the current density distribution effect). On the other hand, during anodic pulses, due to the same effect, the anodic current lines are concentrated at the top of the freshly formed copper nuclei causing their dissolution. The dissolution process of copper nuclei facilitates to freshly formed hydrogen bubbles to find path to coalesce with initially formed hydrogen bubbles leading to their growth with electrolysis time. With the increasing anodic current density, the dissolution of copper nuclei increases and the number of freshly hydrogen bubbles which will coalesce with initially formed hydrogen bubbles is increased. Simultaneously, the number of hydrogen bubbles which will remain trapped among freshly formed copper nuclei and constructing channel structure through the interior of the deposit is decreased. It means that these simultaneous processes will lead to a reduction of irregular channels formed around small copper grain agglomerates and, hence, up to the increase of the compactness of the honeycomb-like structures.

The change in morphology of electrodeposited copper from very disperse agglomerates of copper grains to branchy dendrites (Fig. 5.12d–f) clearly points out the decrease of effectiveness of solution stirring of evolved hydrogen and the increase of the degree of diffusion control of electrodeposition process with the increasing anodic current density. As a consequence, in the hydrogen evolution range, the effective overpotential amplitude, $\eta_{A,eff}$ (Eq. (5.17)), increases with the increasing anodic current density.

5.6.4 Comparison of the Honeycomb-Like Structures Obtained by the Galvanostatic and the Reversing Current Regimes

One of the ways to realize the advantages of use of the RC regime on the improvement of micro- and nanostructural characteristics of the honeycomb-like electrodes is comparison of their characteristics obtained by the appropriate RC regime and by the galvanostatic (DC) regime at the current density equal to the average current density in the RC regime. This comparison is based on the fact that metal electrodeposition by the RC regime in the millisecond range occurs at the average current density [10, 60].

Figure 5.13 shows the honeycomb-like electrodes obtained by the RC and DC regimes. The selected average current density in the RC and the current density in the DC regimes were 0.12 A cm^{-2}. Both honeycomb-like electrodes were obtained with the approximately same quantity of evolved hydrogen which corresponded to $\eta_{I,av}(H_2)$ of 22.0 ± 0.8 % [23, 52]. It was understandable because both electrodeposition processes occurred at the same current density.

The typical elements of which these honeycomb-like electrodes are constructed are shown in Fig. 5.13c–f. The holes obtained by the RC regime (Fig. 5.13c) were about twice smaller than those obtained at the constant current density (Fig. 5.13d). The average size of the holes obtained by the RC regime was about 50 μm, while holes obtained in the DC mode were about 100 μm. The cauliflower-like agglomerates of Cu grains with small dendrites on them are formed by the RC regime (Fig. 5.13e), and they were more compact than those obtained in the DC regime. The nucleation exclusion zones are observed around the cauliflower-like agglomerates of Cu grains obtained in the RC regime. On the other hand, the relatively large cauliflower-like Cu grain agglomerates surrounded by deep irregular channels which the origin is of evolved hydrogen were formed in the DC mode (Fig. 5.13f).

The noticeable increase in the number of holes, and the decrease in both hole size and wall width among holes, is obtained by application of the RC regime in relation to the constant DC regime. In this way, the specific surface of the honeycomb-like electrodes was considerably increased confirming superiority of the regime of reversing current in relation to other current regimes of electrolysis.

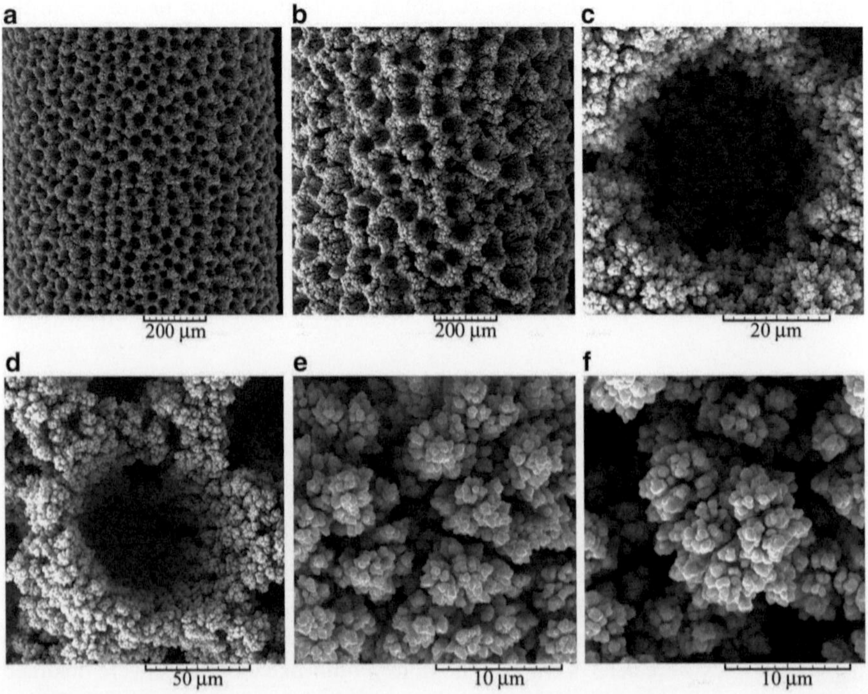

Fig. 5.13 The honeycomb-like electrodes and the elements of which are constructed formed by (**a**, **c**, **e**) the RC regime. The cathodic current density: 440 mA cm^{-2}. The anodic current density: 40 mA cm^{-2}. The cathodic pulse: 10 ms. The anodic pulse: 20 ms and (**b**, **d**, **f**) the constant galvanostatic (DC) regime. The current density: 120 mA cm^{-2} (Reprinted from Ref. [23] with kind permission from Springer)

References

1. Winand R (1994) Electrodeposition of metals and alloys – new results and perspectives. Electrochim Acta 39:1091–1105
2. Nikolić ND, Živković PM, Jokić B, Pavlović MG, Stevanović JS (2014) Comparative analysis of the polarisation and morphological characteristics of electrochemically produced powder forms of the intermediate metals. Maced J Chem Chem Eng 33:169–180
3. Shin HC, Dong J, Liu M (2003) Nanoporous structures prepared by an electrochemical deposition process. Adv Mater 15:1610–1614
4. Nikolić ND, Popov KI, Pavlović LjJ, Pavlović MG (2006) The effect of hydrogen codeposition on the morphology of copper electrodeposits. I the concept of effective overpotential. J Electroanal Chem 588:88–98
5. Nikolić ND, Popov KI (2010) Hydrogen co-deposition effects on the structure of electrodeposited copper. In: Djokić SS (ed) Electrodeposition: theory and practice, vol 48, Series: modern aspects of electrochemistry. Springer, New York, pp 1–70
6. Ibl N (1961) Probleme des konvektiven Stofftransports bei der Elektrolyse. Chemie Ing Techn 33:69–74 (in German)
7. Ibl N (1963) Probleme des Stofftransportes in der angewandten Elektrochemie. Chemie Ing Techn 35:353–361 (in German)

8. Jenssen LJ, Hoogland JG (1970) The effect of electrolytically evolved gas bubbles on the thickness of the diffusion layer. Electrochim Acta 15:1013–1023
9. Bockris JO'M, Reddy AKN, Gamboa–Aldeco M (2000) Modern electrochemistry 2A, Fundamentals of electrodics, 2nd edn. Kluwer Academic/Plenum Publishers, New York
10. Popov KI, Nikolić ND (2012) General theory of disperse metal electrodeposits formation. In: Djokić SS (ed) Electrochemical production of metal powders, vol 54, Series: modern aspects of electrochemistry. Springer, New York, pp 1–62
11. Martins L, Martins JI, Romeira AS, Costa ME, Costa J, Bazzaoui M (2004) Morphology of copper coatings electroplated in an ultrasonic field. Mat Sci Forum 455–456:844–848
12. Nikolić ND, Wang H, Cheng H, Guerrero C, Ponizovskaya EV, Pan G, Garcia N (2004) Magnetoresistance controls arboreous bead-dendritic growth of magnetic electrodeposits: experimental and theoretical results. J Electrochem Soc 151:C577–C584
13. Nikolić ND (2007) The effects of a magnetic field on the morphologies of nickel and copper deposits: the concept of "effective overpotential". J Serb Chem Soc 72:787–797
14. Pavlović MG, Kindlova Š, Roušar I (1992) The initiation of dendritic growth of electrodeposited copper on a rotating disc electrode with changing copper concentration and diffusion layer thickness. Electrochim Acta 37:23–27
15. Aogaki R, Fueki K, Mukaibo T (1975) Application of magnetohydrodynamic effect to the analysis of electrochemical reactions-2. Diffusion process in MHD forced flow of electrolyte solutions. Denki Kagaku 43:509–514
16. Shin HC, Liu M (2005) Three-dimensional porous copper-tin alloy electrodes for rechargeable lithium batteries. Adv Funct Mater 15:582–586
17. Kim R, Han D, Nam D, Kim J, Kwon H (2010) Effects of substrate morphology and postelectrodeposition on structure of Cu foam and their application for Li-Ion batteries. J Electrochem Soc 157:D269–D273
18. Yin J, Jia J, Zhu L (2008) Macroporous Pt modified glassy carbon electrode: preparation and electrocatalytic activity for methanol oxidation. Int J Hydrogen Energy 33:7444–7447
19. Plowman BJ, Jones LA, Bhargava SK (2015) Building with bubbles: the formation of high surface area honeycomb-like films via hydrogen bubble templated electrodeposition. Chem Commun 51:4331–4346
20. Li Y, Jia W-Z, Song Y-Y, Xia X-H (2007) Superhydrophobicity of 3D porous copper films prepared using the hydrogen bubble dynamic template. Chem Mater 19:5758–5764
21. Nikolić ND, Branković G, Pavlović MG, Popov KI (2009) The effects of the pause to pulse ratio in the regime of pulsating overpotential on the formation of honeycomb-like structures. Electrochem Commun 11:421–424
22. Nikolić ND, Branković G, Popov KI (2011) Optimization of electrolytic process of formation of open and porous copper electrodes by the pulsating current (PC) regime. Mater Chem Phys 125:587–594
23. Nikolić ND (2012) Porous copper electrodes formed by the constant and the periodically changing regimes of electrolysis. In: Djokić SS (ed) Electrochemical production of metal powders, vol 54, Series: modern aspects of electrochemistry. Springer, New York, pp 187–249
24. Nikolić ND, Pavlović LjJ, Pavlović MG, Popov KI (2007) Formation of dish-like holes and a channel structure in electrodeposition of copper under hydrogen co-deposition. Electrochim Acta 52:8096–8104
25. Nikolić ND, Branković G, Pavlović MG, Popov KI (2008) The effect of hydrogen codeposition on the morphology of copper electrodeposits. II correlation between the properties of electrolytic solutions and the quantity of evolved hydrogen. J Electroanal Chem 621:13–21
26. Shin HC, Liu M (2004) Copper foam structures with highly porous nanostructured walls. Chem Mater 16:5460–5464
27. Nikolić ND, Popov KI, Pavlović LjJ, Pavlović MG (2007) Determination of critical conditions for the formation of electrodeposited copper structures suitable for electrodes in electrochemical devices. Sensors 7:1–15

28. Nikolić ND, Popov KI, Pavlović LjJ, Pavlović MG (2007) Phenomenology of a formation of a honeycomb – like structure during copper electrodeposition. J Solid State Electrochem 11:667–675
29. Nikolić ND, Maksimović V, Pavlović MG, Popov KI (2009) Cross-section analysis of morphology of electrodeposited copper obtained in hydrogen co-deposition range. J Serb Chem Soc 74:689–696
30. Nikolić ND, Popov KI, Pavlović LjJ, Pavlović MG (2006) Morphologies of copper deposits obtained by the electrodeposition at high overpotentials. Surf Coat Technol 201:560–566
31. Nikolić ND, Pavlović LjJ, Krstić SB, Pavlović MG, Popov KI (2008) Influence of ionic equilibrium in the $CuSO_4$-H_2SO_4-H_2O system on the formation of irregular electrodeposits of copper. Chem Eng Sci 63:2824–2828
32. Casas JM, Alvarez F, Cifuentes L (2000) Aqueous speciation of sulfuric acid-cupric sulfate solutions. Chem Eng Sci 55:6223–6234
33. Nikolić ND, Pavlović LjJ, Branković G, Pavlović MG, Popov KI (2008) The ionic equilibrium in the $CuSO_4$-H_2SO_4-H_2O system and the formation of the honeycomb-like structure during copper electrodeposition. J Serb Chem Soc 73:753–760
34. Nikolić ND, Pavlović LjJ, Pavlović MG, Popov KI (2007) Effect of temperature on electrodeposition of disperse copper deposits. J Serb Chem Soc 72:1369–1381
35. Amadi A, Gabe DR, Goodenough M (1991) An electrochemical mass transport sensor to study agitation in electroplating processes. J Appl Electrochem 21:1114–1119
36. Vogt H, Balzer RJ (2005) The bubble coverage of gas-evolving electrodes in stagnant electrolytes. Electrochim Acta 50:2073–2079
37. Kim J-H, Kim R-H, H-Sang K (2008) Preparation of copper foam with 3-dimensionally interconnected spherical pore network by electrodeposition. Electrochem Commun 10:1148–1151
38. Nam D, Kim R, Han D, Kim J, Kwon H (2011) Effects of $(NH_4)_2SO_4$ and BTA on the nanostructure of copper foam prepared by electrodeposition. Electrochim Acta 56:9397–9405
39. Cherevko S, Xing X, Chung C-H (2011) Hydrogen template assisted electrodeposition of sub-micrometer wires composing honeycomb-like porous Pb films. Appl Surf Sci 257:8054–8061
40. Cherevko S, Xing X, Chung C-H (2010) Electrodeposition of three-dimensional porous silver foams. Electrochem Commun 12:467–470
41. Cherevko S, Chung C-H (2011) Direct electrodeposition of nanoporous gold with controlled multimodal pore size distribution. Electrochem Commun 13:16–19
42. Popov KI, Djokić SS, Grgur BN (2002) Fundamental aspects of electrometallurgy. Kluwer Academic/Plenum Publishers, New York
43. Cherevko S, Chung C-H (2010) Impact of key deposition parameters on the morphology of silver foams prepared by dynamic hydrogen template deposition. Electrochim Acta 55:6383–6390
44. Plowman BJ, O'Mullane AP, Selvakannan PR, Bhargava SK (2010) Honeycomb nanogold networks with highly active sites. Chem Commun 46:9182–9184
45. Ott A, Jones LA, Bhargava SK (2011) Direct electrodeposition of porous platinum honeycomb structures. Electrochem Commun 13:1248–1251
46. Cherevko S, Kulyk N, Chung CH (2012) Nanoporous palladium with sub-10 nm dendrites by electrodeposition for ethanol and ethylene glycol oxidation. Nanoscale 4:103–105
47. Marozzi CA, Chialvo AC (2000) Development of electrode morphologies of interest in electrocatalysis. Part 1: electrodeposited porous nickel electrodes. Electrochim Acta 45:2111–2120
48. Jović VD, Jović BM, Maksimović VM, Pavlović MG (2007) Electrodeposition and morphology of Ni, Co and Ni–Co alloy powders: part II. Ammonium chloride supporting electrolyte. Electrochim Acta 52:4254–4263

49. Nikolić ND, Branković G, Maksimović VM, Pavlović MG, Popov KI (2009) Influence of potential pulse conditions on the formation of honeycomb-like copper electrodes. J Electroanal Chem 635:111–119
50. Popov KI, Nikolić ND, Živković PM, Branković G (2010) The effect of the electrode surface roughness at low level of coarseness on the polarization characteristics of electrochemical processes. Electrochim Acta 55:1919–1925
51. Nikolić ND, Branković G, Maksimović VM, Pavlović MG, Popov KI (2010) Application of pulsating overpotential regime on the formation of copper deposits in the range of hydrogen co-deposition. J Solid State Electrochem 14:331–338
52. Nikolić ND, Branković G (2012) Comparison of open porous copper structures obtained by the different current regimes of electrolysis. Mater Lett 70:11–15
53. Sun BK, O'keefe TJ (1998) Growth of electrolytic copper dendrites and their adhesion to an epoxy resin. Surf Coat Technol 106:44–52
54. Nikolić ND, Branković G, Pavlović MG (2012) Correlate between morphology of powder particles obtained by the different regimes of electrolysis and the quantity of evolved hydrogen. Powder Technol 221:271–277
55. Nikolić ND, Branković G, Maksimović VM (2012) Morphology and internal structure of copper deposits electrodeposited by the pulsating current regime in the hydrogen co-deposition range. J Solid State Electrochem 16:321–328
56. Popov KI, Pavlović MG (1993) Electrodeposition of metal powders with controlled grain size and morphology. In: White RE, Bockris JOM, Conway BE (eds) Modern aspects of electrochemistry, vol 24. Plenum, New York, pp 299–391
57. Nikolić ND, Branković G (2010) Effect of parameters of square-wave pulsating current on copper electrodeposition in the hydrogen co-deposition range. Electrochem Commun 12:740–744
58. Chandrasekar MS, Pushpavanam M (2008) Pulse and pulse reverse plating—conceptual, advantages and applications. Electrochim Acta 53:3313–3322
59. Nikolić ND, Branković G, Maksimović VM (2011) Effect of the anodic current density on copper electrodeposition in the hydrogen co-deposition range by the reversing current (RC) regime. J Electroanal Chem 661:309–316
60. Popov KI, Maksimović MD (1989) Theory of the effect of electrodeposition at periodically changing rate on the morphology of metal deposition. In: Conway BE, Bockris JO'M, White RE (eds) Modern aspects of electrochemistry, vol 19. Plenum press, New York, pp 193–250

Chapter 6
Electrochemically Produced Metal Powders

6.1 Introduction

Powders are finely divided solids, smaller than 1000 μm in its maximum dimension. A particle is defined as the smallest unit of a powder. The particles of powder may assume various forms and sizes, whereas the powders, as an association of such particles, exhibit, more or less, the same characteristics as if they were formed under identical conditions and if the manipulation of the deposits after removal from the electrode was the same [1, 2]. The size of particles of many metal powders can vary in a quite wide range from a few nanometers to several hundreds of micrometers. The most important properties of a metal powder are the specific surface, the apparent density, the flowability, and the particle grain size distribution. These properties, called decisive properties, characterize the behavior of a metal powder.

Different methods for the production of metal powders including mechanical commuting, chemical reaction, electrolysis, and liquid metal atomization are used in practice [1]. Powders of about 60 metals can successfully be produced by electrolysis. The majority of metallic powders are obtained by molten salt electrolysis. However, due to technological advantages and various industrial applications, most of the practically useful powders, *e.g.*, copper, iron, nickel, *etc.*, are produced from aqueous solutions [3]. Formation of powders by electrolysis is an economical processing method with a low capital investment and operational cost. The main advantages of this method in relation to other methods of powder production are high purity of the produced powder which can be easily pressed and sintered and low oxygen content [1, 2]. It has environmentally friendly way of powder production which enables working in a closed circuit [4].

Dendrites are the most common shape of electrochemically produced powder particles [5]. Aside from dendrites, powder particles can be also obtained in the form of flakes, fibrous, spongy, wires, cauliflower-like, and the many other irregular forms. The powder particles can spontaneously fall off or can be removed from the

K.I. Popov et al., *Morphology of Electrochemically and Chemically Deposited Metals*, DOI 10.1007/978-3-319-26073-0_6

electrode by tapping or other similar techniques. In this chapter, all these electrodeposited forms before their removal from the electrode surface are also treated as powder particles.

Metal powders can be formed by both potentiostatic and galvanostatic regimes of electrolysis [2, 5–7]. In the galvanostatic regime of electrolysis, all metals which can be electrodeposited exhibit a tendency to appear in the form of powders at current densities larger than a certain critical value which is equal to the limiting diffusion current density [7, 8]. In the potentiostatic regime of electrolysis, metals are electrodeposited in the powder form at overpotentials inside the plateau of the limiting diffusion current density, as well as at the higher ones [7]. In addition, metal powders can be produced under periodically changing regimes of electrolysis, such as pulsating overpotential (PO), pulsating current (PC), and reversing current (RC) [5–7]. Application of electrodeposition at a periodically changing rate offers a great possibility in the electrochemical production of metal powders.

6.2 Morphology of Powder Particles in the Dependence of the Exchange Current Density and Hydrogen Overpotential

The shape of electrochemically formed powder particles strongly depends on the electrodeposition conditions, such as the electrolysis regime, the composition of solution, the type of the working electrode, and the temperature and the nature of the metal. Regarding the values of the exchange current density, i_0, and melting points, T_m, metals are classified into three classes [9, 10]:

(a) *Class I, so-called normal metals*: Pb, Sn, Tl, Cd, Hg, Zn, Ag (simple electrolytes). These metals are characterized by low T_m and high i_0 ($i_0 > 1$ A dm^{-2}) values. Additionally, they show high overpotentials for hydrogen discharge.
(b) *Class II, intermediate metals*: Cu, Au, Ag (complex electrolytes). These metals are characterized by moderate T_m, medium i_0 (i_0 in the interval from 10^{-2} to 1 A dm^{-2}), and lower hydrogen overpotentials.
(c) *Class III, inert metals*: Fe, Co, Ni, Mn, Cr, Pt. These metals have high T_m, low i_0, and very low hydrogen overpotentials. For this class of metals, i_0 is between 10^{-2} and 10^{-12} A dm^{-2}.

6.2.1 *Class I: So-Called Normal Metals*

Due to the high exchange current density values, i_0, the processes of electrodeposition of the normal metals belong to the fast electrochemical processes which enable formation of powder at the relatively low overpotentials [5]. The dendrites of normal metals are mainly 2D (two dimensional), and the shape of the dendrites of

Fig. 6.1 Pb, Ag, and Cd powders electrodeposited from (**a**) Pb; 0.10 M $Pb(NO_3)_2 + 2.0$ M $NaNO_3$; overpotential, $\eta = 100$ mV, (**b**) Pb; 0.10 M $Pb(CH_3COO)_2 + 1.5$ M $NaCH_3COO + 0.15$ M CH_3COOH; $\eta = 100$ mV, (**c**) Ag; 0.060 M $AgNO_3 + 1.2$ M $NaNO_3 + 0.050$ M HNO_3; $\eta = 150$ mV, (**d**) Cd; 0.10 M $CdSO_4 + 0.50$ M H_2SO_4; $\eta = 110$ mV, $\times$ 400, (**e**) Pb; 0.30 M $Pb(NO_3)_2 + 2.0$ M $NaNO_3$; $\eta = 55$ mV, and (**f**) Ag; 0.10 M $AgNO_3 + 2.0$ M $NaNO_3$; $\eta = 100$ mV (Reprinted from Refs. [11–13] with kind permission from Springer, Ref. [14] with permission from Elsevier and Ref. [15] with permission from Electrochemical Society)

this group of metals follows the classical Wranglen definition of a dendrite (Fig. 6.1). According to Wranglen [16], a dendrite is a skeleton of a monocrystal and consists of a stalk and branches, thereby resembling a tree. The dendrite consisted only of the stalk, and primary branches is referred as primary (P) dendrite. If the primary branches in turn develop secondary branches, the dendrite is called secondary (S). The two-dimensional (2D) dendrite refers to dendrites with branches that lie in the same plane as that of the primary stalk [17].

The shape of dendrites depends not only on the nature of metals but also on the type of used electrolyte. For example, the 2D Pb dendrites of the P type are obtained from the basic (nitrate) electrolyte (Fig. 6.1a), while the 2D Pb dendrites of the S type are obtained from the complex (acetate) electrolyte (Fig. 6.1b) [11, 14, 18]. As far as the shape of dendrites of the other metals from this group, they also belong to either P (Ag; Fig. 6.1c) [12, 15] or S (Cd; Fig. 6.1d) types [5, 12, 13, 19]. The branchy Pb and Cd dendrites of the S type are usually referred as the fern-like ones [13, 14]. Aside from the (P) and (S) dendrites, these metals are also electrodeposited in the needle-like form (Fig. 6.1e, f). The appearance of tertiary branches in branchy Pb dendrites is also possible [11].

In contrast to the very different morphology of the Pb powder particles obtained from the different electrolytes, the crystallographic structure of powder particles does not depend on the type of electrolyte [11, 18]. Irrespective of the type of electrolytes, the dominant presence of Pb crystallites oriented in the (111) plane was observed in the powder particles obtained from both basic and complex electrolytes.

6.2.2 *Class II: Intermediate Metals*

The shape of dendrites of the group of the intermediate metals is completely different from those of the group of the normal metals. They are the 3D (three dimensional) pine-like constructed from a long trunk or stem with corncob-like branches oriented in all directions (Fig. 6.2a–c) [7, 12, 15, 20]. There is no any difference in the shape of dendrites among metals from this group (Fig. 6.2a, c – Ag (ammonium electrolyte) and Fig. 6.2b – Cu). The similar shape of dendrites is also obtained by the processes of electrodeposition of Au [21] and Ag from the other types of complex electrolytes (*i.e.*, in the presence of tungstosilicate [22] and citric [23] acids). This similarity in the form of dendrites at the macrolevel clearly indicates that the pine-like shape of dendrites represents the typical form of a dendrite characterizing the group of intermediate metals.

The corncob-like forms were mutually parallel and oriented at an approximate angle of about 60° in relation to the stalk or trunk, indicating the existence of high structural symmetry of this group of dendrites. The corncob-like forms were constructed from small agglomerates of metal grains indicating on the similarity the pine-like dendrites, not only on the macro- but also on the microlevel. The treatment of copper dendrites using ultrasound [24, 25] showed that the corncob-like forms represent the basic element of which these dendrites were constructed.

The pine-like Ag and Cu dendrites shown in Fig. 6.2a, b were electrodeposited inside the plateaus of the limiting diffusion current density [20]. The Ag dendrites were also formed outside the plateau of the limiting diffusion current density in the zone of a sharp increase in the current density with an increase in the overpotential (Fig. 6.2c).

However, due to the fact that hydrogen evolution reaction on copper electrode is somewhat faster than on silver electrode (in the case of Ag, there is no hydrogen evolution even at an overpotential of 1000 mV *vs.* Ag reference electrode) [15, 20], hydrogen evolution as a parallel reaction to Cu electrodeposition at high overpotentials commencing at the overpotential inside the plateau of the limiting diffusion current density causes an inhibition of dendritic growth. In the dependence of concentration of Cu^{2+} ions, the honeycomb-like structures are usually formed outside the plateau of the limiting diffusion current density (Fig. 6.2d), and the particles of cauliflower-like shape are obtained after a removal of deposits from

Fig. 6.2 Ag and Cu powders electrodeposited from (**a**) Ag; 0.10 M $AgNO_3 + 0.50$ M $(NH_4)_2SO_4 + NH_4OH$ to dissolve the silver sulfate precipitate; $\eta = 650$ mV, (**b**) Cu; 0.10 M $CuSO_4 + 0.50$ M H_2SO_4; $\eta = 650$ mV, part in circles: holes formed of detached hydrogen bubbles, (**c**) Ag; 0.10 M $AgNO_3 + 0.50$ M $(NH_4)_2SO_4 + NH_4OH$ to dissolve the silver sulfate precipitate; $\eta = 1000$ mV, (**d–f**) Cu; 0.10 M $CuSO_4 + 0.50$ M H_2SO_4; $\eta = 1000$ mV (Reprinted from Ref. [20] with permission from Society of Chemists and Technologists of Macedonia)

them (Fig. 6.2e, f). The cauliflower-like particles were constructed from small agglomerates of Cu grains, indicating on the similarity of microstructure of these particles with dendritic ones [7].

6.2.3 *Class III: Inert Metals*

Due to the low overpotentials for hydrogen evolution reaction for metals from this group, electrodeposition of inert metals occurs parallel with hydrogen evolution reaction in the whole range of current densities and potentials. Dendritic growth is mainly inhibited by generated hydrogen, and the spongy-like, cauliflower-like, and globular particles, very similar to each other, were predominately formed by electrodepositions of Co, Ni, and Fe (Fig. 6.3a–c) [12, 26, 27, 29, 30]. The shape of 3D fern-like dendrites was strongly affected by generated hydrogen, and the

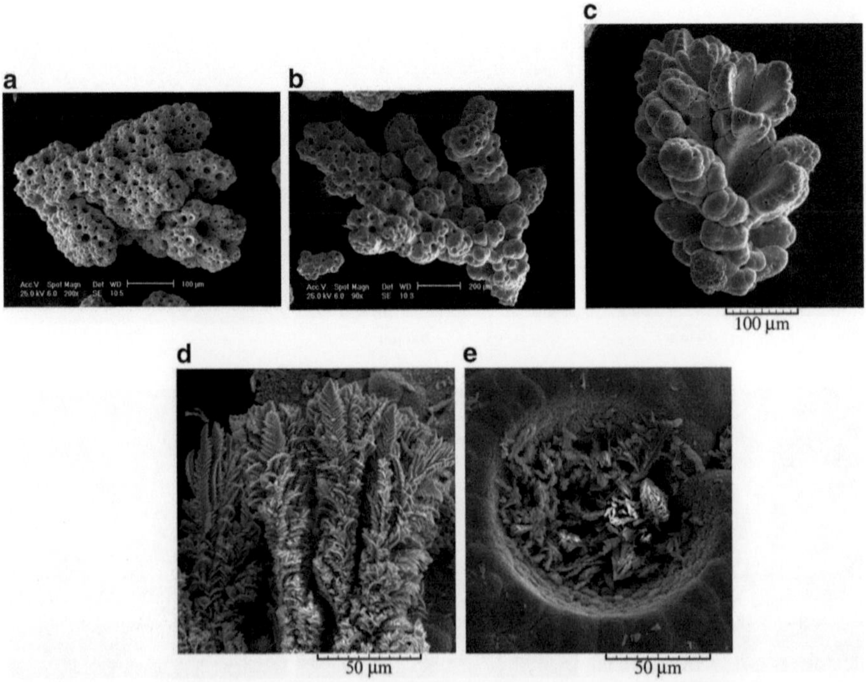

Fig. 6.3 Ni, Co, and Fe powders electrodeposited from (**a**) Ni: 0.10 M $NiCl_2$ + 1.0 M NH_4Cl + 0.70 M NH_4OH under the condition of limiting current density, (**b**) Co: 0.10 M $CoCl_2$ + 1.0 M NH_4Cl + 0.70 M NH_4OH under the condition of limiting current density, (**c**) Fe: 1.0 M NH_4Cl + 0.20 M $Na_3C_6H_5O_7$ + 0.10 M $FeCl_2$, pH = 4.0, (**d**, **e**) Co: 0.10 M $CoSO_4$ + 0.70 M NH_4OH + 1.0 M $(NH_4)_2SO_4$; $i = 500$ mA cm^{-2} (Reprinted from Ref. [12] with kind permission from Springer, Refs. [26, 27] with permission from Elsevier, and Ref. [28] with permission from the Serbian Chemical Society)

dendrites are formed at the lower current densities, as well as in the interior of holes where there is no the effect of evolved hydrogen on a solution stirring in the near-electrode layer (Fig. 6.3d, e) [28].

6.3 The Characteristics of Electrochemically Produced Powder: A General Discussion

Comparative survey of powders obtained in potentiostatic and galvanostatic regimes of electrolysis is shown in Fig. 6.4. In spite of the fact that particles obtained by galvanostatic deposition are less dendritic than those obtained by potentiostatic deposition due to smaller overpotential at the end of the galvanostatic than the potentiostatic regimes [5], there is no any substantial difference in the morphology of powder particles obtained between these two constant regimes of

Fig. 6.4 Pb particles electrodeposited from 0.50 M $Pb(NO_3)_2 + 2.0$ M $NaNO_3$: (**a**) the potentiostatic regime: $\eta = 150$ mV, (**b**) the galvanostatic (DC) regime: $i = 160$ mA cm^{-2}, and Cu particles electrodeposited from 0.10 M $CuSO_4 + 0.50$ M H_2SO_4: (**c**) $\eta = 650$ mV, (**d**) $i = 16$ mA cm^{-2}, (**e**) $\eta = 1000$ mV, and (**f**) $i = 440$ mA cm^{-2} (Reprinted from Refs. [12, 31] with kind permission from Springer)

electrolysis (Pb powder (Fig. 6.4a – the potentiostatic regime and Fig. 6.4b – the galvanostatic regime) and Cu powder (Fig. 6.4c, e – the potentiostatic regime and Fig. 6.4d, f – the galvanostatic regime)).

As previously mentioned, the shape of powder particles depends on both the exchange current density and hydrogen evolution reaction. The lowering of the exchange current density causes a branching of dendrites and their transformation from the 2D into 3D ones. The range of overpotentials corresponding to electrodeposition of metals in the powder form is shifted toward the higher overpotentials, and the increase of the diffusion part at the polarization curve was observed with the decrease of the exchange current density.

The process of complex formation also decreases the exchange current density causing both the increase of the diffusion part and the branching of dendrites [14]. Comparative analysis of the polarization and morphological characteristics of lead for the basic (nitrate) and complex (acetate) electrolytes is presented in Fig. 6.5. In the case of silver, this lowering of the exchange current density is so strong to cause transfer of Ag from the group of normal ($i_0 \rightarrow \infty$ for the nitrate

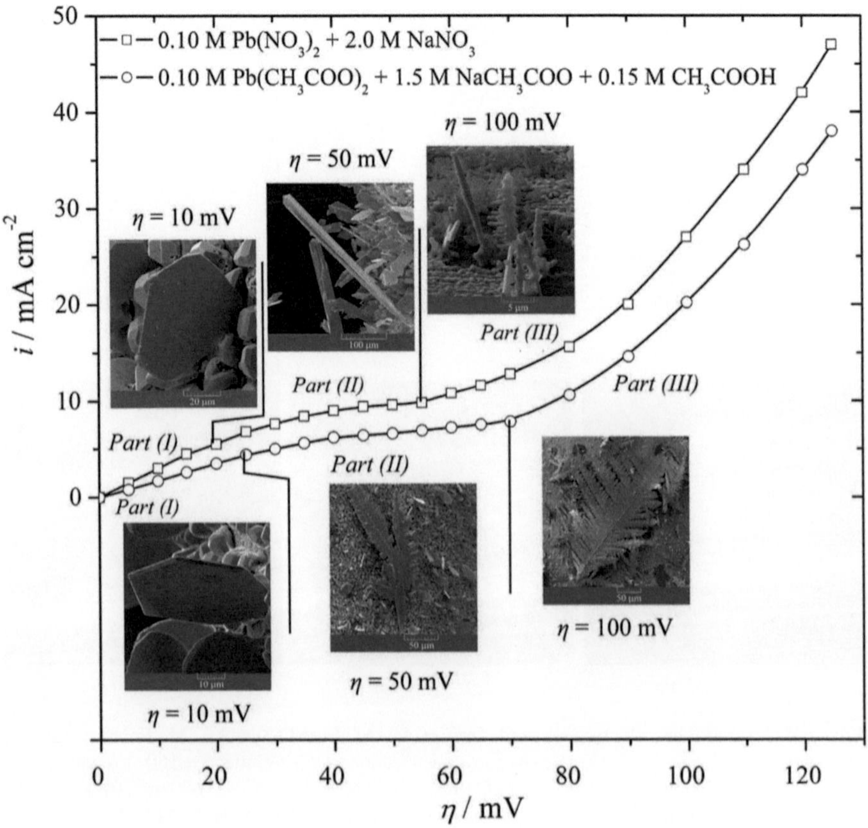

Fig. 6.5 The comparative survey of the polarization curves and the typical morphologies of lead deposits electrodeposited from 0.10 M $Pb(NO_3)_2 + 2.0$ M $NaNO_3$ and 0.10 M Pb $(CH_3COO)_2 + 1.5$ M $NaCH_3COO + 0.15$ M CH_3COOH (Reprinted from Ref. [11] with kind permission from Springer and Ref. [14] with permission from Elsevier)

electrolytes; Fig. 6.1c, f) to the group of intermediate metals ($i_0 \ll i_L$ for the ammonium electrolyte; Fig. 6.2a, c).

The second factor which strongly affects morphology of powder particles is hydrogen evolution reaction as parallel reaction to metal electrodeposition. The effect of this parallel reaction already becomes visible during electrodeposition of Cu at the high overpotentials (Fig. 6.2d–f), and evolution of hydrogen just defines the transitional character of the intermediate metals between the normal and the inert metals. Thanks to this reaction, the same shape of the polarization curves for ammonium Ag (the high hydrogen overpotential for hydrogen evolution reaction) and Cu (the lower hydrogen evolution overpotential than for Ag) gives a completely different morphology of Ag and Cu powders in the zone of fast increase of the current density with increasing the overpotential (Fig. 6.6).

Fig. 6.6 Comparative survey of the polarization and morphological characteristics of Cu and Ag (Reprinted from Ref. [20] with permission from Society of Chemists and Technologists of Macedonia)

Due to the ionic equilibrium of the species in the $CuSO_4$-H_2SO_4-H_2O system (the most often employed electrolytes for the electrodeposition of copper are sulfate ones), the shape of powder particles depends on $CuSO_4$ and H_2SO_4 concentrations [32–34]. For a constant H_2SO_4 concentration, the H^+ ion concentration decreases with increasing Cu^{2+} concentration that is manifested by the change of shape of powder particles from the cauliflower-like to dendritic ones.

Of course, hydrogen evolution affects mechanism of formation of powder particles. The dendrites are formed without, as well as with, a quantity of evolved hydrogen (the case of Cu) which was insufficient to achieve any effect on the hydrodynamic conditions in the near-electrode layer. Then, the electrodeposition process was primarily controlled by the diffusion of ions to the electrode surface, rather than the kinetic of the electrodeposition [5, 6]. The cauliflower-like particles are formed in the conditions of vigorous hydrogen evolution with the strong effect of evolved hydrogen on the hydrodynamic conditions in the near-electrode layer, and the concept of "effective overpotential" is proposed to explain formation of these particles [25, 35].

Finally, hydrogen evolution becomes crucial factor determining the shape of powder particles of the group of the inert metals, and the concept of "effective overpotential" is applicable to explain the formation of these powder particles. Due to vigorous hydrogen evolution, dendritic growth is almost completely inhibited. Analysis of the polarization curves for Co and Ni [12, 26, 30] showed that their shape is a consequence of the so-called electrode effect which is connected to the bubble formation at high current densities; as a result, the evolution of bubbles became the rate-limiting step of the electrochemical process. The current for hydrogen evolution at more negative potentials reached 60–70 % of the total current.

6.4 Application of Periodically Changing Regimes of Electrolysis on Formation of Metal Powders

Powder particles with different grain size and morphology can be obtained by varying the wave of periodically changing current or overpotential [6]. In the dependence of the shape of square-wave pulsating overpotential (PO), it can be obtained either dendritic (Fig. 6.7a) or cauliflower-like (Fig. 6.7b) Cu particles [36]. Please note that both particles are obtained with same overpotential amplitude of 1000 mV. The applied parameters of the PO regime have influence on both hydrogen evolution reaction and metal electrodeposition rate. Generally, application of square-wave PO with deposition pulses, t_c, shorter than pause durations, t_p, suppresses the hydrogen evolution reaction and favors the growth of dendrites, while the applied square-wave PO with t_c longer than t_p favors formation of cauliflower-like particles. In this way, the effect of shortening of deposition pulse is equivalent to the decrease of the overpotential of electrodeposition in the potentiostatic regime of electrolysis. It is clear that dendrite obtained by the PO regime (Fig. 6.7a) is very similar to the one obtained at a constant overpotential of 650 mV (Fig. 6.2b). On the other hand, morphology of cauliflower-like particle shown in Fig. 6.7b approaches to the one obtained at an overpotential of 1000 mV (Fig. 6.2d).

The similar consideration is also valid for the regime of pulsating current (PC) [37]. The branchy dendrites are formed with deposition pulses, t_c, shorter than pause duration, t_p, (Fig. 6.7c), and cauliflower-like particles are formed with t_c longer than t_p (Fig. 6.7d). Effects attained by the choice of appropriate parameters of square-wave PC were equivalent to those obtained by electrodeposition at the constant overpotential from solutions of different concentrations of $CuSO_4$ and H_2SO_4. Increasing concentration of Cu^{2+} ions was equivalent to the effect of the decreasing deposition pulses, while the constant pause duration was equivalent to H_2SO_4 concentration used. The effect of increasing H_2SO_4 concentration was equivalent to the effect to the decreasing pause duration, while the constant $CuSO_4$ concentration is equivalent to the constant deposition pulse used [37]. Simultaneously, the constant overpotential used corresponds to the amplitude current density used.

The periodically changing regimes also affect strongly the shape of powder particles of metals with high overpotentials for hydrogen discharge (*Class I*). Unlike of the fern-like dendrites obtained at the constant overpotential

Fig. 6.7 Cu particles electrodeposited from 0.15 M $CuSO_4$ in 0.50 M H_2SO_4 by the regime of (**a**) pulsating overpotential (PO); the overpotential amplitude, η_A: 1000 mV; pause duration, t_p: 10 ms; deposition pulse, t_c: 3 ms, (**b**) PO; η_A: 1000 mV; t_p: 10 ms; t_c: 20 ms, (**c**) pulsating current (PC); the current density amplitude, i_A: 0.20 A cm^{-2}; t_p: 10 ms; t_c: 1 ms, (**d**) PC; i_A: 0.20 A cm^{-2}; t_p: 10 ms; t_c: 50 ms, and Pb particles obtained from 0.10 M $Pb(CH_3COO)_2$ + 1.5 M $NaCH_3COO$ + 0.15 M CH_3COOH: (**e**) the constant overpotential: 75 mV; $t = 50$ s; × 500, and (**f**) PO; η_A: 75 mV; t_c: 0.1 ms; pause to pulse ratio: 3; × 1000 (Reprinted from Ref. [7] with kind permission from Springer and Refs. [36–38] with permission from Elsevier)

(Fig. 6.7e), Pb particles of irregular shape (denoted as irregular crystals or precursors of dendrites) (Fig. 6.7f) were obtained by the appropriate square-wave PO with the overpotential amplitude equal to the overpotential of electrodeposition in the constant potentiostatic regime of electrolysis [5, 38]. As a remainder, in the constant potentiostatic regimes of electrolysis, irregular crystals are formed at lower overpotentials than the fern-like dendrites.

6.5 Analysis of Decisive Properties of Powders and Their Mutual Relations

The properties of metal powder, denoted as the decisive properties, characterize the behavior of metal powders. The most important of them are the specific surface, the apparent density, the flowability, the particle grain size, and the particle size distribution [1]. These properties were analyzed by Popov et al. [39–54] for Cu powder that is taken as a model system. It is shown that some of properties of Cu powders can be mutually related, as well as that the specific surface of copper powder can be related to overpotential of electrodeposition [39, 40]. In this way, correlation between properties of metal powders and conditions of electrodeposition was enabled.

6.5.1 Correlation Between Specific Surface and Overpotential of Electrodeposition

The specific surface of a powder and a powdered deposit is determined as the surface per unit of the mass of powder.

It is well known [55, 56] that the surface coarseness during potentiostatic electrodeposition in the mixed activation–diffusion control increases with the increase of the current density of electrodeposition. The surface coarseness also increases during potentiostatic electrodeposition at the limiting diffusion current density with an increase of overpotential [57, 58], resulting in the formation of dendrites. In metal electrodeposition in the limiting diffusion current density range, the real current density remains constant regardless of overpotential used. Simultaneously, the larger overpotential of electrodeposition is used, the more disperse deposit characterized by the increased specific surface is formed.

The last effect can be qualitatively discussed as follows. In the Eq. (1.31), the activation part of electrodeposition overpotential required for the charge transfer, η_{act}, is given by Eq. (2.89), while the rest of the overpotential is due to mass transfer limitations, η_{diff}, given by Eq. (6.1):

$$\eta_{diff} = \frac{b_c}{2.3} \ln \frac{1}{1 - \frac{i}{i_L}} \tag{6.1}$$

Hence, if $i \to i_L$,

$$\eta_{act} = \frac{b_c}{2.3} \ln \frac{i_L}{i_0} \tag{6.2}$$

being equal to the critical overpotential for the initiation of dendritic growth [5, 59]. Simultaneously, for $i \to i_L$

$$\eta_{diff} \to \infty \tag{6.3}$$

It is obvious that a very small increase of current density in the limiting diffusion current density range causes a large increase in deposition overpotential. Hence, the charge transfer overpotential and the ohmic drop in the solution remain the same for all overpotentials belonging to the limiting diffusion current density plateau,

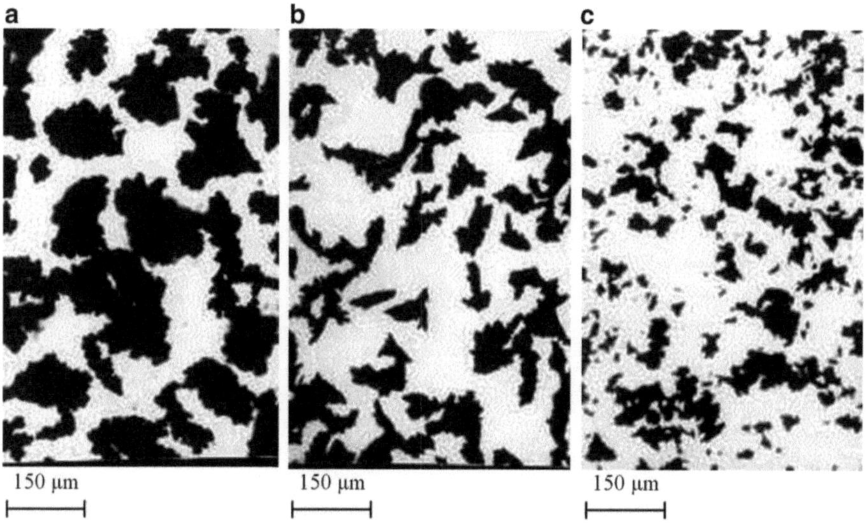

Fig. 6.8 Copper powder particles obtained at different overpotentials from 0.10 M $CuSO_4$ in 0.50 M H_2SO_4 onto stationary platinum wire electrodes painted with shellac. Deposition time: 15 min. Overpotentials of electrodeposition: (**a**) 500 mV, (**b**) 600 mV, and (**c**) 700 mV (Reprinted from Ref. [5] with kind permission from Springer)

regardless of electrodeposition overpotential. This is due to the fact that both the charge transfer overpotential and the ohmic drop only depend on the current density. On the other hand, an increase of the deposition overpotential in the limiting diffusion current density range causes a strong increase of the dispersity, as illustrated by Fig. 6.8, and leads to an increase of the specific surface area of metal deposits. Hence, it seems reasonable to assume that the difference in overpotential of electrodeposition can be related to the increase of the surface of the deposited metal by Eq. (6.4) [39, 40]:

$$\gamma_s(S_2 - S_1) = (\eta_2 - \eta_1)\int_0^t I\mathrm{d}t \tag{6.4}$$

where I is a current of electrodeposition, t is a time of electrodeposition, $\int_0^t I\mathrm{d}t$ is the quantity of passed electricity, η_2 and η_1 are overpotentials of electrodeposition belonging to the limiting diffusion current density range, S_2 and S_1 are the surface area of disperse deposits at the overpotentials η_2 and η_1, respectively, and γ_s is the surface energy of disperse deposit formation.

Equation (6.4) can be rewritten in the form:

$$\gamma_s = \frac{(\eta_2 - \eta_1)\int_0^t I\mathrm{d}t}{(S_2 - S_1)} \tag{6.5}$$

for the direct determination of the energy of disperse solid copper surface formation in sulfate solutions.

The energy of disperse solid copper surface formation, γ_s, calculated by Eq. (6.5) includes all energetic loses during electrodeposition in the range of the limiting diffusion current densities.

For the estimation of the surface energy of disperse deposits formation, γ_s, according to Eq. (6.5), it is necessary to determine surface area of deposits, S_2 and S_1, at overpotentials η_2 and η_1 belonging to the limiting diffusion current density range. Obviously, the surface area of a deposit, S_1, corresponds to an overpotential η_1 at the beginning of the limiting diffusion current density plateau, while the surface area of a deposit, S_2, corresponds to an overpotential η_2 at the end of the limiting diffusion current density plateau. During the depositions, $I - t$ dependences were recorded and the quantity of electricity, $\int_0^t I\mathrm{d}t$ was determined by a graphical integration.

Due to very high surface areas of powdered (or disperse) deposits, the determination of real surface area of this deposit type can represent serious problem [52]. The determination of the real surface area of disperse metal deposits by some common methods, such as the use of STM and AFM techniques (using the option *surface area diff.*, in the accompanying software package), is not possible, because these techniques are only suitable for the determination of the real surface area of compact and relatively smooth surface area [60]. For that reason, the new and relatively easy way for the estimation of the real surface area of disperse deposits was proposed by Popov et al. [39, 40, 52]. For copper solution containing 0.15 M $CuSO_4$ in 0.50 M H_2SO_4, this estimation can be presented in the following way: working electrode (copper or platinum of the surface area $S_{0,i}$) is covered by a thin copper film by electrodeposition at an overpotential of 300 mV during 2 min. After a relaxation of the diffusion layer for 15 min, current at an overpotential of 50 mV, I_0, is recorded, being proportional to the original electrode surface area $S_{0,i}$. The overpotential is then adjusted to the desired value and electrodeposition is carried out. After the determined quantity of electricity had been reached, the overpotential is decreased to 50 mV, and after relaxation of the diffusion layer for 15 min, the current, I_η, corresponding to the surface area S_f generated during electrodeposition, is determined. The surface area of the deposit is then calculated using the Eq. (6.6):

$$S_f = S_{0,i}\frac{I_\eta}{I_0} - S_{0,i} = S_{0,i}\left(\frac{I_\eta}{I_0} - 1\right) \tag{6.6}$$

It is well known [57] that dendrites are not formed at overpotentials of electrodeposition lower than a critical overpotential for dendritic growth initiation, η_i, and that powdered deposits characterized by a very large surface area are formed at overpotentials higher than some critical value, η_c.

For $\eta_2 \geq \eta_c$, $\eta_1 \geq \eta_i$, and $S_2 \gg S_1$, Eq. (6.5) can be rewritten in the form:

$$S_2 = \frac{(\eta_2 - \eta_1)\int_0^t I\mathrm{d}t}{\gamma_s} \tag{6.7}$$

On the other hand, the quantity of electrodeposited metal m_{sp} is given by:

$$m_{sp} = \frac{M}{nF}\int_0^t I\mathrm{d}t \tag{6.8}$$

assuming the current efficiency for metal electrodeposition, $\eta_I(M)$ to be 1, and M is the atomic mass of deposited metal. From Eqs. (6.7) and (6.8), the specific powder (or powdered deposit) surface S_{sp} is:

$$S_{sp} = \frac{S_2}{m} = \frac{(\eta_2 - \eta_1)nF}{\gamma_s M} \tag{6.9}$$

If the current efficiency for metal electrodeposition, $\eta_I(M)$, is lower than 1, Eq. (6.8) becomes:

$$m_{sp} = \eta_I(M)\frac{M}{nF}\int_0^t I\mathrm{d}t \tag{6.10}$$

and Eq. (6.9) becomes:

$$S_{sp} = \frac{(\eta_2 - \eta_1)nF}{\eta_I(M)\gamma_s M} \tag{6.11}$$

Equations (6.10) and (6.11) are valid in the hydrogen evolution range at overpotentials lower than the critical one for the change of the growth of dendrites. The situation is dramatically different in galvanostatic electrodeposition of powder. In this case, due to the increase of the surface coarseness, the low increase of the limiting diffusion current density caused by the increase of the surface area of a deposit leads to the strong decrease of overpotential of the electrodeposition in the limiting diffusion current density range even with the short deposition times. The internal structure of powder particles is not changed with a time during potentiostatic electrodeposition, and it only depends on the presence or the absence of hydrogen evolution. In the galvanostatic case, the structure of particles becomes more compact with the increasing time of electrodeposition, and it can be changed from dendrites at the beginning of the electrodeposition process to the compact one with the longer electrodeposition time, as shown in Fig. 6.9a, b [41, 47, 50]. In the hydrogen evolution range, overpotential of electrodeposition is determined by the hydrogen reduction and at the sufficiently large initial current densities, $i \gg i_L$, as well as with enough short electrodeposition times, formation of copper powder in a galvanostatic deposition becomes similar to the one in potentiostatic electrodeposition, as shown in Fig. 6.9c, d.

The described method for the determination of the specific surface of electrodeposited copper is applicable if some kind of a Faradaic cage is not formed on the surface of deposit, *i.e.*, when the formed structure is "open" to the bulk of electrolyte solution in potentiostatic deposition.

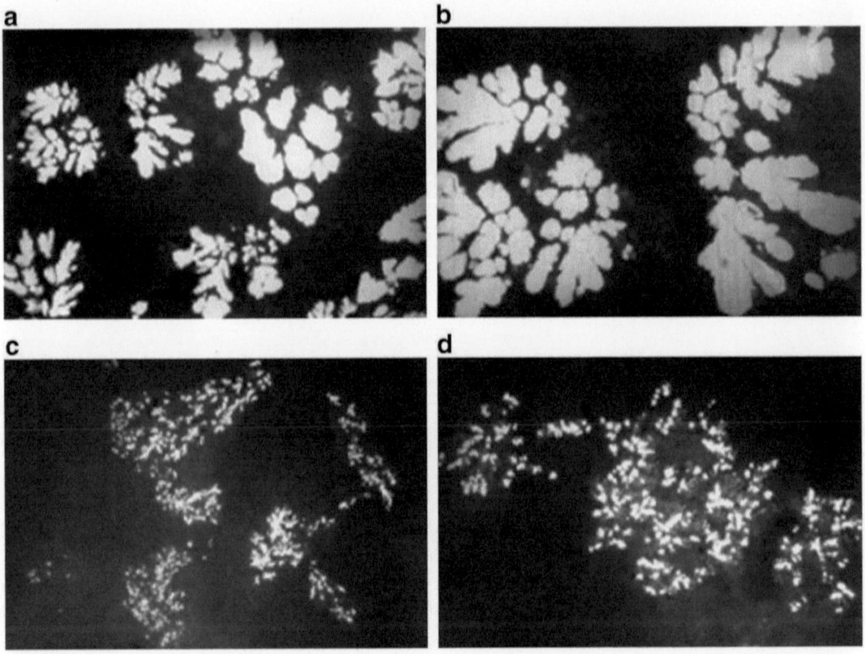

Fig. 6.9 SEM photomicrographs of copper powder particles obtained in constant current deposition. $c(Cu^{2+}) = 15$ g dm^{-3}; $c(H_2SO_4) = 140$ g dm^{-3}; electrolyte circulation rate: 0.11 dm^3 min^{-1}; temperature: (50 ± 2) °C; fraction: (149–177) μm; time of powder removal by brush: 15 min: (**a**, **b**) $i = 1800$ A m^{-2}, apparent density: 1.122 g cm^{-3}, and (**c**, **d**) $i = 3600$ A m^{-2}, apparent density: 0.524 g cm^{-3} (Reprinted from Ref. [50] with permission from the Serbian Chemical Society)

According to Calusaru [3], the specific surface of copper powder is 500–3000 $cm^2\ g^{-1}$ depending on the electrodeposition conditions. The critical overpotential for the dendritic growth initiation in copper sulfate acid solution is about 0.30 V, and the critical overpotential for the instantaneous dendritic growth initiation, which is equal to the critical overpotential of copper powder formation, is about 0.70 V [57]. Assuming that electrodeposition is carried out at an overpotential of the instantaneous dendritic growth initiation, it can be shown with the following: using the difference in the actual deposition overpotential and the overpotential corresponding to the beginning of the plateau of the limiting diffusion density as 0.40 V, γ_s determined in the presented way as 2.7 J cm^{-2} [39, 40], and $\eta_I(Cu) = 1$, the minimum specific powder surface, S_{sp} of 500 $cm^2\ g^{-1}$ is calculated, that is in good agreement with the findings of Calusaru [3].

In this case, the number of electrons involved in the electrodeposition reaction, n, is 2, the atomic mass of Cu, M, is 63.55 g mol^{-1}, and Faraday constant, F, is 96485 A s mol^{-1}.

6.5.2 The Apparent Density as a Function of Specific Surface and Overpotential of the Electrodeposition

The apparent density or volumetric mass is defined as the mass per unit volume of powder [1]. It is well known that copper powders characterized with high values of specific surface exhibit low apparent density. Powder particles from the same fraction of different powders occupy approximately the same volume, but the structure of the metallic copper can be considerably different with apparent densities, as well as the different specific surfaces [42, 43]. Obviously, the more disperse the powder particles are, the smaller is the apparent density of the copper powder, and the larger is the specific surface.

Using the data of Calusaru [3], it can be established that correlation between the apparent density and the specific surface of Cu powder is:

$$\rho_{ad} = \frac{K'}{S_{sp}} \tag{6.12}$$

where K' is constant, ρ_{ad} is the apparent density, and S_{sp} is the specific surface of Cu powder. The apparent density of powder decreases exponentially with increasing the specific surface, and a constant K' can be determined from the dependence of ρ_{ad} on S_{sp} [48].

The dependence of ρ_{ad} on deposition overpotential can be easily obtained by substitution of S_{sp} from Eq. (6.11) in Eq. (6.12) as:

$$\rho_{ad} = \frac{K' \gamma_s M \eta_I(M)}{(\eta_2 - \eta_i) n F} \tag{6.13}$$

6.5.3 The Size of Representative Particle and Particle Size Distribution

As told already, a copper powder is not formed of particles of identical size and morphology; the individual particles may assume various forms and have very different surface areas for the same average size of granule [3]. As a result of this, it is not possible to relate the powder properties with the deposition process parameters and the deposition conditions, and, hence, a representative particle of the metal powder, having at least one property the same as the powder, should be determined. Obviously, a powder can then be considered as a group of identical particles.

The representative powder particle can be defined as follows: the specific surface is the common property of a metal powder and an individual powder particle. The specific surface of powder, S_{sp}, measured in cm^2g^{-1} [3], can be determined by the method available in Ref. [61]. On the other hand, the specific surface of an individual powder particle can be calculated from its known regular geometric form. Hence, a representative powder particle is that one which is characterized by the same specific surface as a powder consisting of a mixture of different particles. The specific surface of a powder particle, $S_{sp,p}$, is given by:

$$S_{sp,p} = \frac{S_{par}}{G} \tag{6.14}$$

where S_{par} and G are the surface and the mass of particle, respectively. For a spherical particle, the specific surface $S_{sp,spher}$ is then:

$$S_{sp,spher} = \frac{3}{\rho_m r_{par}} \tag{6.15}$$

where ρ_m is the density of the metal and r_{par} is the radius of the particle. The size of a representative metal powder particle, $r_{par,r}$, can be determined by substitution of S_{sp} instead of $S_{sp,spher}$ into Eq. (6.15) which after further rearrangement gives:

$$r_{par,r} = \frac{3}{\rho_m S_{sp}} \tag{6.16}$$

In this way, the size of spherical representative powder particle is related to the property of the powder which can be determined experimentally and even calculated, as already shown in Refs. [39, 40]. The above discussion is valid if the representative particle is made of compact metal.

Assuming that copper powder can be treated as being made of some homogenous material of the density which is equal to the apparent density of the powder, ρ_{ad}, Eq. (6.16) can be written as:

$$r_{\text{par,r}} = \frac{3}{S_{\text{sp}}\rho_{\text{ad}}} \tag{6.17}$$

The shape of the particle size distribution curves can also be discussed assuming that particles are made of a compact metal [48].

The representative particle of a copper powder has the same specific surface as the powder, meaning that mass of powder should have the same surface as mass of the representative powder particles. It follows from the above fact that two equal portions of two fractions of the particles can have the same surface as the same quantity of representative powder particles. Hence,

$$S_{1,\text{m}} + S_{2,\text{m}} = 2S_{\text{r,m}} \tag{6.18}$$

or

$$S_{\text{sp1}} + S_{\text{sp2}} = 2S_{\text{sp}} \tag{6.19}$$

where $S_{1,\text{m}}$, $S_{2,\text{m}}$, and $S_{\text{r,m}}$ are the surfaces of particles 1, 2, and representative ones, respectively, and S_{sp1}, S_{sp2}, and S_{sp} are the corresponding values of specific surface. Taking into account, Eqs. (6.15) and (6.19) can be rewritten in the form:

$$\frac{1}{r_{\text{par,1}}} + \frac{1}{r_{\text{par,2}}} = \frac{2}{r_{\text{par,r}}} \tag{6.20}$$

where $r_{\text{par,1}}$ and $r_{\text{par,2}}$ are the radii of the particles 1 and 2 and $r_{\text{par,r}}$ is the radius of the representative particle. It follows from Eq. (6.20) that:

$$r_{\text{par,1}} = \frac{r_{\text{par,2}} \cdot r_{\text{par,1}}}{2r_{\text{par,2}} - r_{\text{par,r}}} \tag{6.21}$$

which is valid for $r_{\text{par,2}} > r_{\text{par,r}}$, if

$$r_{\text{par,1}} < r_{\text{par,r}} < r_{\text{par,2}} \tag{6.22}$$

Assuming that the largest fraction of particles corresponds to the representative ones, it is now possible to calculate the shape of the particle size distribution curves, which is in perfect agreement with literature data [48].

6.5.3.1 Real Systems

In real conditions, the size and shape of powder particles depend on a regime of electrolysis, a composition of solution, deposition time, cathodic material, temperature, hydrodynamic regime, *etc.*, as illustrated by Fig. 6.10.

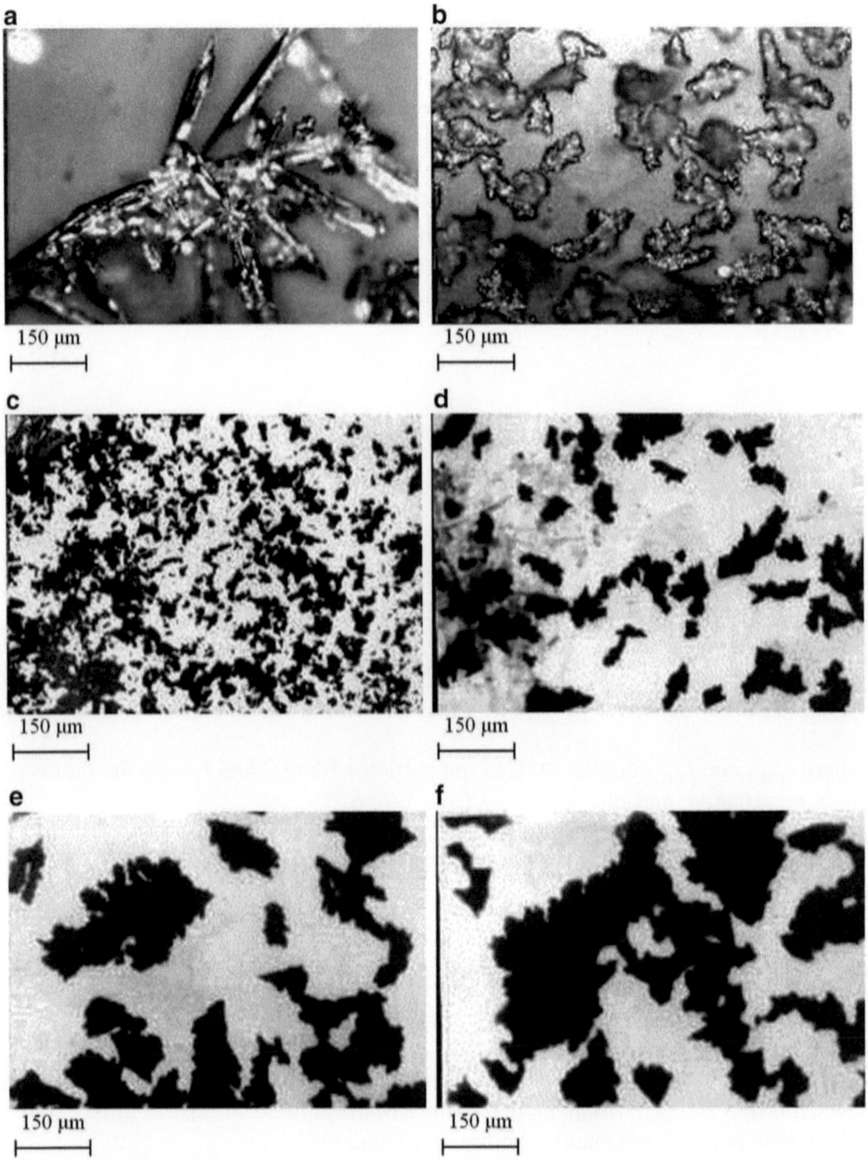

Fig. 6.10 (**a**) Silver powder particles obtained at a constant overpotential of 150 mV from 0.10 M $AgNO_3$ + 0.50 M HNO_3. Deposition time: 15 min, (**b**) silver powder particles obtained at a constant overpotential of 600 mV from 0.10 M $AgNO_3$ + 0.50 M $(NH_4)_2SO_4$ + NH_4OH to dissolve the silver sulfate precipitate. Deposition time: 15 min. The depositions were carried out on platinum electrodes at the room temperature and copper powder particles obtained at a constant overpotential of 700 mV on Cu electrodes. Deposition time: 15 min: (**c**) 0.10 M $CuSO_4$ + 0.50 M H_2SO_4, the room temperature, (**d**) 0.10 M $CuSO_4$ + 0.10 M H_2SO_4, the room temperature, (**e**) 0.50 M $CuSO_4$ + 0.50 M H_2SO_4, the room temperature, and (**f**) 0.10 M $CuSO_4$ + 0.50 M H_2SO_4, temperature: 50 ° C (Reprinted from Ref. [5] with kind permission from Springer)

During electrodeposition of Cu at overpotentials belonging to the plateau of the limiting diffusion current density, the size and hence distribution of powder particles strongly depend on the applied overpotential of electrodeposition. The reason for it is a relatively wide range of overpotentials belonging to the plateau of the limiting diffusion current density. As already mentioned, there is no a vigorous hydrogen evolution inside this plateau. The typical particle size distribution curves for Cu powders obtained by electrodeposition at overpotentials inside (600 and 700 mV) and outside (800 mV) the plateau of the limiting diffusion current density from 0.10 M $CuSO_4$ in 0.50 M H_2SO_4 on Pt electrodes are shown in Fig. 6.11a [62]. From Fig. 6.11a, it can be seen that the increase in overpotential leads to the formation of smaller particles and to narrower particle size distribution curves.

The type of used cathodic materials had the strong effect on the shape of particle size distribution curves [6, 7, 62]. For example, at overpotentials of 600 and 700 mV (*i.e.*, at overpotentials belonging to the plateaus of the limiting diffusion current density), smaller particles and narrower distribution curves were obtained for the electrodeposition on Pt electrodes than on Al ones. This is due to fact that Al is covered with a relatively thick oxide film which causes an enlarged ohmic resistance of the electrode–solution interface of Al in relation to the one for Pt [62]. At an overpotential of 800 mV (this overpotential is outside the plateau of the limiting diffusion current density), there was no any difference in particle size distribution curves obtained on these electrodes. The reason for it is the fact that an overpotential of 800 mV is situated in the hydrogen evolution range where the process is dominantly controlled by hydrogen evolution reaction.

Anyway, increasing overpotential leads to the formation of a more disperse deposit characterized by the decreased particle size. This can be explained by the fact that increasing overpotential leads to the decrease of the height of protrusion at which dendrites start to grow instantaneously. Hence, increasing overpotential means a larger number of growth sites suitable for growth of dendrites. On the other hand [63], the velocity of dendrite growth has maximum for some optimal value of the dendrite tip radius. The optimal tip radius decreases with the increasing overpotential. With the dendrite tip radii larger than the optimal value, the difference between maximal and actual velocities of dendrite growth increases with the increasing overpotential. Hence, smaller particles and narrower particle size distribution curves are expected with the increasing overpotential of powder formation [5, 62].

The particle size distribution curves for Cu powders obtained on Pt electrodes in a galvanostatic regime at currents of 28.6, 52.0, and 133 mA are shown in Fig. 6.11b. The selected currents corresponded to average currents recorded in the potentiostatic electrodepositions at overpotentials of 600, 700, and 800 mV, respectively [62]. Formation of larger particles and less narrow distribution curves in the galvanostatic regime than those formed in the potentiostatic regime (Fig. 6.11a) can be considered as follows: during electrodeposition in the galvanostatic regime in the hydrogen evolution range, overpotential is determined by hydrogen reduction, and for the difference of overpotentials of the order 100 mV, ten times larger current of electrodeposition is required. Hence, in real

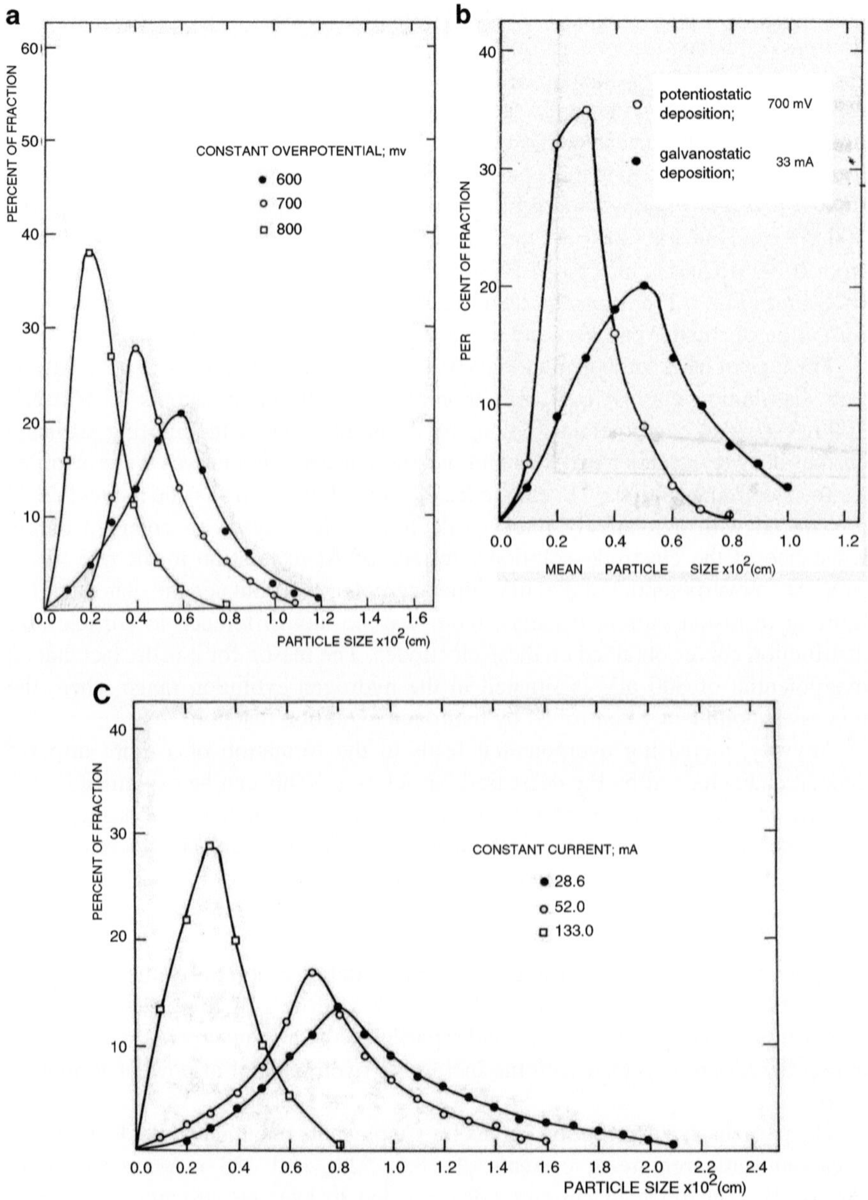

Fig. 6.11 (**a**) Particle size distribution curves for Cu powders obtained by potentiostatic electrodeposition on Pt electrodes, (**b**) particle size distribution curves for Cu powders obtained by galvanostatic electrodeposition on Pt electrodes. The surface area of the electrode: 0.63 cm^2 and (**c**) particle size distribution curves for Cu powders obtained by the potentiostatic and galvanostatic (the average current in the potentiostatic regime) electrodepositions on Cu electrodes. The surface area of the electrode: 0.63 cm^2 (Reprinted from Refs. [6, 7, 62] with kind permission from Springer)

conditions, the smaller differences in a size of the particles are expected during electrodeposition by different current densities, as well as less narrow particle size distribution curves relative to ones obtained in potentiostatic electrodeposition. The similar situation was observed when Cu was used as cathode material (Fig. 6.11c).

Anyway, the effect of the increasing current density in the galvanostatic electrodeposition is qualitative same as the increase of overpotential in potentiostatic electrodeposition, and the essence of the particle size distribution curve formation is the same in both cases.

6.5.4 The Effect of the Particle Shape and Structure on the Flowability of Electrolytic Copper Powder

The flowability of a copper powder depends on the interparticle friction, which is dominated by the surface area and surface roughness of the particles. As the surface area and surface roughness increase, the amount of friction in the powder mass increases and the powder exhibits less efficient flow. The same appears with the shape of particle.

The more irregular the particle shape is, the less efficient is the powder flow. Resistance to flow is the main feature of friction, which decreases as the particles approach a smooth spherical shape. The effect of particle size distribution on the powder flowability is also important. If the powder consists of monosized particles, which are more or less in point contact with one another, making the contact surface as low as possible, powdered deposits can flow. If the powder consists of different particles, the interstitial voids of the larger particles can be filled by the smaller ones, the contact surface area increases, and the flow of the powder is less efficient [1]. Hence, the best conditions for the free flow of the powder are fulfilled if the powder consists of monosized particles of spherical shape with a surface structure approaching to the structure of a smooth metal surface. This happens when the surface parts of the particles corresponding to the metal segments are larger than, or equal to, the pores between them [45], as illustrated by Fig. 6.12, and can be discussed quantitatively as follows.

It is well known that flowability of copper powders [3, 64] mainly depends on the apparent density of a copper powder. It was shown by Peissker [64] that free flow of a copper powder can be only expected if the apparent density of the powder is larger than 2.2–2.3 g cm^{-3}, while poor flow is possible at lower densities also. This can be explained in the following way. It was shown [44] that copper powder can be treated as a continuous medium, the density of which is equal to the apparent density of the powder ρ_{ad}. The density of compact metal is ρ_{m}. As shown in Fig. 6.13, the continuous medium can be divided into equal cubes with edge height a_{c}. It is obvious that spheres in the radius of which is $a_{\mathrm{c}}/2$ occupy effectively the same volume and that a powder consisting of spherical particles equal to each other will be characterized by free flow. Hence, the representative particle of a copper

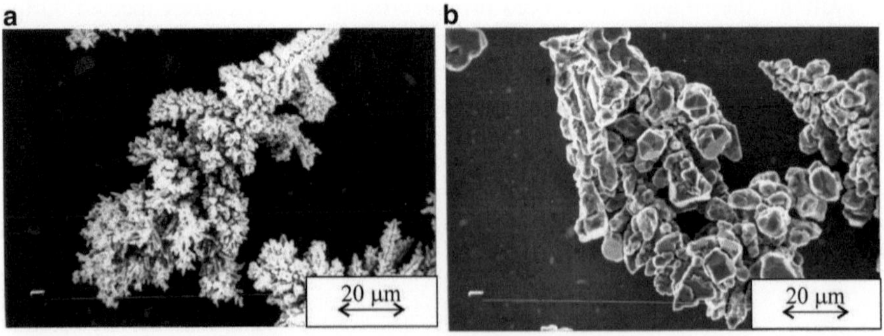

Fig. 6.12 SEM photomicrographs of Cu powder particles obtained (**a**) the constant galvanostatic regime: $i = 3600$ A m^{-2}; fraction: (149–177) μm; apparent density: 0.524 g cm^{-3}, and (**b**) the regime of reversing current: Amplitude current density: 3600 A m^{-2}; cathodic to anodic time ratio: 2.5; cathodic pulse duration: 1 s; apparent density: 1.624 g cm^{-3}. The conditions of electrolysis: $c(Cu^{2+}) = 15$ g dm^{-3}, $c(H_2SO_4) = 140$ g dm^{-3}, electrolyte circulation rate: 0.11 dm^3 min^{-1}, temperature: (50 ± 2) °C, time of powder removal by brush: 15 min (Reprinted from Ref. [7] with kind permission from Springer and Ref. [43] with permission from the Serbian Chemical Society)

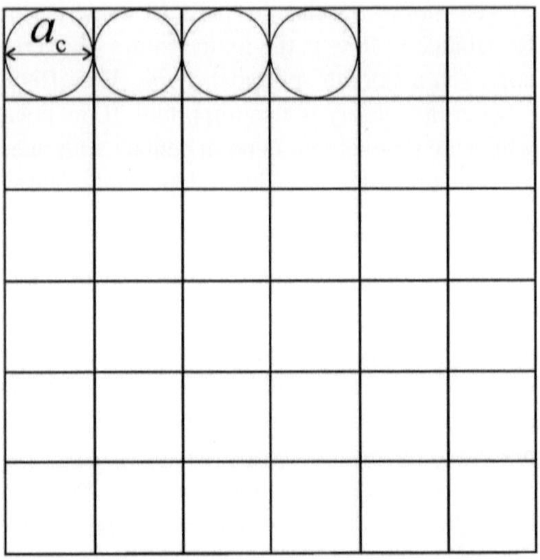

Fig. 6.13 The cross section of the part of the continuous medium divided in equal cubes and of the same volume as cubes (Reprinted from Ref. [45] with permission from the Serbian Chemical Society)

powder in respect to free flow must be spherical. The relationship between the density of the spherical representative powder particle and the apparent density of the powder can be established in the following way. It can be seen from Fig. 6.13 that instead of cubes with edge height a_c and a density ρ_{ad}, spherical particles with a radius $a_c/2$ and a density $\rho_{ad,c}$ can be considered.

Hence,

$$\frac{4}{3}\left(\frac{a_c}{2}\right)^3 \pi \rho_{ad,c} = {a_c}^3 \rho_{ad} \tag{6.23}$$

and

$$\rho_{ad,c} = \frac{6}{\pi} \rho_{ad} \tag{6.24}$$

This means that the density of the spherical particles of which a powder is composed is about twice as big as the apparent density of the powder. On the other hand, it is necessary to bear in mind that copper powder particles are dendritic by nature [42], hence they are porous. It is obvious that the free flow of a powder consisting of spherical particles can be expected only if the surface parts of the particles corresponding to the metal segments are larger than, or equal to, the pores between them.

Hence, the critical density of a particle itself $\rho_{ad,c}$ is then given by:

$$\rho_{ad,c} = \frac{1}{2} \rho_m \tag{6.25}$$

or taking into account Eq. (6.24) after further rearranging one obtains:

$$\rho_{ad} = \frac{\pi}{12} \rho_m \tag{6.26}$$

Using $\rho_m = 8.9$ g cm^{-3} for Cu, it follows from Eq. (6.26) that the critical value of the apparent density resulting in free flow of copper powder is 2.32 g cm^{-3}. This is in good agreement with the findings of Peisseker [64] for non-sieved copper powders.

References

1. German RM (1994) Powder metallurgy science, 2nd edn. Metal Powder Industries Federation, Princeton
2. Pavlović MG, Popov KI (2005) Electrochem encyclopedia. http://electrochem.cwru.edu/ed/encycl/
3. Calusaru A (1979) Electrodeposition of metal powders, Materials science monographs. Elsevier, New York
4. Orhan G, Hapci G (2010) Effect of electrolysis parameters on the morphologies of copper powder obtained in a rotating cylinder electrode cell. Powder Technol 201:57–63
5. Popov KI, Djokić SS, Grgur BN (2002) Fundamental aspects of electrometallurgy. Kluwer Academic/Plenum Publishers, New York
6. Popov KI, Pavlović MG (1993) Electrodeposition of metal powders with controlled grain size and morphology. In: White RE, Bockris JO'M, Conway BE (eds) Modern aspects of electrochemistry, vol 24. Plenum Press, New York, pp 299–391

7. Nikolić ND, Popov KI (2012) Electrodeposition of copper powders and their properties. In: Djokić SS (ed) Electrochemical production of metal powders, vol 54, Series: modern aspects of electrochemistry. Springer, New York, pp 125–185
8. Ibl N (1962) Advances in electrochemistry and electrochemical engineering, vol 2. Interscience, New York
9. Winand R (1994) Electrodeposition of metals and alloys – new results and perspectives. Electrochim Acta 39:1091–1105
10. Kozlov VM, Peraldo Bicelli L (1999) Influence of the nature of metals on the formation of the deposit's polycrystalline structure during electrocrystallization. J Cryst Growth 203:255–260
11. Nikolić ND, Popov KI (2014) A new approach to the understanding of the mechanism of lead electrodeposition. In: Djokić SS (ed) Electrodeposition and surface finishing, vol 57, Series: modern aspects of electrochemistry. Springer, New York, pp 85–132
12. Jović VD, Nikolić ND, Lačnjevac UČ, Jović BM, Popov KI (2012) Morphology of different electrodeposited pure metal powders. In: Djokić SS (ed) Electrochemical production of metal powders, vol 54, Series: modern aspects of electrochemistry. Springer, New York, pp 63–123
13. Popov KI, Krstajić NV, Čekerevac MI (1996) The mechanism of formation of coarse and disperse electrodeposits. In: White RE, Conway BE, Bockris JO'M (eds) Modern aspects of electrochemistry, vol 30. Plenum Press, New York, pp 261–312
14. Nikolić ND, Vaštag DjDj, Živković PM, Jokić B, Branković G (2013) Influence of the complex formation on the morphology of lead powder particles produced by the electrodeposition processes. Adv Powder Technol 24:674–682
15. Djokić SS, Nikolić ND, Živković PM, Popov KI, Djokić NS (2011) Electroless deposition and electrodeposition of metallic powders: a comparison. ECS Trans 33:7–31
16. Wranglen G (1960) Dendrites and growth layers in the electrocrystallization of metals. Electrochim Acta 2:130–146
17. Liaw HM, Faust JW Jr (1973) Effect of growth parameters on habit and morphology of electrodeposited lead dendrites. J Cryst Growth 18:250–256
18. Nikolić ND, Maksimović VM, Branković G, Živković PM, Pavlović MG (2013) Influence of the type of electrolyte on morphological and crystallographic characteristics of lead powder particles. J Serb Chem Soc 78:1387–1395
19. Popov KI, Čekerevac MI (1989) Dendritic electrocrystallization of cadmium from acid sulphate solution. II. The effect of the geometry of dendrite precursors on the shape of dendrites. Surf Coat Technol 37:435–440
20. Nikolić ND, Živković PM, Jokić B, Pavlović MG, Stevanović JS (2014) Comparative analysis of the polarisation and morphological characteristics of electrochemically produced powder forms of the intermediate metals. Maced J Chem Chem Eng 33:169–180
21. Lv Z-Y, Li A-Q, Fei Y, Li Z, Chen J-R, Wang A-J, Feng J-J (2013) Facile and controlled electrochemical route to three-dimensional hierarchical dendritic gold nanostructures. Electrochim Acta 109:136–144
22. Han J, Liu J (2012) Electrodeposition of crystalline dendritic silver in 12-tungstosilicate acid system. J Nanoeng Nanomanuf 2:171–174
23. Mandke MV, Han S-H, Pathan HM (2012) Growth of silver dendritic nanostructures via electrochemical route. CrystEngComm 14:86–89
24. Nikolić ND, Pavlović LjJ, Pavlović MG, Popov KI (2008) Morphologies of electrochemically formed copper powder particles and their dependence on the quantity of evolved hydrogen. Powder Technol 185:195–201
25. Nikolić ND, Popov KI (2010) Hydrogen co-deposition effects on the structure of electrodeposited copper. In: Djokić SS (ed) Electrodeposition: theory and practice, vol 48, Series: modern aspects of electrochemistry. Springer, New York, pp 1–70
26. Jović VD, Jović BM, Maksimović VM, Pavlović MG (2007) Electrodeposition and morphology of Ni, Co and Ni–Co alloy powders: part II. Ammonium chloride supporting electrolyte. Electrochim Acta 52:4254–4263

27. Lačnjevac UČ, Jović BM, Jović VD (2009) Morphology and composition of the Fe–Ni powders electrodeposited from citrate containing electrolytes. Electrochim Acta 55:535–543
28. Maksimović VM, Nikolić ND, Kusigerski VB, Blanuša JL (2015) Correlation between morphology and magnetic properties of electrochemically produced cobalt powder particles. J Serb Chem Soc 80:197–207
29. Jović VD, Maksimović VM, Pavlović MG, Popov KI (2006) Morphology, internal structure and growth mechanism of electrodeposited Ni and Co powders. J Solid State Electrochem 10:373–379
30. Jović VD, Jović BM, Pavlović MG (2006) Electrodeposition of Ni, Co and Ni–Co alloy powders. Electrochim Acta 51:5468–5477
31. Nikolić ND, Branković G, Lačnjevac U (2012) Formation of two-dimensional (2D) lead dendrites by application of different regimes of electrolysis. J Solid State Electrochem 16:2121–2126
32. Lowenheim FA (1978) Electroplating. McGraw-Hill Book Company, New York/St. Louis
33. Nikolić ND, Pavlović LjJ, Krstić SB, Pavlović MG, Popov KI (2008) Influence of ionic equilibrium in the $CuSO_4$-H_2SO_4-H_2O system on the formation of irregular electrodeposits of copper. Chem Eng Sci 63:2824–2828
34. Nikolić ND, Pavlović LjJ, Branković G, Pavlović MG, Popov KI (2008) The ionic equilibrium in the $CuSO_4$-H_2SO_4-H_2O system and the formation of the honeycomb-like structure during copper electrodeposition. J Serb Chem Soc 73:753–760
35. Nikolić ND, Popov KI, Pavlović LjJ, Pavlović MG (2006) The effect of hydrogen codeposition on the morphology of copper electrodeposits. I. The concept of effective overpotential. J Electroanal Chem 588:88–98
36. Nikolić ND, Branković G, Pavlović MG (2012) Correlate between morphology of powder particles obtained by the different regimes of electrolysis and the quantity of evolved hydrogen. Powder Technol 221:271–277
37. Nikolić ND, Branković G (2010) Effect of parameters of square-wave pulsating current on copper electrodeposition in the hydrogen co-deposition range. Electrochem Commun 12:740–744
38. Popov KI, Stojilković ER, Radmilović V, Pavlović MG (1997) Morphology of lead dendrites electrodeposited by square-wave pulsating overpotential. Powder Technol 93:55–61
39. Popov KI, Nikolić ND, Rakočević Z (2002) An estimation of the interfacial energy of the copper-copper sulphate solution interface and of the specific surface of copper powder. J Serb Chem Soc 67:635–638
40. Popov KI, Nikolić ND, Rakočević Z (2002) The estimation of solid copper surface tension in copper sulfate solutions. J Serb Chem Soc 67:769–775
41. Nikolić ND, Krstić SB, Pavlović LjJ, Pavlović MG, Popov KI (2008) The mutual relation of decisive characteristics of electrolytic copper powder and effect of deposition conditions on them. In: Hayashi K (ed) Electroanalytical chemistry research trends. NOVA Publishers, New York, pp 185–209
42. Pavlović MG, Pavlović LjJ, Ivanović ER, Radmilović V, Popov KI (2001) The effect of particle structure on apparent density of electrolytic copper powder. J Serb Chem Soc 66:923–933
43. Popov KI, Pavlović LjJ, Ivanović ER, Radmilović V, Pavlović MG (2002) The effect of reversing current deposition on the apparent density of electrolytic copper powder. J Serb Chem Soc 67:61–67
44. Popov KI, Nikolić ND, Rakočević Z (2002) A representative particle of copper powder and branching of copper dendrites. J Serb Chem Soc 67:861–866
45. Popov KI, Krstić SB, Pavlović MG (2003) The critical apparent density for the free flow of copper powder. J Serb Chem Soc 68:511–513
46. Popov KI, Krstić SB, Obradović MČ, Pavlović MG, Pavlović LjJ, Ivanović ER (2003) The effect of the particle shape and structure on the flamability of copper powder I. Modelling of a representative powder particle. J Serb Chem Soc 68:771–777

47. Popov KI, Pavlović MG, Pavlović LjJ, Ivanović ER, Krstić SB, Obradović MČ (2003) The effect of the particle shape and structure on the flowability of copper powder II. The experimental verification of the model of the representative powder particle. J Serb Chem Soc 68:779–783
48. Popov KI, Živković PM, Krstić SB (2003) The apparent current density as a function of the specific surface of copper powder and the shape of the particle size distribution curve. J Serb Chem Soc 68:903–907
49. Popov KI, Krstić SB, Obradović MČ, Pavlović MG, Pavlović LjJ, Ivanović ER (2004) The effect of the particle shape and structure on the flamability of electrolytic copper powder III. A model of the surface of a representative particle of flowing copper powder electrodeposited by reversing current. J Serb Chem Soc 69:43–51
50. Popov KI, Krstić SB, Pavlović MG, Pavlović LjJ, Maksimović VM (2004) The effect of the particle shape and structure on the flamability of electrolytic copper powder. IV. The internal structure of the powder particles. J Serb Chem Soc 69:817–825
51. Popov KI, Nikolić ND, Krstić SB, Pavlović MG (2006) Physical modelling of representative copper powder particles. J Serb Chem Soc 71:397–400
52. Nikolić ND, Popov KI, Pavlović LjJ, Pavlović MG (2007) New method of the determination of specific surface of copper obtained by the electrodeposition. Mater Prot 48:3–8
53. Pavlović MG, Nikolić ND, Popov KI (2003) The current efficiency during the cathodic period of reversing current in copper powder deposition and the overall current efficiency. J Serb Chem Soc 68:649–656
54. Pavlović MG, Pavlović LjJ, Doroslovački ID, Nikolić ND (2004) The effect of benzoic acid on the corrosion and stabilisation of electrodeposited copper powder. Hydrometallurgy 73:155–162
55. Popov KI, Pavlović LjJ, Pavlović MG, Čekerevac MI (1988) Electrode surface coarsening in potentiosttic copper electrodeposition. Surf Coat Technol 35:39–45
56. Popov KI, Pavlović MG, Pavlović LjJ, Čekerevac MI, Remović GŽ (1988) Electrode surface coarsening in pulsating overpotential copper electrodeposition. Surf Coat Technol 34:355–363
57. Popov KI, Maksimović MD, Trnjančev JD, Pavlović MG (1981) Dendritic electrocrystallization and the mechanism of powder formation in the potentiostatic electrodeposition of metals. J Appl Electrochem 11:239–246
58. Popov KI, Maksimović MD, Pavlović MG, Lukić DT (1980) The mechanism of copper powder formation in potentiostatic deposition. J Appl Electrochem 10:299–308
59. Despić AR, Popov KI (1972) Transport controlled deposition and dissolution of metals. In: Conway BE, Bockris JO'M (eds) Modern aspects of electrochemistry, vol 7. Plenum Press, New York, pp 199–313
60. Nikolić ND, Rakočević Z, Popov KI (2005) Nanostructural analysis of bright metal surfaces in relation to their reflectivities. In: Conway BE, Vayenas CG, White RE, Gamboa-Adelco ME (eds) Modern aspects of electrochemistry, vol 38. Kluwer Academic/Plenum Publishers, New York, pp 425–474
61. Schatt W, Wierters KP (1997) Powder metallurgy – processing and materials. European Powder Metallurgy Association, Technical University Dresden, Shrewsbury
62. Popov KI, Pavlović MG, Maksimović MD, Krstajić SS (1978) The comparison of galvanostatic and potentiostatic copper powder deposition on platinum and aluminium electrodes. J Appl Electrochem 8:503–514
63. Barton JL, Bockris JO'M (1962) The electrolytic growth of dendrites from ionic solutions. Proc Roy Soc A268:485–505
64. Peissker E (1984) Production and properties of electrolytic copper powder. Int J Powder Metallurgy Powder Technol 20:87–102

Chapter 7
Electrodeposited Alloys and Multilayered Structures

7.1 Introduction

It is general experience in materials science that alloy can exhibit qualities that are unobtainable with parent metals. This is particularly true for electrodeposited alloys, mainly due to formation of metastable phases and intermediate layers. Some important properties of materials, such as hardness, ductility, tensile strength, Young's modulus, corrosion resistance, solderability, wear resistance, antifriction service, etc., may be enhanced. At the same time, some properties that are not characteristic for parent metals, such as high magnetic permeability, other magnetic and electrical properties, amorphous structure, etc., can also be obtained. In some cases, alloy coatings may be more suitable for subsequent electroplate overlayers and conversion chemical treatments [1].

Some alloys may be more easily obtained by electrodeposition than by metallurgical processes. This is particularly true for alloys composed of metals having large differences in melting temperatures or metals that cannot be mixed in a liquid state. Such metals can very often be codeposited from the solutions (e.g., alloys Ag–Ni, Ag–Co, and Cd–Co). Taking into account that some metals cannot be electrodeposited from the aqueous solutions (Ti, V, W, Nb, Zr, etc.), they could be electrodeposited from the melts of their salts. In recent times, the processes of metal and alloy electrodeposition from the room-temperature molten salts were also investigated and developed (cf. electrodeposition of Al–Cu, Al–Co, Al–Ni alloys from $AlCl_3$–MeEtImCl melt).

The fast-growing requirements of modern industry for materials with special qualities in the last century have given rise to increasing interest in electrodeposition of alloys, particularly in corrosion protection and in the modern electronic industry [1].

From 1842 until the end of the nineteenth century, over 180 alloys involving 40 elements have been electrodeposited [2]. An excellent review of the achievements up to 1962 is given in the book by Brenner [3], while from practical point of

K.I. Popov et al., *Morphology of Electrochemically and Chemically Deposited Metals*, DOI 10.1007/978-3-319-26073-0_7

view, it is recommended to consider the book of Bondar, Grimina, and Pavlov [4], which contains recipes and references for more than 1100 baths for alloy electrodeposition.

Concerning scientific approach of the electrodeposition of alloys, all the results obtained until 1995, mainly connected with the thermodynamics and kinetics of alloy electrodeposition, are summarized in the chapter by Despić and Jović [1] and Jović et al. [5]. In the present chapter, the morphology of electrodeposited alloys will be the main subject.

7.2 Electrodeposition of Alloys from Aqueous Solutions

7.2.1 Conditions for Electrodeposition of Alloys

The metals immersed in the solution of their simple salts establish the reversible potential. The values of the reversible potentials for different metals could differ for about 3 V. Electrodeposition of metals could take place only at potentials more negative than the reversible ones. Accordingly, in the solution of ions of two metals (cf. Cu^{2+} and Zn^{2+}) with one being on the positive side of the potential scale (vs. SHE) (Cu) and another one being on the negative side of the potential scale (Zn), intensive electrodeposition of Cu could take place at potentials at which Zn would not electrodeposit at all. Taking into account that the reversible potentials of metals could change with the presence of different anions in the solution (complexation of metal ions) and that the rates of electrodeposition of different metals are usually different, it is possible to achieve conditions for simultaneous electrodeposition of these two metals [1, 5].

For simultaneous electrodeposition of two metals, A and B, their electrodeposition potentials (E) must be identical, $E(\mathrm{A}) = E(\mathrm{B})$, i.e.,

$$E_{\mathrm{r}}(\mathrm{A}) + \eta(\mathrm{A}) = E_{\mathrm{r}}(\mathrm{B}) + \eta(\mathrm{B}) \tag{7.1}$$

where $E_{\mathrm{r}}(\mathrm{A})$ and $E_{\mathrm{r}}(\mathrm{B})$ are reversible potentials of metals A and B, while $\eta(\mathrm{A})$ and $\eta(\mathrm{B})$ correspond to the overpotentials needed for the electrodeposition of these two metals. The reversible potential could be changed by the change of metal ion concentration in the solution and by the temperature of the solution and is defined by the Nernst equation:

$$E_{\mathrm{r}}(\mathrm{A}) = E^{\Theta}(\mathrm{A}) + \frac{RT}{pF}\ln a(\mathrm{A}^{\mathrm{p}+}) \tag{7.2}$$

$$E_{\mathrm{r}}(\mathrm{B}) = E^{\Theta}(\mathrm{B}) + \frac{RT}{qF}\ln a(\mathrm{A}^{\mathrm{q}+}) \tag{7.3}$$

where $E^{\ominus}(\mathrm{A})$ and $E^{\ominus}(\mathrm{B})$ are standard potentials of metals A and B, a activities of corresponding metal ions in the solution, and p and q numbers of electrons to be exchanged during the process of metal electrodeposition.

The condition defined by Eq. (7.1) could be accepted only as a first approximation, since the potential of the metal electrodeposition is undefined quantity if the value of corresponding current density is not known. It appears that a better definition of the conditions for simultaneous electrodeposition of two metals would be current density (i) at which both metals electrodeposit with approximately the same current density. More precisely, for two-component alloy to be electrodeposited with the molar ratio of the more noble metal x and the less noble metal $(1-x)$, assuming that Faraday's law is obeyed, the following relations should be fulfilled:

$$x = \frac{n_{\mathrm{A}}}{n_{\mathrm{A}} + n_{\mathrm{B}}} = \frac{\frac{i_{\mathrm{A}}}{p}}{\frac{i_{\mathrm{A}}}{p} + \frac{i_{\mathrm{B}}}{q}} \tag{7.4}$$

and

$$(1-x) = \frac{n_{\mathrm{B}}}{n_{\mathrm{A}} + n_{\mathrm{B}}} = \frac{\frac{i_{\mathrm{B}}}{q}}{\frac{i_{\mathrm{A}}}{p} + \frac{i_{\mathrm{B}}}{q}} \tag{7.5}$$

where n_{A} and n_{B} are numbers of moles of components A and B. Hence, the current density ratio for the electrodeposition of these two metals should be defined as

$$\frac{i_{\mathrm{A}}}{i_{\mathrm{B}}} = \frac{p}{q} \frac{x}{(1-x)} \tag{7.6}$$

The condition defined by Eq. (7.6) could be achieved by proper adjustment of three essential variables: the concentration of the electrodepositing ions at the electrode/solution interface (where the discharge occurs), the electrode potential, and the temperature [1, 5].

For better understanding of the conditions defined by Eqs. (7.1, 7.2, 7.3, 7.4, 7.5, and 7.6), it is important to present polarization curves (current density vs. potential relationships) for electrodeposition of each metal. The characteristic cases are presented in Figs. 7.1 and 7.2. The first case is presented in Fig. 7.1a: the overpotential for electrodeposition of the more noble metal A is higher than that for the less noble metal B. From the potentials $E_{\mathrm{r}}(\mathrm{A})$ to $E_{\mathrm{r}}(\mathrm{B})$, only more noble metal electrodeposition occurs, while the electrodeposition of alloy commences at the potential E_1. In the potential range from $E_{\mathrm{r}}(\mathrm{B})$ to E_1, metal A electrodeposits with higher current density than metal B (the alloy contains more metal A than B). At the potential E_1, both metals electrodeposit with the same current density, and the alloy contains the same amount of both metals. At the potentials more negative than E_1, the metal B electrodeposits with higher current density and, accordingly, the alloy contains more metal B than metal A. The second case is presented in

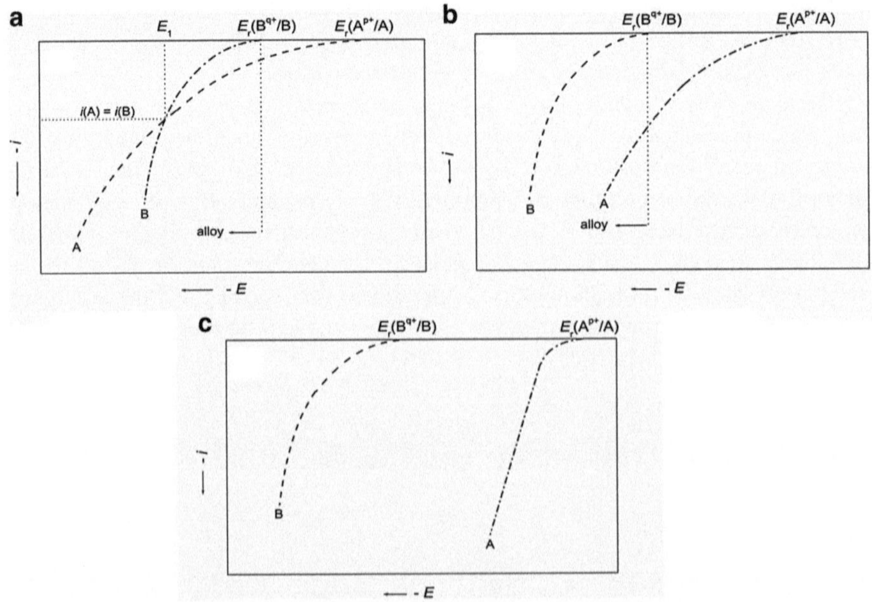

Fig. 7.1 Schematic presentation of the characteristic cases for alloy electrodeposition. (**a**) The overpotential for electrodeposition of the more noble metal A is higher than that for the less noble metal B; (**b**) The overpotential for electrodeposition of metal A is slightly lower than that for metal B; (**c**) Alloy electrodeposition is impossible (Reprinted from Ref. [5] with kind permission from Springer)

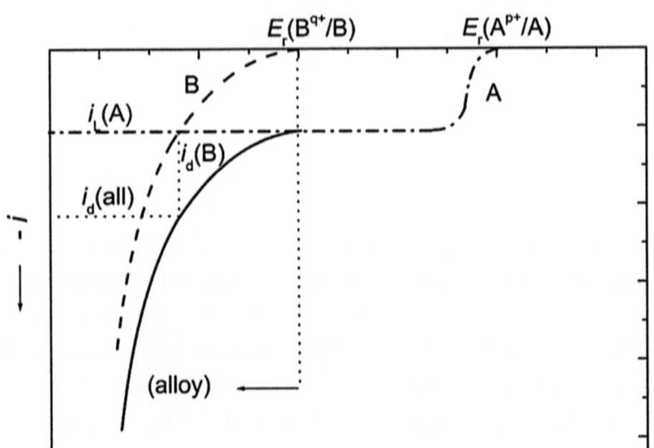

Fig. 7.2 Polarization curves for the electrodeposition of more noble metal (A) and less noble metal (B): $i_L(A)$ diffusion limiting current density for the electrodeposition of metal (A), $i_d(B)$ current density for the electrodeposition of metal (B), $i_d(all)$ current density for the electrodeposition of alloy (Reprinted from Ref. [5] with kind permission from Springer)

Fig. 7.1b: the overpotential for electrodeposition of metal A is slightly lower than that for metal B, i.e., the polarization curves are almost parallel. Hence, the electrodeposition of alloy commences at the potential $E_r(B)$, while the alloy contains more metal A than B. If the difference between $E_r(A)$ and $E_r(B)$ is high and the overpotential for electrodeposition of the more noble metal A is lower than that for the less noble metal B, the third case presented in Fig. 7.1c applies: in such a case, alloy electrodeposition is impossible. The difference between the reversible potentials of two metals could be changed (lowered) by the change of metal ion concentration (activity), and in most cases, this is achieved by the complexation.

Simultaneous electrodeposition of two metals is possible even if the difference in their reversible potentials is high, if the applied current density for alloy electrodeposition is higher than the diffusion limiting current density for the electrodeposition of the more noble metal. Such a case is schematically presented in Fig. 7.2.

If $p = q = 2$, the molar ratios of metals (A) and (B) in the alloy are defined by the following relation:

$$\begin{aligned} x(\mathrm{A}) &= \frac{i_{\mathrm{L}}(\mathrm{A})}{i_{\mathrm{L}}(\mathrm{A}) + i_{\mathrm{d}}(\mathrm{B})} = \frac{i_{\mathrm{L}}(\mathrm{A})}{i_{\mathrm{d}}(\mathrm{all})} \\ x(\mathrm{B}) &= \frac{i_{\mathrm{d}}(\mathrm{B})}{i_{\mathrm{L}}(\mathrm{A}) + i_{\mathrm{d}}(\mathrm{B})} = \frac{i_{\mathrm{d}}(\mathrm{B})}{i_{\mathrm{d}}(\mathrm{all})} \end{aligned} \tag{7.7}$$

7.2.2 Types of Electrodeposition of Alloys

Classification of different types of alloy electrodeposition was made by Brenner [3] in 1962, by defining five groups: equilibrium, irregular, regular, anomalous, and induced codeposition. More detailed explanations including samples for each type were given in Ref. [5].

7.2.2.1 Equilibrium Codeposition

Equilibrium codeposition implies a common reversible potential for both metal constituents so that the reduction of both metal ions would take place at potentials more negative than the reversible ones. To close the gap between the reversible potentials of electrodepositing metals, it is necessary to make the concentration of simple salts (undergoing complete dissociation) of the more noble metal impractically low and of the less noble metal impractically high. The best way to overcome this problem could result from complexation of metal ions with different ligands. Complexation usually changes the activity of the resulting species in solution by many orders of magnitude, while keeping the total amount of one or other metal in solution sufficiently high for a good supply of plating material to the cathode. It is very often case that the ions of both metals form complexes with one and the same ligand with similar values of the stability (formation) constants, so that the change of the potential of the electrodeposition of each metal is the same (or similar) value. Usually, in such a case, the complexation with two different ligands could result in a more pronounced change of the electrodeposition potentials of two metals.

Electrodeposition of the Ni–Sn Alloy

In a further text, an example for the electrodeposition of Ni–Sn alloy from the solution containing pyrophosphate and glycine ligands is presented [6], where Sn belongs to the normal metals while Ni belongs to the inert metals (see Sect. 6.2). According to the literature [7], standard potential of the Ni electrodeposition is −0.23 V versus SHE, while that for Sn is −0.1364 V versus SHE and accordingly Ni is less noble metal. Taking into account that the overvoltage for Ni electrodeposition [8] is much higher than that for Sn electrodeposition [6], the difference between the potentials of electrodeposition of these two metals should be larger than that of their standard potentials. In the data presented in Dean's Handbook of Chemistry [9], Ni forms two pyrophosphate complexes, $[Ni(P_2O_7)]^{2-}$ and $[Ni(P_2O_7)_2]^{6-}$ as well as three glycine complexes, $[Ni(NH_2CH_2COO)]^{+}$, $[Ni(NH_2CH_2COO)_2]$, and $[Ni(NH_2CH_2COO)_3]^{-}$. All data about different complexes of Ni and Sn in the pyrophosphate and glycine solutions and corresponding reactions for their formation and their formation (stability) constants are given in the paper of Duffield et al. [10]. All species and their stability constants used for the calculation of the distribution of different complexes in the solution containing Sn, Ni, pyrophosphate, and glycine ions are listed in Table 7.1.

The calculation of the distribution of complexes in the solution containing pyrophosphate and glycine showed that $[Sn(P_2O_7)_2]^{6-}$ is dominant complex with Sn at pH 8.0, while two complexes of Ni dominate at a given pH: $[Ni(P_2O_7)_2]^{6-}$ and $[Ni(NH_2CH_2COO)_3]^{-}$. This is shown in Fig. 7.3.

The values of the equilibrium potentials (E_{eq}) of prevailing complexes, calculated using explanations based on the Gibbs energy change for reaction of certain complex formation [11] (assuming that the ions activities are equal to their concentrations), are also presented in Table 7.1. As can be seen, the equilibrium potential for electrodeposition of Sn by the reduction of $[Sn(P_2O_7)_2]^{6-}$ complex is −0.847 V versus SCE, while the equilibrium potentials for the reduction of $[Ni(P_2O_7)_2]^{6-}$ and $[Ni(NH_2CH_2COO)_3]^{-}$ complexes are more positive, being about −0.716 V versus SCE and situation becomes opposite to that for electrodeposition from the solution of simple ions. After the complexation Ni becomes more noble metal, while Sn becomes less noble one. Hence, it could be concluded that at pH 8.0 Sn would electrodeposit from the complex $[Sn(P_2O_7)_2]^{6-}$, while Ni would electrodeposit simultaneously from two complexes, $[Ni(P_2O_7)_2]^{6-}$ and $[Ni(NH_2CH_2COO)_3]^{-}$, in the presence of both complexing anions. The equilibrium potentials for electrodeposition of Sn and Ni still differ for 0.131 V. As already stated, because of high overpotential for Ni electrodeposition [6], it could be expected that two metals possess identical, or similar, potential of electrodeposition. This is exactly the case for these two metals in the pyrophosphate–glycine solution.

Table 7.1 All complexes present in the solution containing 0.1 M $SnCl_2$ + 0.1 M $NiCl_2$ + 0.6 M $K_4P_2O_7$ + 0.3 M NH_2CH_2COOH, their concentrations, stability constants, and equilibrium potentials of prevailing complexes

Complexes	log β	Conc./M	E_{eq}/V vs. SCE
$[H(P_2O_7)]^{3-}$	8.14		
$[H_2(P_2O_7)]^{2-}$	14.01		
$[H_3(P_2O_7)]^{-}$	15.78		
$[H_4(P_2O_7)]$	16.63		
$[H(NH_2CH_2COO)]$	9.64		
$[H_2(NH_2CH_2COO)]^{+}$	12.05		
$[Sn\ (NH_2CH_2COO)H]^{2+}$	12.78		
$[Sn(NH_2CH_2COO)]^{+}$	10.02		
$[Sn(P_2O_7)]^{2-}$	13.05	0.007	−0.847
$[Sn(P_2O_7)H]^{-}$	15.92		
$[Sn(P_2O_7)H_2]$	17.47		
$[Sn(P_2O_7)_2]^{6-}$	16.27	0.093	−0.847
$[Sn(P_2O_7)_2H]^{5-}$	22.31		
$[Sn(P_2O_7)_2H_2]^{4-}$	26.79		
$[Sn(P_2O_7)_2H_3]^{3-}$	30.07		
$[Sn(P_2O_7)_2H_4]^{2-}$	31.58		
$[Sn(P_2O_7)OH]^{3-}$	5.32		
$[Sn(P_2O_7)(OH)_2]^{2-}$	−4.77		
$[Sn(P_2O_7)_2OH]^{5-}$	7.04		
$[Ni(NH_2CH_2COO)]^{+}$	5.60		
$[Ni(NH_2CH_2COO)_2]$	10.40	0.009	−0.716
$[Ni(NH_2CH_2COO)_3]^{-}$	13.80	0.057	−0.716
$[Ni(P_2O_7)]^{2-}$	5.80	0.005	−0.716
$[Ni(P_2O_7)_2]^{6-}$	7.40	0.029	−0.716

The polarization curve for Ni–Sn alloy electrodeposition onto Ni electrode [12] is shown in Fig. 7.4. The electrodeposition process commences at about −0.83 V versus SCE being activation controlled down to about −0.95 V versus SCE, while in the potential range from about −0.95 V versus SCE to about −1.20 V versus SCE, well-defined diffusion limiting current density (−10 mA cm^{-2}) is established. In the region of the activation control (squares marked in Fig. 7.4a), Ni–Sn alloy coatings were electrodeposited at the current densities of −2, −4, and −6 mA cm^{-2}. Flat and compact electrodeposits of the thickness of about 5 μm were obtained in all cases, as shown in Fig. 7.4b. The composition of the coatings changed with the increase of cathodic current density from about 37 at.% Ni (for sample obtained at −2 mA cm^{-2}) to about 45 at.% Ni (for sample obtained at −6 mA cm^{-2}) [12], but in all cases, both metals were present in the coating, indicating a good example for equilibrium codeposition.

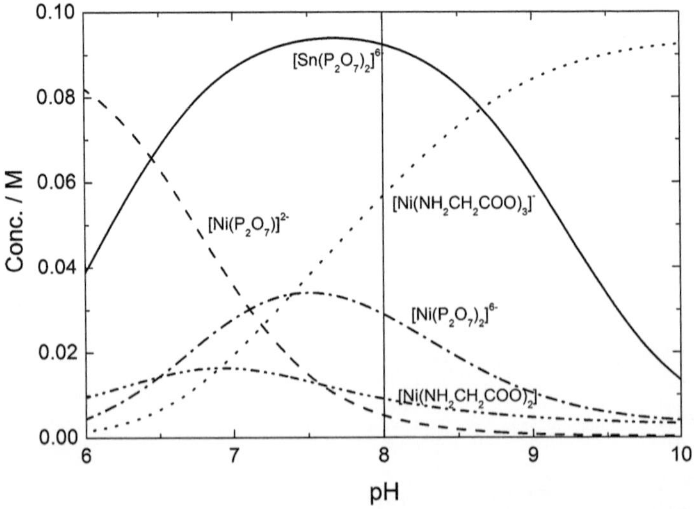

Fig. 7.3 Distribution of different complexes in the solution containing 0.1 M $SnCl_2 + 0.1$ M $NiCl_2 + 0.6$ M $K_4P_2O_7 + 0.3$ M NH_2CH_2COOH as a function of the solution pH (Reprinted from Ref. [5] with kind permission from Springer)

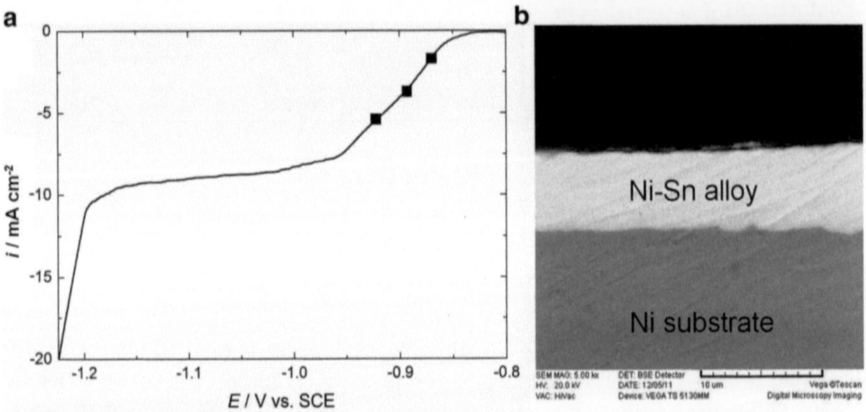

Fig. 7.4 (**a**) Polarization curve for electrodeposition of the Ni–Sn alloy onto Ni electrode. (**b**) Typical cross section of coatings obtained at different current densities marked in (**a**) with *solid squares* (■) (Reprinted from Ref. [5] with kind permission from Springer)

Morphology of the Ni–Sn Alloy

In this case, two electrodepositing metals are characterized by different mechanisms of electrodeposition: Sn nucleation is very fast with nuclei growing normal to the electrode surface, while nucleation of Ni is slow with lateral growth of the electrodeposit. In Fig. 7.5 are shown pure Sn electrodeposits obtained in two

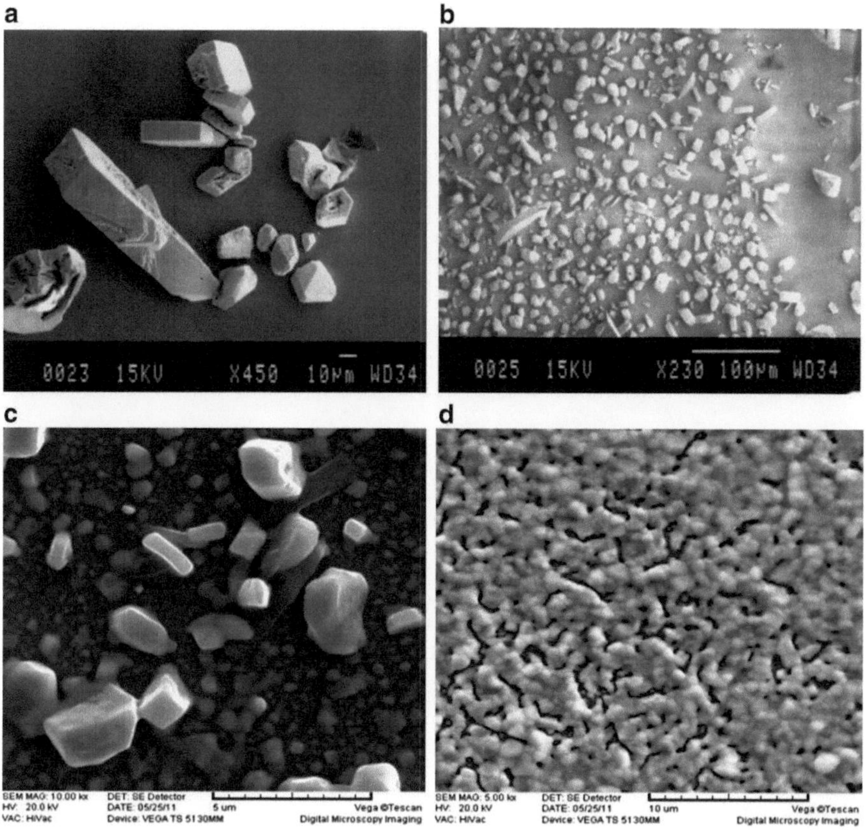

Fig. 7.5 SEM micrographs of the Sn electrodeposit obtained at the overpotentials of −20 mV (**a**) and −60 mV (**b**) on the 40 nm thick Cu layer evaporated onto Si(111) in the solution containing 0.5 M Sn–MSA + 1.0 M MSA. SEM micrographs of the Sn electrodeposit obtained onto Ni electrode during the potentiostatic pulse $E = -1.15$ V versus SCE for 1000 s in the solution 0.1 M $SnCl_2$ + 0.6 M $K_4P_2O_7$ + 0.3 M NH_2CH_2COOH of the pH 8.0, (**c**, **d**): (**c**) part of the surface with less dense electrodeposit; (**d**) part of the surface with more dense electrodeposit (Reprinted from Ref. [6] with the permission of The Electrochemical Society)

different solutions: SEM micrographs of the Sn electrodeposit obtained in the solution containing 0.5 M Sn–MSA + 1.0 M MSA (MSA – methane sulfonic acid) at the overpotentials of −20 mV (a) and −60 mV (b) on the 40 nm thick Cu layer evaporated onto Si(111); SEM micrographs of the Sn electrodeposit obtained onto Ni electrode during the potentiostatic pulse $E = -1.15$ V versus SCE for 1000 s in the solution 0.1 M $SnCl_2$ + 0.6 M $K_4P_2O_7$ + 0.3 M NH_2CH_2COOH of the pH 8.0, (c) and (d); (c) part of the surface with less dense electrodeposit; and (d) part of the surface with more dense electrodeposit [6]. Hence, under the conditions of constant potential (overpotential), it is practically not possible to obtain compact and flat Sn electrodeposit from the investigated solutions (without additives for brightening and leveling) independently of the materials and type of

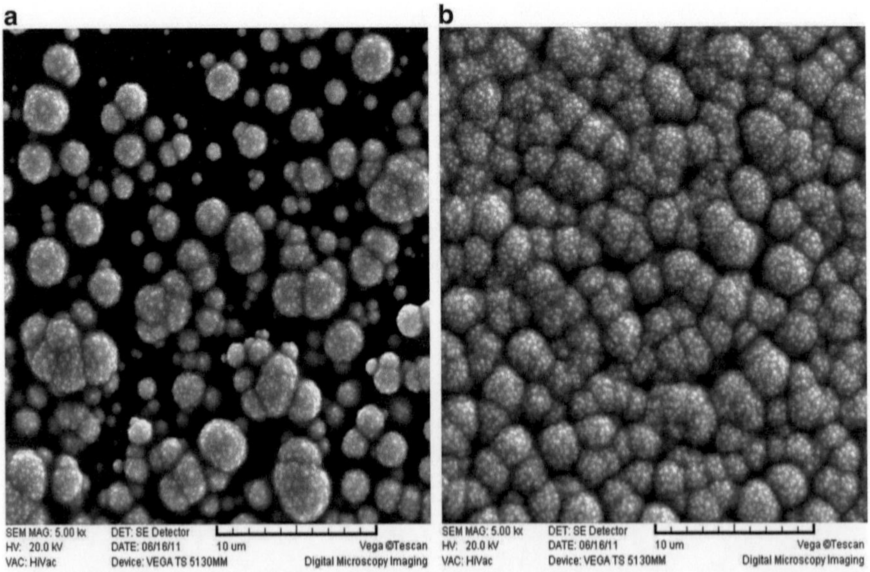

Fig. 7.6 SEM micrographs of the Ni–Sn electrodeposit obtained onto GC electrode during the potentiostatic pulse $E = -1.00$ V versus SCE for 1000 s in the solution 0.1 M $SnCl_2 + 0.1$ M $NiCl_2 + 0.6$ M $K_4P_2O_7 + 0.3$ M NH_2CH_2COOH of the pH 8.0: (**a**) less dense electrodeposit, average composition 54 at.% Ni–46 at.% Sn; (**b**) more dense electrodeposit, average composition 47 at.% Ni–53 at.% Sn

the electrode surface. Rectangular type crystals, growing normally to the electrode surface, are obtained in both cases.

During the initial period of Ni–Sn alloy electrodeposition, similar behavior could be detected with less and more dense parts of the electrodeposit, as shown in Fig. 7.6. In the case of alloy electrodeposition, the shape of crystals is different. Ball-like crystals with the average composition 54 at.% Ni–46 at.% Sn (the composition was obtained by the EDS analysis) could be detected at the less dense part of the electrodeposit (a), while the more dense part of the electrodeposit of the average composition 47 at.% Ni–53 at.% Sn is characterized with distorted balls (b). After prolonged electrodeposition in the range of activation-controlled electrodeposition ($i \leq 10$ mA cm^{-2}, Fig. 7.4a), compact electrodeposit is obtained [6], as shown in Fig. 7.4b.

At higher cathodic current densities (potentials more negative than -1.2 V versus SCE, Fig. 7.4a), sudden increase of current density indicates simultaneous evolution of hydrogen. In such a case, the morphology of electrodeposited Ni–Sn alloy coatings [12] changes from the flat into nodular one, with the appearance of large, micron-sized pores, as shown in Fig. 7.7 for the Ni–Sn electrodeposit obtained at the current density of -75 mA cm^{-2}.

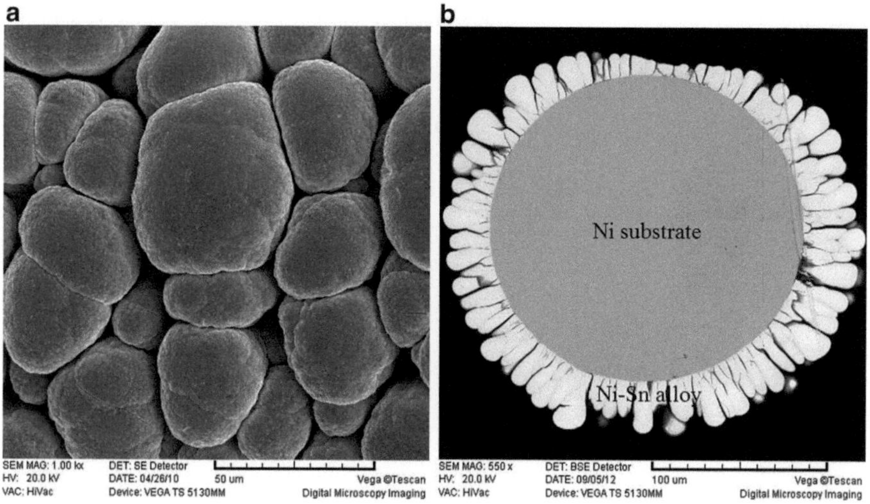

Fig. 7.7 SEM micrographs of the Ni–Sn electrodeposit obtained onto Ni 40 mesh electrode at $i = -75$ mA cm^{-2} in the solution 0.1 M $SnCl_2$ + 0.1 M $NiCl_2$ + 0.6 M $K_4P_2O_7$ + 0.3 M NH_2CH_2COOH of the pH 8.0: (**a**) surface of the electrodeposit; (**b**) cross section of the electrodeposit

7.2.2.2 Irregular Codeposition

The irregular type of codeposition is very often characterized by simultaneous influence of cathodic potential and diffusion phenomena, i.e., it mainly occurs under the activation and/or mixed control of the electrodeposition processes. The rate of electrodeposition in such a case is expressed by the Butler–Volmer equation which is usually used for the kinetics of electrochemical processes [1, 5]:

$$\frac{i}{i^{\mathrm{o}}} = \frac{i_{\mathrm{o}}}{i^{\mathrm{o}}}\left[\exp\left(\frac{\alpha_a F}{RT}\eta\right) - \left(\frac{c}{c_{\mathrm{o}}}\right)\exp\left(\frac{-\alpha_c F}{RT}\eta\right)\right] \tag{7.8}$$

where $\eta = E - E_{\mathrm{r}}$ corresponds to the overpotential needed for the electrodeposition of metals (see Eq. (7.1)), c and c_{o} represent concentrations of the discharging species at the surface of the electrode and in the bulk of the solution, respectively, i_{o} is "exchange current density" related to the rate constant of the electrodeposition process, and α_{a} and α_{c} are anodic and cathodic transfer coefficients related to the mechanism of discharge (note that for a cathodic process, both η and i acquire negative signs. Also note that the current densities are divided by arbitrarily chosen unit current i^{o} in order to obtain dimensionless values for further use).

If the discharge of electrodepositing species is sufficiently slow so that their supply to the electrode surface occurs without difficulty, the concentration c virtually does not deviate from c_{o}, and such a case is termed "activation-controlled" electrodeposition with the rate-determining step being the activation energy of the discharge process. At any cathodic overpotential larger than −40 mV, the first term in Eq. (7.8) becomes negligible, so that this equation can be transformed into a simpler one, known as the Tafel equation:

$$\eta = a - b\log\left(-\frac{i}{i^{o}}\right) \tag{7.9}$$

where the Tafel constant a is

$$a = \frac{2.3RT}{\alpha_c F}\log\left(\frac{i_o}{i^{o}}\right) \tag{7.10}$$

while the slope of the linear dependence obtained from a plot η versus $\log(-i)$ (Tafel slope) is

$$b = \frac{2.3RT}{\alpha_c F} \tag{7.11}$$

The above reasoning applies equally and independently to both metals (A) and (B), i_A and i_B, and the total current density being $i_{alloy} = i_A + i_B$.

It should be stated here that the concept of overpotential is related to the reversible potential of a pure metal in a given solution. In the case of codeposition of two metals and the formation of a phase $A_xB_{(1-x)}$, this potential has no physical meaning since it represents an arbitrary point to which i_o is related.

Electrodeposition of the Ag–Cd Alloy

An example for irregular codeposition is presented in Fig. 7.8 for the system Ag–Cd (two intermediate metals). The Ag, Cd, and Ag–Cd alloy coatings were electrodeposited onto stationary Pt electrode from cyanide electrolytes [13], using the following solutions respectively: 0.14 M $CdSO_4{\cdot}8/3H_2O$ + 0.56 M KCN, 0.032 M $KAg(CN)_2$ + 0.56 M KCN, and 0.032 M $KAg(CN)_2$ + 0.14 M $CdSO_4{\cdot}8/3H_2O$ + 0.56 M KCN. Figure 7.8 shows polarization curves recorded at a sweep rate of 1 mV s^{-1} in the electrolyte containing both metals separately or together. The electrodeposition of Ag (dashed line) is characterized by the cathodic shoulder, which is most likely indication of the diffusion limiting current density (taking into account small concentration of Ag) at a potential of about −0.82 V versus Ag/AgCl, with the electrodeposition starting at about −0.79 V versus Ag/AgCl. The increase of the cathodic current density at potentials more negative than −1.2 V versus Ag/AgCl during the Ag electrodeposition is the consequence of simultaneous hydrogen evolution. Pure Cd electrodeposition (dotted line) starts at the same potential as that of Ag. At potentials more negative than −1.6 V versus Ag/AgCl current oscillations with amplitude higher than 10 mA cm^{-2} have been observed. In the case of Ag–Cd alloy electrodeposition, a shoulder representing diffusion-controlled electrodeposition of Ag (solid line) appears at about −0.51 V versus Ag/AgCl (for about 0.3 V

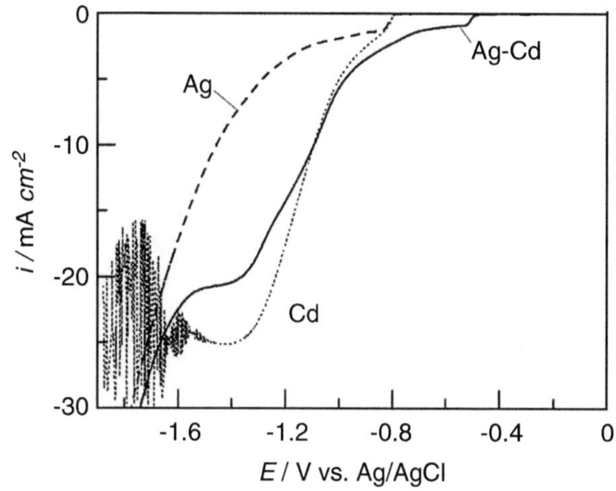

Fig. 7.8 Polarization curves recorded at a sweep rate of 1 mV s^{-1} in the electrolyte containing 0.032 M $KAg(CN)_2$ + 0.56 M KCN (Ag), 0.14 M $CdSO_4{\cdot}8/3H_2O$ + 0.56 M KCN (Cd), and 0.032 M $KAg(CN)_2$ + 0.14 M $CdSO_4{\cdot}8/3H_2O$ + 0.56 M KCN (Ag–Cd) (Reprinted from Ref. [5] with kind permission from Springer)

Table 7.2 Concentration of different Ag complexes and their equilibrium potentials (E_{eq})

Solution composition	Concentration / %		E_{eq}/V versus Ag/AgCl		
	$[Ag(CN)_2]^-$	$[Ag(CN)_3]^{2-}$	$[Ag(CN)_4]^{3-}$	$[Ag(CN)_3]^{2-}$	$[Ag(CN)_2]^-$
Pure Ag electrodeposition	35.9	61.2	2.6	−0.737	
Ag–Cd alloy electrodeposition	98.1	1.9	–		−0.479

Reprinted from Ref. [5] with kind permission from Springer

more positive than the shoulder of pure Ag). Hence, in the solution containing both metal ions, electrodeposition of Ag commences at more positive potential. In order to find out the reason for such behavior, the analysis of Ag complexes with cyanide has been performed. The results are presented in Table 7.2. As can be seen, when pure Ag is present in the KCN solution the dominant complex is $[Ag(CN)_3]^{2-}$, with the equilibrium potential of −0.737 V versus Ag/AgCl. In the presence of Cd ions, most of the CN^- anions are consumed in Cd–CN complexes ($[Cd(CN)]^+$, $[Cd(CN)_2]$, $[Cd(CN)_3]-$, and $[Cd(CN)_4]^{2-}$) and the dominant Ag–CN complex becomes $[Ag(CN)_2]^-$, with the equilibrium potential of −0.479 V versus Ag/AgCl. Hence, from the presented analysis, it is obvious that the first shoulder on a solid curve in Fig. 7.8 for Ag–Cd alloy electrodeposition corresponds to the electrodeposition of pure Ag from $[Ag(CN)_2]^-$ complex [13, 14].

Morphology of the Ag–Cd Alloy

Five alloy samples were electrodeposited to the total charge of −1.3 C cm^{-2} at different potentials (−1.0, −1.4, −1.6, −1.8, and −2.0 V versus Ag/AgCl). The morphology of the coating electrodeposited at −1.0 V versus Ag/AgCl is presented in Fig. 7.9. The morphology of this coating is typical to those where the gradual increase in crystallite size with increase in Cd content is observed [13, 14]. At more negative potential (−1.4 V versus Ag/AgCl), the coating becomes very heterogeneous (see Fig. 7.10a) and reflections of Cd, $AgCd_3$, AgCd, and Ag phases were registered on the XRD spectra [13]. The heterogeneity of the coating could be the consequence of two factors: the alloy composition and the influence of simultaneous hydrogen evolution, since the natural convection must be significantly disturbed by the hydrogen evolution at this potential.

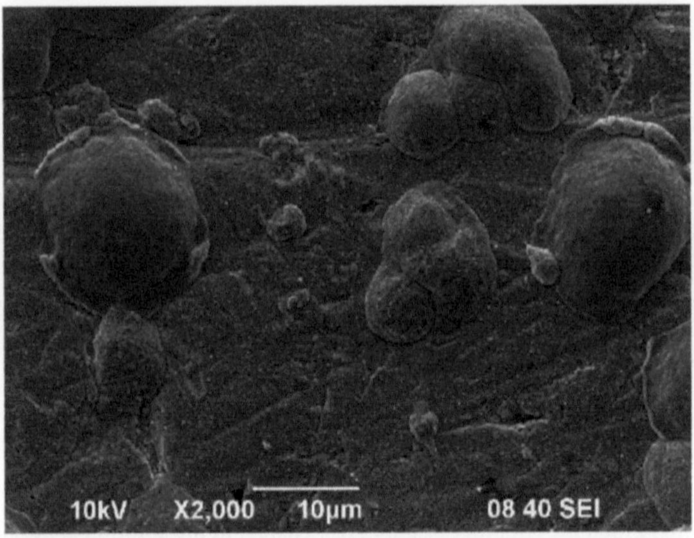

Fig. 7.9 Surface morphology (SEM) of the alloy coating electrodeposited at a potential $E = -1.0$ V versus Ag/AgCl (Reprinted from Ref. [13] with permission from Elsevier)

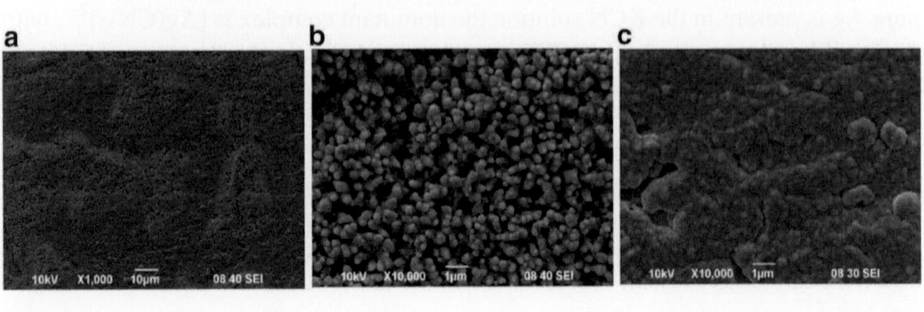

Fig. 7.10 (**a**) Surface morphology (SEM) of the alloy coating electrodeposited at $E = -1.4$ V versus Ag/AgCl. (**b**, **c**) different areas of the same coating at higher magnification (Reprinted from Ref. [13] with permission from Elsevier)

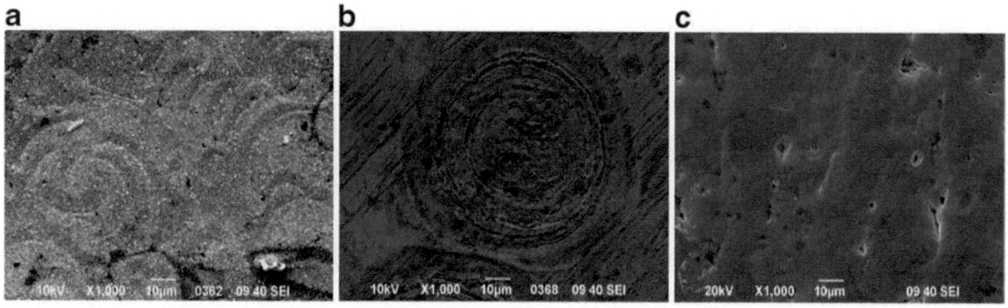

Fig. 7.11 Surface morphology (SEM) of alloy coatings electrodeposited at different (more negative) potentials: (**a**) $E = -1.6$ V versus Ag/AgCl; (**b**) $E = -1.8$ V versus Ag/AgCl; (**c**) $E = -2.0$ V versus Ag/AgCl (Reprinted from Ref. [13] with permission from Elsevier)

The heterogeneity of this coating is well visible – its morphology is presented in Fig. 7.10a. In the rough areas of the coating, the Cd content is about 40 at.% (Fig. 7.10b), while in the smooth areas, it reaches 58 at.% (Fig. 7.10c). The electrodeposition at higher cathodic potentials, −1.6 or −1.8 V versus Ag/AgCl, leads to the appearance of some periodic structured areas on the surface (Fig. 7.11a, b, special structures observed in alloy electrodeposits only; see Sect. 7.2.3), and the reflections of pure Cd disappear on the XRD, which could be connected with the changes in the current efficiency and the enhanced formation of Ag–Cd alloy phases. At the most negative potential of −2.0 V versus Ag/AgCl, the phase $AgCd_3$ is preferentially formed and mainly the reflections corresponding to this phase are registered on the XRD spectra. The morphology of this coating is shown in Fig. 7.11c. The surface is smooth and shiny, indicating the presence of mainly one phase in the electrodeposit ($AgCd_3$) [13].

7.2.2.3 Regular Codeposition

Regular codeposition assumes transport-controlled codeposition in which diffusion of metal ions of both metals is a rate-determining step in the overall codeposition reaction [1, 5].

Under steady-state conditions of electrodeposition, the diffusion is governed by Fick's first law [15].

$$\frac{c}{c_o} = \frac{i_L + i}{i_L} \tag{7.12}$$

with

$$i_L = \frac{zFD}{\delta} c_o \tag{7.13}$$

where D is the diffusion coefficient of the electrodepositing species and δ is the Nernst diffusion layer thickness.

Introducing Eq. (7.12) into Eq. (7.8) and rearranging following relation is obtained:

$$i = \frac{i_o\left[\exp\left(\frac{\alpha_a F}{RT}\eta\right) - \exp\left(-\frac{\alpha_c F}{RT}\eta\right)\right]}{1 + \frac{i_o}{i_L}\exp\left(-\frac{\alpha_c F}{RT}\eta\right)} \tag{7.14}$$

At increasing values of overpotential, the second term in the denominator becomes overwhelming and the current density tends to a potential-independent diffusion limiting one, i_L.

Under the conditions of diffusion control, a plot of E versus log $[(-i_L/i)-1]$ should be made instead of the regular Tafel plot and the relationship should be linear with the slope of $(2.3RT/\alpha_c F)$.

Electrodeposition of the Ag–Pd Alloy

One of the examples of regular codeposition is the electrodeposition of Ag–Pd alloy from high concentration chloride (12 M LiCl) containing bath [16] (both metals belong to intermediate metals due to complexation with Cl^- ions). Polarization curves for the electrodeposition of Ag–Pd alloy from the solution containing 0.005 M $PdCl_2$ + 0.05 M AgCl + 12 M LiCl + 0.1 M HCl (t = 80 °C), recorded onto rotating glassy carbon electrode under the conditions of stationary (RPM = 0)

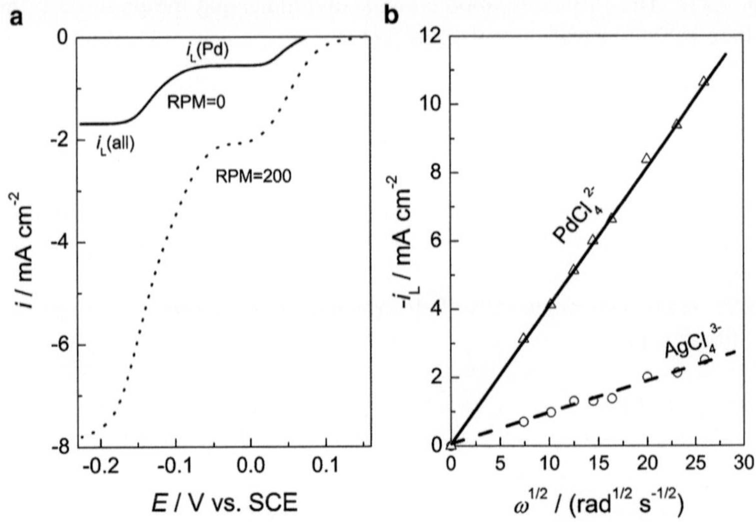

Fig. 7.12 (**a**) Polarization curves for the electrodeposition of Ag–Pd alloy from the solution containing 0.005 M $PdCl_2$ + 0.05 M AgCl + 12 M LiCl + 0.1 M HCl (t = 80 °C), recorded onto rotating glassy carbon electrode. (**b**) The i_L versus $\omega^{1/2}$ dependences for pure Ag and pure Pd electrodeposition onto rotating glassy carbon electrode from the solution containing 0.005 M $PdCl_2$ + 12 M LiCl + 0.1 M HCl and 0.005 M AgCl + 12 M LiCl + 0.1 M HCl, respectively (t = 80 °C) (Reprinted from Ref. [5] with kind permission from Springer)

and convective (RPM = 200) diffusion, are presented in Fig. 7.12a. The Pd electrodeposition commences at about 0.07 V versus SCE, while the codeposition of Ag begins at about −0.07 V versus SCE. A well-defined diffusion limiting current densities for Pd (i_L(Pd)) and alloy (i_L(alloy)) electrodeposition, representing actually the sum of i_L(Pd) and i_L(Ag), could be detected on the polarization curves, indicating that the i_L(Ag) is lower than that of Pd (taking into account that the concentration of AgCl is ten times higher than that of $PdCl_2$) and that the diffusion coefficient for Ag electrodeposition is lower than that for Pd. This is confirmed in Fig. 7.12b. The electrodeposition of Pd occurs by the reduction of $[PdCl_4]^{2-}$, while the electrodeposition of Ag occurs by the reduction of $[AgCl_4]^{3-}$.

The linear i_L versus $\omega^{1/2}$ dependences [16] (Fig. 7.12b) confirm that the diffusion of both species obeys Levich's Eq. (7.15) and from their slopes the diffusion coefficients are obtained. Corresponding values of D for both species amounts to $D([AgCl_4]^{3-}) = 5.5 \times 10^{-6}\ cm^2{\cdot}s^{-1}$ and $D([PdCl_4]^{2-}) = 1.8 \times 10^{-5}\ cm^2{\cdot}s^{-1}$:

$$i_L = 0.62zFD^{2/3}\nu^{-1/6}c_o\omega^{1/2} \tag{7.15}$$

(ν – kinematic viscosity, ω – rotation speed).

In the case of Ag–Pd electrodeposition from ammoniacal electrolyte, the same type of electrodeposition (regular) is valid, but the potentials of electrodeposition of individual metals are opposite [17] to those in chloride-containing electrolyte: Ag is more noble metal with its electrodeposition commencing at about −0.05 V versus Ag/AgCl being characterized by well-defined diffusion limiting current density of about −0.95 mA cm^{-2}, while Pd is less noble metal starting to electrodeposit at about −0.40 V versus Ag/AgCl, as shown in Fig. 7.13. Such behavior is the result of

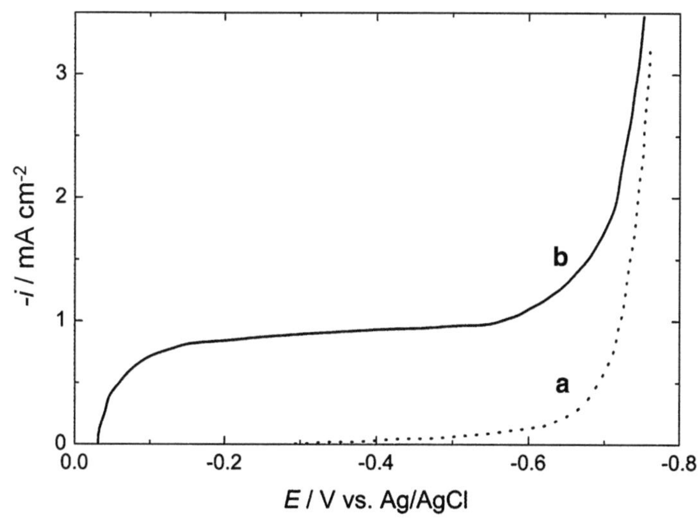

Fig. 7.13 Polarization curves for Pd and Ag–Pd electrolytes ($v = 1$ mV s^{-1}): (**a**) 20 g dm^{-3} Pd; (**b**) 2 g dm^{-3} Ag + 20 g dm^{-3} Pd; pH 11.5 adjusted by addition of NH_3 (Reprinted from Ref. [17] with the permission of Johnson Matthey Plc)

much stronger complexation of Pd^{2+} with NH_3 causing more negative potential for Pd electrodeposition (stability constant for complex $[Pd(NH_3)_4]^{2+}$ amounts to 6.3×10^{32}, while that for $[Ag(NH_3)_2]^+$ amounts to 1.6×10^7).

Morphology of the Ag–Pd Alloy

Morphology of Ag–Pd alloys electrodeposited from the ammonium-containing electrolyte [17] is shown in Fig. 7.14. At small current density, (a) matt electrodeposit was obtained, while at high current density, (b) bright electrodeposit was obtained, due to decrease of Ag content in the alloy.

Morphology of the Ag–Pd alloy of approximate composition 50 at.% Ag – 50 at. % Pd obtained from the chloride-containing electrolyte [16] is similar to that presented in Fig. 7.14 for $i = -15$ mA cm^{-2} (semi-bright).

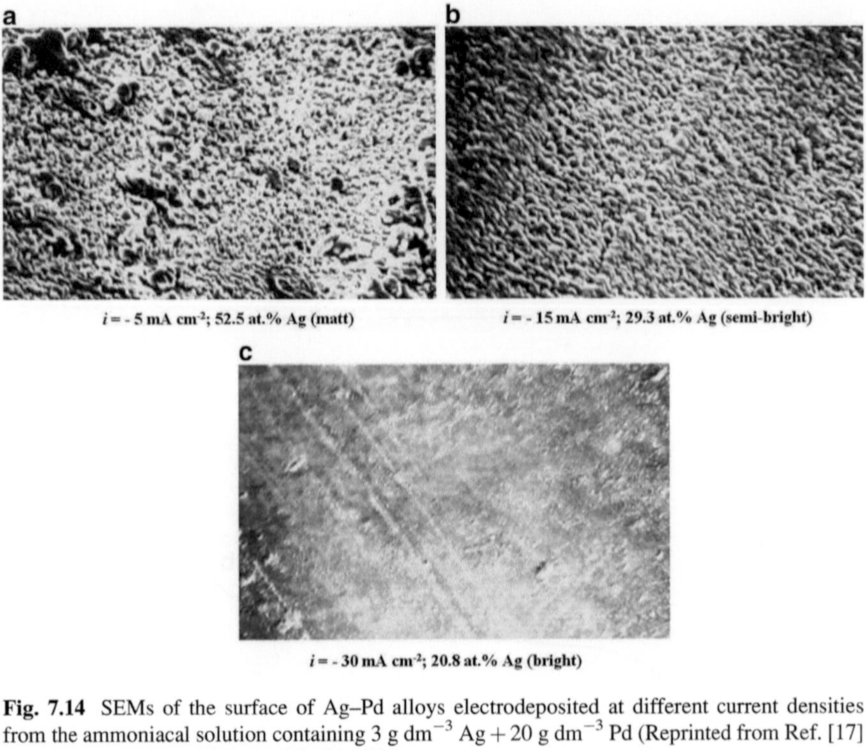

Fig. 7.14 SEMs of the surface of Ag–Pd alloys electrodeposited at different current densities from the ammoniacal solution containing 3 g dm^{-3} Ag + 20 g dm^{-3} Pd (Reprinted from Ref. [17] with the permission of Johnson Matthey Plc)

Electrodeposition of the Ag–Sn Alloy

Two types of solutions were examined for Ag–Sn alloy electrodeposition [18]: sulfate solution containing thiourea as a complexing agent for Ag^+ ions and pyrophosphate and iodide solution which form a stable complex with both Ag^+ and Sn^{2+} ions. In sulfate solution, Sn was a normal metal, while Ag was intermediate one due to formation of complexes with thiourea and iodide ions. In pyrophosphate solution, both metals belonged to intermediate ones due to formation of complexes with pyrophosphate and iodide ions. The polarization curves for alloy electrodeposition measured by the potential sweep method ($v = 0.5$ mV s^{-1}) in sulfate and pyrophosphate–iodide solutions are shown in Fig. 7.15. A current density rapidly increased at about −0.07 V versus NHE with the current density plateau up to about −0.27 V versus NHE corresponding to the pure Ag electrodeposition in the sulfate solution. Additional current density increase and plateau at more negative potentials correspond to the alloy electrodeposition (①). Similar behavior is detected for pyrophosphate–iodide solution (②). In both electrolytes, electrodeposition of both metals was suppressed due to complexes formation. The content of Ag in both cases abruptly decreased with the increase of electrodeposition current density.

Morphology of the Ag–Sn Alloy

Morphology of Ag–Sn alloy electrodeposits obtained from sulfate solution at different current densities is presented in Fig. 7.16, while the morphology of the same alloy electrodeposits obtained from pyrophosphate–iodide solution at different current densities is presented in Fig. 7.17.

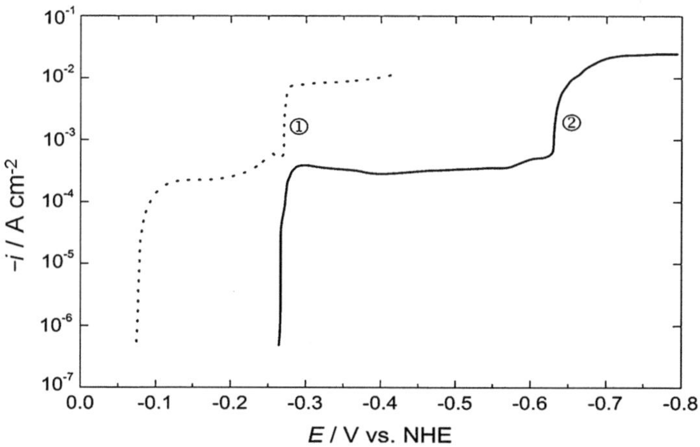

Fig. 7.15 Polarization curves for Ag–Sn alloy electrodeposition from sulfate (①) and pyrophosphate–iodide (②) solutions (Reprinted from Ref. [18] with the permission of the Japan Institute of Metals and Materials)

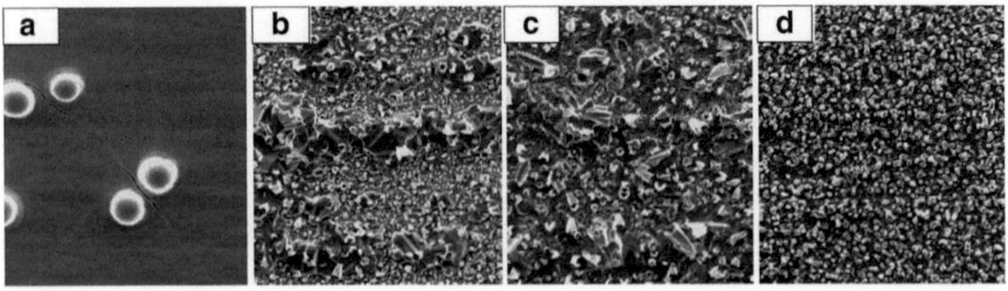

Fig. 7.16 Morphology of Ag–Sn electrodeposits obtained from sulfate solution at different current densities. (**a**) $i = -0.2$ mA cm^{-2}, 98 mass % Ag; (**b**) $i = -1.0$ mA cm^{-2}, 45 mass % Ag; (**c**) $i = -2.0$ mA cm^{-2}, 20 mass % Ag; (**d**) $i = -10.0$ mA cm^{-2}, 13 mass % Ag (Reprinted from Ref. [26] with the permission of Journal of Chemical Sciences)

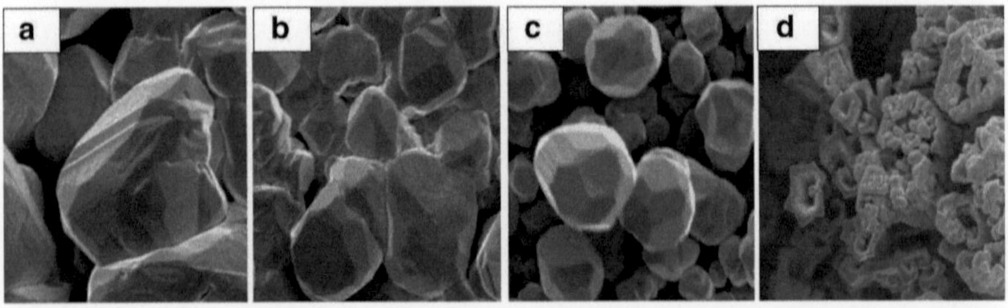

Fig. 7.17 Morphology of Ag–Sn electrodeposits obtained from pyrophosphate–iodide solution at different current densities. (**a**) $i = -0.2$ mA cm^{-2}, 100 mass % Ag; (**b**) $i = -0.4$ mA cm^{-2}, 81 mass % Ag; (**c**) $i = -0.5$ mA cm^{-2}, 77 mass % Ag; (**d**) $i = -0.6$ mA cm^{-2}, 62 mass % Ag (Reprinted from Ref. [26] with the permission of Journal of Chemical Sciences)

At a current density of −0.2 mA cm^{-2} (a), Fig. 7.16, the electrodeposit of practically pure Ag (98 mass % Ag) showed a smooth surface despite partial grains and cracks in the smooth area. At a current density of −1.0 mA cm^{-2} (b), the content of Ag decreased to 45 mass %, the smooth area disappeared, and the electrodeposit consisted of grains over the entire surface. Further increase in the cathodic current density to −2.0 mA cm^{-2} (c) and −10.0 mA cm^{-2} (d) caused smaller grains and significant decrease of the Ag content.

The surface morphology of Ag–Sn alloys electrodeposited from a pyrophosphate–iodide solution was completely different. At the smallest electrodeposition current density ($i = -0.2$ mA cm^{-2}), it consisted of large blocks, while at $i = -0.6$ mA cm^{-2}, the electrodeposit became coarse and spongy (Fig. 7.17) [18].

7.2.2.4 Anomalous Codeposition

According to Brenner's classification [3], anomalous codeposition is characterized by the fact that the less noble metal electrodeposits before the more noble one as the potential is driven cathodic. As a consequence, the content of the less noble metal in the alloy is higher than its content in the solution. In all cases of anomalous codeposition, both metals belong to the group of intermediate metals [1, 5].

Electrodeposition of the Co–Ni Alloy

An excellent example [19] is found in the case of Co–Ni alloy electrodeposition, where both metals belong to the inert metals. Pure Co and pure Ni were electrodeposited at RPM = 1000 onto a gold disc electrode from the solutions containing 0.2 M $CoSO_4$ + 0.2 M $Na_3C_6H_5O_7$ and 1 M $NiSO_4$ + 0.2 M $Na_3C_6H_5O_7$ respectively. For electrodeposition of their alloys of different compositions, only the concentration of Co^{2+} has been changed, being 0.005 M, 0.01 M, 0.025 M, 0.05 M, and 0.2 M, respectively. Alloy layers of different compositions of the thickness of approximately 0.34 μm were obtained by galvanostatic electrodeposition at two different current densities ($i = -2.5$ mA cm^{-2} (○) and $i = -28.5$ mA cm^{-2} (□)) to the same amount of charge ($Q_{dep} = -1$ C cm^{-2}). Polarization curves for pure metal electrodeposition (Co and Ni) and Co–Ni alloy electrodeposition (alloy) from the solution containing 0.025 M $CoSO_4$ + 1 M $NiSO_4$ + 0.2 M $Na_3C_6H_5O_7$ are presented in Fig. 7.18a (all other polarization curves for Co–Ni alloy electrodeposition are placed between that for Ni and that for alloy – shaded area). As can be seen, all polarization curves for Co–Ni alloy electrodeposition are placed at more positive potentials than either of pure metals, which is clear indication for the anomalous codeposition of these two metals.

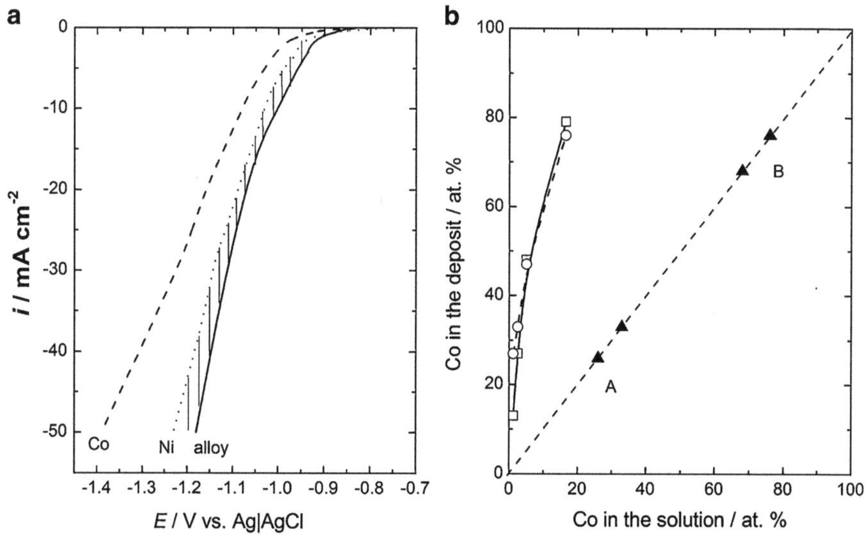

Fig. 7.18 (**a**) Polarization curves (corrected for IR drop) for the electrodeposition of pure metals (Co and Ni) and Co–Ni alloy at the gold disc electrode (RPM = 1000) obtained from the solutions: 0.2 M $CoSO_4$ + 0.2 M $Na_3C_6H_5O_7$ (Co); 1 M $NiSO_4$ + 0.2 M $Na_3C_6H_5O_7$ (Ni); 0.025 M $CoSO_4$ + 1 M $NiSO_4$ + 0.2 M $Na_3C_6H_5O_7$ (alloy). (**b**) Brenner's diagram for the system Co–Ni: samples electrodeposited at $i = -2.5$ mA cm^{-2} (○) and at $i = -28.5$ mA cm^{-2} (□) (Reprinted from Ref. [5] with kind permission from Springer)

Based on the chemical analysis, the content of Co in the electrodeposit increases within the range ~8 at. % to ~80 at. % with increasing concentration of Co^{2+} ions in the solution. Brenner's diagram presented in Fig. 7.18b clearly shows pronounced anomalous codeposition.

Different explanations for such behavior are offered in the literature [3]. The most likely one appears to be "the hydroxide suppression mechanism" [20– 23]. According to this concept, coevolution of hydrogen during the electrodeposition causes an increase of pH at the electrode/solution interface, producing hydrolysis of less noble metal species and their precipitation as a layer of solid hydroxide. Formed hydroxide layer provides a good supply of ions of the less noble metal for their discharge and electrodeposition but suppresses the transport of species of the more noble metal to the cathode surface, causing anomalous codeposition.

Morphology of the Co–Ni Alloy

Co–Ni alloys were electrodeposited from a typical Watts-type electrolyte, containing $NiSO_4$ (200 g dm^{-3}), NaCl (20 g dm^{-3}), H_3BO_3 (30 g dm^{-3}), sodium lauryl sulfate (0.1 g dm^{-3}), and $CoSO_4$ (0–80 g dm^{-3}) [24]. Alloys were electrodeposited on AISI-1045 steel substrates at a current density of -30 mA cm^{-2}, bath temperature of 45 °C, and pH 4.0. The dependence of the morphology on the composition is shown in Fig. 7.19.

Typical morphology of a Watts Ni electrodeposit is shown in Fig. 7.19a, characterized with relatively large grain size (3–10 μm) and polyhedral crystallites. With the increase of Co content from 7 to 49 wt. % (Fig. 7.19b–d), a gradual decrease in the grain size of the Co–Ni alloy down to a submicron grain size is observed. When the Co content reached the 49 wt. %, close observation of SEM morphology at high magnification (Fig. 7.19g) revealed that the Co–Ni alloys have spherical cluster surface piled with a large number of equally sized grains with

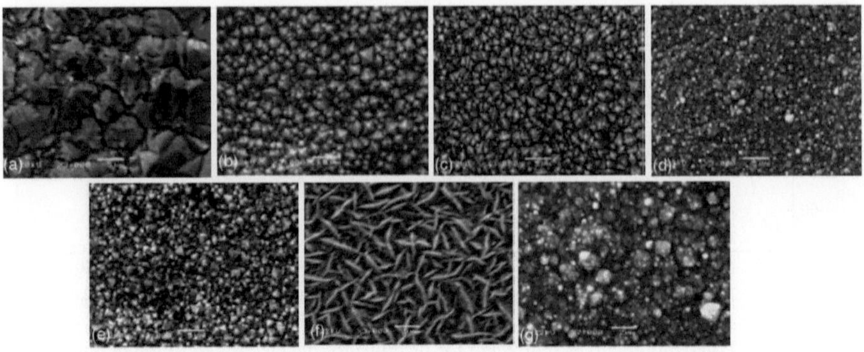

Fig. 7.19 SEM morphologies of Co–Ni alloy electrodeposits with their Co contents of (**a**) 0 wt.%, (**b**) 7 wt.%, (**c**) 27 wt.%, (**d**) 49 wt.%, (**e**) 66 wt.%, (**f**) 81 wt.%, (**g**) high magnification of Ni–49 wt.% Co alloy (Reprinted from Ref. [24] with the permission of Elsevier)

spherical shape. At above 49 wt. % Co, the grain size of Co–Ni electrodeposits, however, increased with the increase of Co content in alloys. When increasing Co content up to 81 wt. %, the morphology of the Co–Ni alloys changed dramatically, and with less compact structure, the Co–Ni alloy showed a rather regularly branched structure with extended acicular 3–6 μm length crystallites (Fig. 7.19f). Such a change of the morphology is related to the change in the phase composition of electrodeposited Co–Ni alloys. XRD investigations revealed that the phase structure of Co–Ni alloys gradually changed from *fcc* into *hcp* with the increase of Co content [24].

The influence of saccharin addition on the morphology and nanostructure of Co–Ni electrodeposits containing 80 at. %·Co was investigated by SEM and TEM analysis [25]. Alloys were electrodeposited from the bath containing 30 g dm^{-3} $NiSO_4 \times 6H_2O$, 30 g dm^{-3} $CoSO_4 \times 7H_2O$, 15 g dm^{-3} $NiCl_2 \times 6H_2O$, 15 g dm^{-3} H_3BO_3, and 10 g dm^{-3} NaCl at $i = -$ 5.1 mA cm^{-2}, while the amount of added saccharin varied from 0 to 12 g dm^{-3}. Figure 7.20 shows that already at 2 g dm^{-3} flat and nanocrystalline electrodeposit was obtained. The grain size of these electrodeposits sharply decreased from 50 to 5 nm with the increase of the amount of added saccharin from 0.0 to 0.05 g dm^{-3}, while slight increase of the grain size to about 10 nm at saccharin concentrations higher than 1 g dm^{-3} was detected by XRD.

TEM analysis of the electrodeposit with the lowest grain size of 5 nm (amount of saccharin 0.05 g dm^{-3}) is shown in Fig. 7.21a, while that of the electrodeposit obtained at saccharin concentration of 2.0 g dm^{-3} is shown in Fig. 7.22b. Dark-field image (Fig. 7.21a) shows grains of 30 nm size, containing a high density of nanoscale twins. In the electrodeposit obtained from the bath containing 2.0 g dm^{-3} of saccharin [25] (Fig. 7.21b), nanocrystalline grains are smaller (about 20 nm). Nanoscale twins are also present, although the twin density is somewhat lower than that observed in Fig. 7.21a.

Electrodeposition of the Fe–Zn Alloy

In this case, alloy contains normal (Zn) and inert (Fe) metal. Electrodeposition of Fe–Zn alloy was carried out from the sulfate bath [26]. The optimized composition

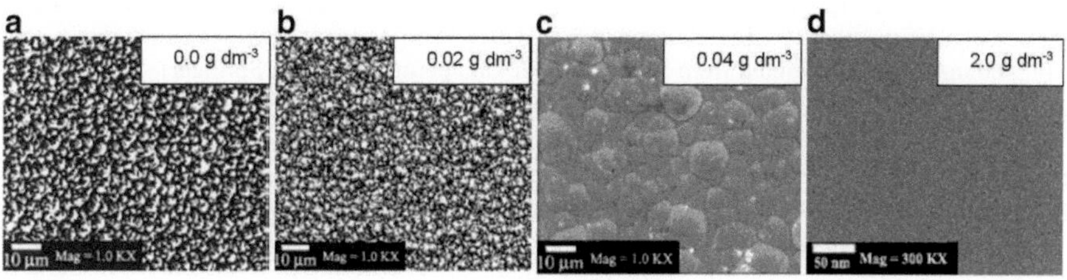

Fig. 7.20 The effect of saccharin addition on the topography of electrodeposited Co 80 at.% -Ni 20 at.% *hcp* alloys. (**a**) 0.0 g dm^{-3} saccharin; (**b**) 0.02 g dm^{-3} saccharin; (**c**) 0.04 g dm^{-3} saccharin; (**d**) 2.0 g dm^{-3} saccharin (Reprinted from Ref. [25] with kind permission from Springer)

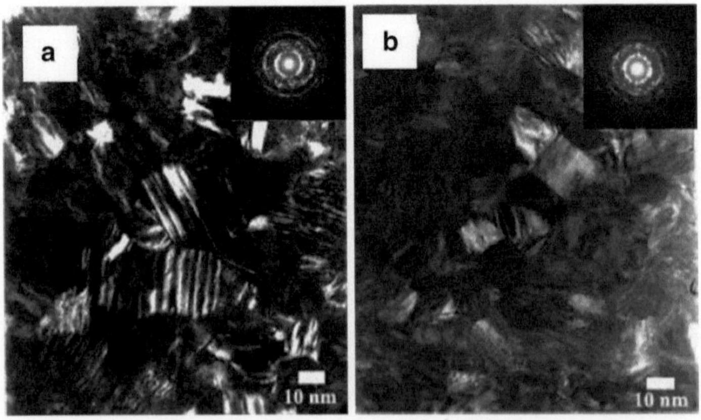

Fig. 7.21 Dark-field TEM images of the Co–Ni electrodeposits containing 80 at. % Co, electrodeposited with 0.05 g dm^{-3} (**a**) and 2.0 g dm^{-3} (**b**) of saccharine (Reprinted from Ref. [25] with kind permission from Springer)

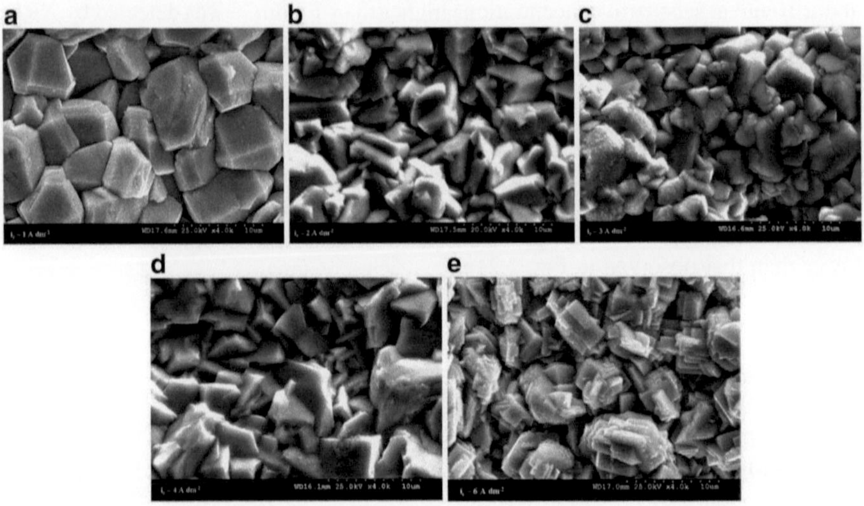

Fig. 7.22 SEM morphologies of Fe–Zn alloy electrodeposits obtained at different current densities: (**a**) −10 mA cm^{-2}, (**b**) −20 mA cm^{-2}, (**c**) −30 mA cm^{-2}, (**d**) −40 mA cm^{-2}, and (**e**) −60 mA cm^{-2} (Reprinted from Ref. [26] with kind permission from Springer)

of the bath used for this study was $ZnSO_4 \cdot 7H_2O$, 0.28 M; $FeSO_4(NH_4)_2SO_4 \cdot 6H_2O$, 0.41 M; and KCl, 1.07 M. Alloys were electrodeposited at different current densities from −10 mA cm^{-2} to −80 mA cm^{-2} and the amount of Fe changed from 6 to 38 wt. % Fe.

Morphology of the Fe–Zn Alloy

SEM images of electrodeposits obtained at different current densities are presented in Fig. 7.22. It could be observed that the electrodeposits morphology was influenced by the alloy composition. Three types of morphologies were observed. For Zn-rich alloy (<10 wt. % Fe content, $i = -10$ mA cm^{-2}), distorted hexagonal morphology was detected. For an alloy with optimum (10–25 wt. %) Fe content, triangular pyramidal morphology ($i = -20$ to -40 mA cm^{-2}) was observed. Relatively smooth and compact electrodeposit with fine-grained morphology (with an Fe content of ≈17 wt. %) was observed at $i = -30$ mA cm^{-2}. Coarser electrodeposits with stacked platelet columnar morphology were produced at cathodic current densities above $i = -50$ mA cm^{-2} (Fe content >25 wt. %). Reduction of crystallite size at $i = -30$ mA cm^{-2} indicates that the nucleation rate was higher than the grain growth rate. This indicates that the overall electrodeposition rate was predominantly controlled by charge-transfer processes rather than mass-transfer processes. But at higher cathodic current densities (above -50 mA cm^{-2}), the overall electrodeposition rate was controlled by mass-transfer limitations particularly for Zn. So the grain growth rate was higher than the nucleation rate and this leads to the formation of coarser morphology enriched with Fe [26].

Electrodeposition of the Fe–Ni Alloy

Fe–Ni alloys (both metals belong to inert ones) were electrodeposited under galvanostatic conditions on the steel substrate for durations 25, 50, and 100 min [27]. The amount of Fe in the electrodeposit was varied by changing the Ni^{2+}/Fe^{2+} ratio in the solution containing constant concentration of $NiCl_2 \cdot 6H_2O$ (23.5 g dm^{-3}), different concentrations of $FeCl_2 \cdot 6H_2O$ (1.0, 1.5, and 2.1 g dm^{-3}), and 20.0 g dm^{-3} H_3BO_3. The plating took place at a constant stirring rate (RPM = 500), by applying current density $i = -20$ mA cm^{-2}. The pH 2 was kept by adding a few drops of sulfuric acid.

Morphology of the Fe–Ni Alloy

Morphology of the Fe–Ni alloys electrodeposited at different Ni^{2+}/Fe^{2+} mass ratios is presented in Fig. 7.23. Figure 7.23a, for sample of Ni^{2+}/Fe^{2+} mass ratio 20.7, displayed a typical elliptical particles shape, with the larger particles growing preferentially out of the surface. By increasing the Fe content (Fig. 7.23b), electrodeposits showed clusters of spherical fine particles embedded in elongated coarse ones with appearance of grain boundaries. For samples prepared at the lowest Ni^{2+}/Fe^{2+} mass ratio of 9.8 (Fig. 7.23c), the microstructure was made of flattened areas containing very fine particles distributed all over the surface. The increase of electrodeposition time had no influence on the appearance of the Fe–Ni alloy electrodeposits.

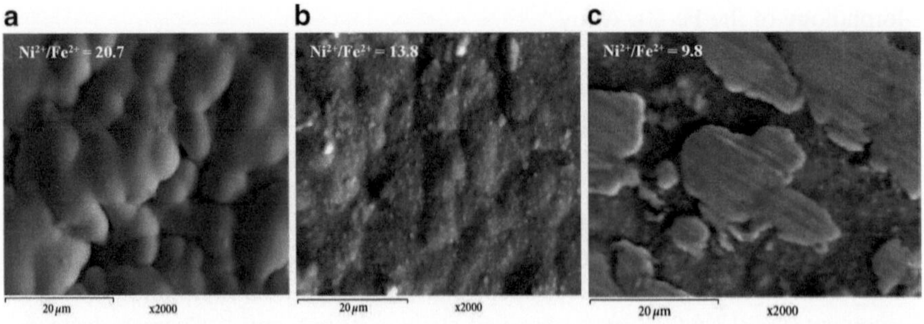

Fig. 7.23 SEM of electrodeposited Ni base layers at current density -20 mA cm^{-2} and deposition time 25 min. as a function of Ni^{2+}/Fe^{2+} mass ratio in the electrolytic bath (Reprinted from Ref. [27] with permission from Hindawi Publishing Corporation)

7.2.2.5 Induced Codeposition

Although it has been shown that Mo, W, Ti, and Ge could not be separately electrodeposited from aqueous solutions, it was discovered that they could be codeposited with the iron-group metals (Fe,Ni,Co) in the presence of appropriate complexing agents. This type of alloy electrodeposition was defined by Brenner [3] as induced codeposition.

Electrodeposition of the Mo–Ni Alloy

In most of the papers concerning the process of induced codeposition of Mo and Ni, where Mo was present in the form of molybdate ion (MoO_4^{2-}), it was assumed that both metals were reduced to a metallic state, producing a Mo–Ni alloy coating. The most probable mechanism of alloy electrodeposition was proposed by Podlaha and Landolt [28–30] and Marlot et al. [31]. Their investigations were performed under controlled mass-transport conditions (rotating cylinder electrode). It was shown that mass-transport control is an important factor for induced codeposition process. In a Ni-rich electrolyte, Mo deposition was mass-transport limited and the alloy composition was strongly influenced by the electrode rotation rate, while in a Mo-rich electrolyte, the rate of Mo electrodeposition was limited by the flux of nickel ions and alloy composition was independent of hydrodynamic effects [28]. The model assumed that Ni electrodeposition occurred on the surface not covered by the molybdate reaction intermediate, by direct reduction of Ni species (all of them being complex of Ni^{2+} cations with the citrate anions) independently on the molybdate reaction, which could occur only in the presence of Ni species [28–31]. The electrodeposition of Mo–Ni alloy was described by following reduction reactions [29]:

$$[Ni(C_6H_5O_7)]^- + 2e^- \rightarrow Ni(s) + (C_6H_5O_7)^{-3} \qquad (7.16)$$

$$MoO_4^{-2} + [Ni(C_6H_5O_7)]^- + 2H_2O + 2e^- \rightarrow [Ni(C_6H_5O_7)MoO_2]_{ads}^- + 4OH^- \quad (7.17)$$

$$[Ni(C_6H_5O_7)MoO_2]_{ads}^- + 2H_2O + 4e^- \rightarrow Mo(s) + [Ni(C_6H_5O_7)]^- + 4OH^- \quad (7.18)$$

$$H_2O + e^- \rightarrow \tfrac{1}{2} H_2 + OH^- \quad (7.19)$$

These reactions were postulated with an assumption that the alloy electrodeposition was always accompanied by the simultaneous hydrogen evolution (reaction (7.19)). This model has been confirmed by in situ surface Raman spectroscopic studies, by revealing existence of adsorbed intermediate $[Ni(C_6H_5O_7)MoO_2]_{ads}^-$ at the electrode surface [32].

It was later shown for the Mo–Ni–O powder electrodeposition [33] that this mechanism is only partially correct and that among metallic Ni and Mo, the MoO_3 phase is also electrodeposited. Taking into account that NiO, MoO_3, and $MoNi_4$ phases were detected in as-deposited samples by TEM, it was obvious that the mechanism of alloy electrodeposition similar to the one presented by Eqs. (7.16, 7.17, 7.18, and 7.19) could be applied for the formation of $MoNi_4$ phase only, since it assumed complete reduction of both metal ions. By the analogy with the mechanism proposed by Podlaha and Landolt [28–30], it was suggested that the $MoNi_4$ phase could be formed by the following mechanism [34]:

$$[Ni(NH_3)_3]^{2+} + 2e^- \rightarrow Ni(s) + 3NH_3 \quad (7.20)$$

$$MoO_4^{2-} + [Ni(NH_3)_3]^{2+} + 2H_2O + 2e^- \rightarrow [Ni(NH_3)_3MoO_2]_{ads}^{2+} + 4OH^- \quad (7.21)$$

$$[Ni(NH_3)_3MoO_2]_{ads}^{2+} + 2H_2O + 4e^- \rightarrow Mo(s) + [Ni(NH_3)_3]^{2+} + 4OH^- \quad (7.22)$$

with both metals being reduced to a metallic state, Mo(s) and Ni(s).

The most probable mechanism for the MoO_3 phase formation was [34].

$$MoO_4^{2-} + [Ni(NH_3)_3]^{2+} + 2H_2O + 2e^- \rightarrow MoO_3 + Ni(s) + 3NH_3 + 2OH^- \quad (7.23)$$

Morphology of the Mo–Ni Alloy

Typical nodular morphology of the Mo–Ni alloy surface [35] is shown in Fig. 7.24a, while cross section (Fig. 7.24b) revealed the presence of large cracks in the electrodeposit [36]. Such behavior was characteristic for all Mo–Ni alloys independently of the solution composition and applied current density.

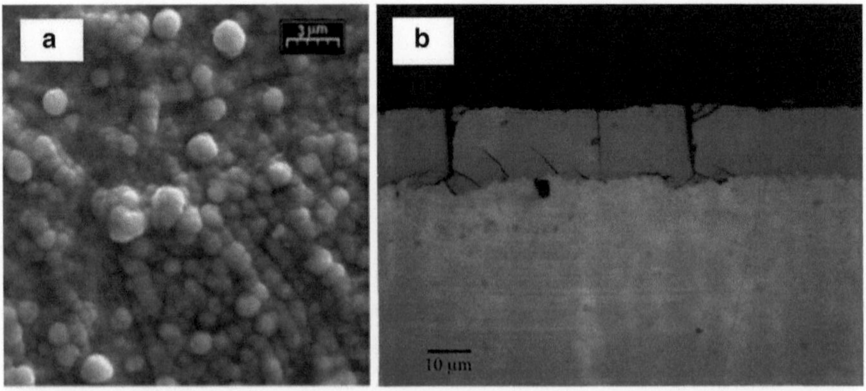

Fig. 7.24 Typical surface morphology (**a**) and cross section (**b**) of the Ni–Mo electrodeposits (**a** – reprinted from Ref. [35] with the permission of Elsevier; **b** – reprinted from Ref. [36] with permission from the International Association of Hydrogen Energy)

Electrodeposition of the Fe–Mo Alloy

The Fe–Mo alloy coatings were electrodeposited onto mild steel substrate from the solution containing $FeCl_3$, Na_2MoO_4, $Na_4P_2O_7$, and $NaHCO_3$, pH 9.3 [37]. Depending on the concentrations of $FeCl_3$, $Na_4P_2O_7$, and Na_2MoO_4, as well as the applied current density, the wt. % of Mo in the coatings (approximate thickness 20 μm) changed from about 45 wt. % to about 70 wt. %, while the current efficiency (η_i) varied from about 10 to 50 %, as shown in Fig. 7.25. It should be stated that in all samples certain (in some cases significant) amount of oxygen has been detected, but this was neglected assuming that the mechanism for induced codeposition defined by Eqs. (7.16, 7.17, 7.18, and 7.19) is operative and, accordingly, only percentages of Fe and Mo were taken into account [37].

Morphology of the Fe–Mo Alloy

Typical morphology of the coatings electrodeposited at different current densities is presented in Fig. 7.26, characterized with nodular surface and large cracks. By the EDS analysis of the Fe–Mo alloy surfaces, it was found that their composition depends on the electrodeposition current density, as shown in Table 7.3. Considering these results, it appeared that the amount of MoO_3 in the electrodeposit decreased with the increase of the electrodeposition current density (lower percentage of oxygen), indicating that the mechanism predicted by Eqs. (7.16, 7.17, and 7.18) and (7.20, 7.21, and 7.22) for Ni–Mo prevailed at higher current densities, while the mechanism for MoO_3 formation (Eq. (7.23) prevailed at lower current densities of Fe–Mo alloy electrodeposition [38].

The cross section of the Fe–Mo alloy electrodeposited at $i = -100$ mA cm^{-2} is shown in Fig. 7.27. All characteristics of the morphology are seen (nodules and cracks) on the cross section (a, b). Distribution of the Fe and Mo in the

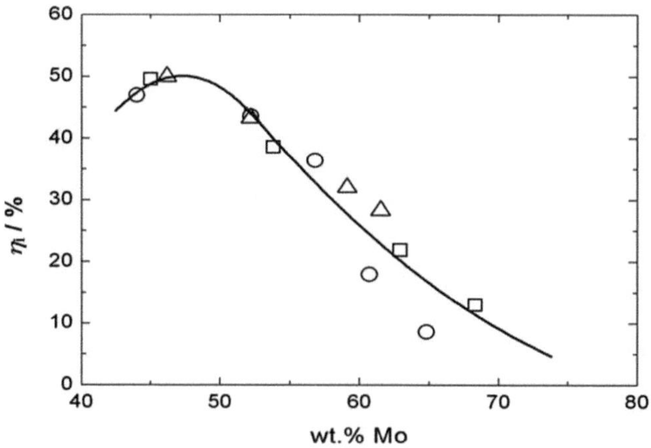

Fig. 7.25 Dependence of η_i on Fe–Mo alloy composition (wt. % Mo), (○) the influence of the $Na_4P_2O_7$ concentration, (□) the influence of the $FeCl_3$ concentration, and (Δ) the influence of the current density for alloy electrodeposition (Reprinted from Ref. [5] with kind permission from Springer)

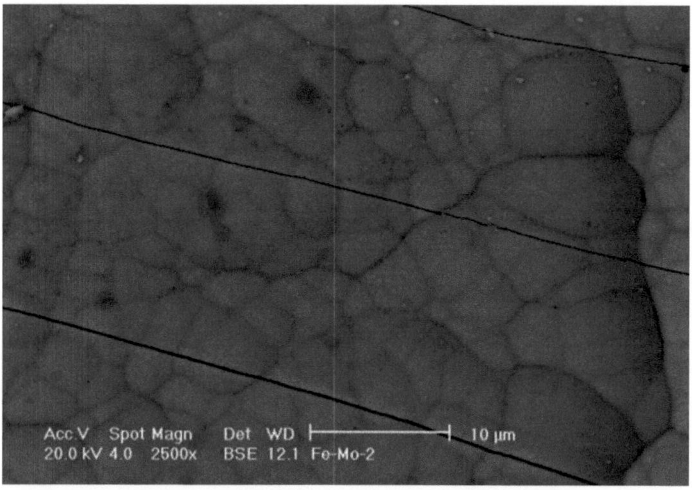

Fig. 7.26 Typical morphology of the Fe–Mo alloy coatings electrodeposited at different current densities (Reprinted from Ref. [5] with kind permission from Springer)

Table 7.3 Composition of the Fe–Mo alloy surfaces, obtained by the EDS analysis, as a function of the electrodeposition current density

i_d/mA·cm^{-2}	at. % O	at. % Mo	at. % Fe
−20	61.20	17.51	21.29
−50	34.05	27.08	38.87
−100	25.80	25.24	48.96

Reprinted from Ref. [5] with kind permission from Springer

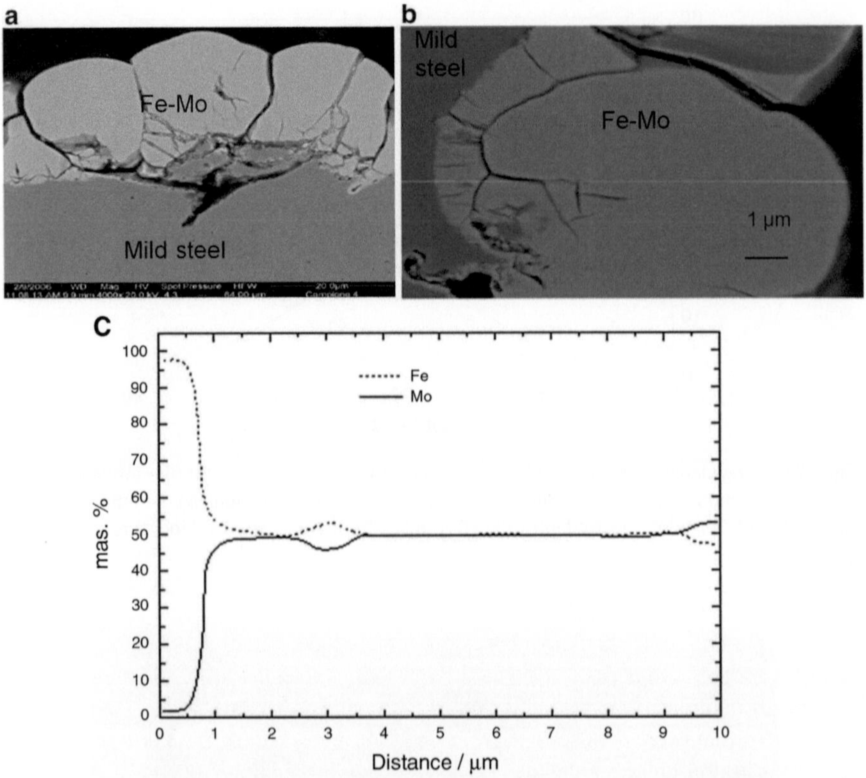

Fig. 7.27 Cross section of the Fe–Mo alloy coating electrodeposited at $i = -100$ mA cm^{-2}, (**a**, **b**). Distribution of the Fe and Mo metals over the electrodeposit obtained by linear EDS analysis (**c**) for sample shown in (**b**) (Reprinted from Ref. [39] with the permission of the Institute for Multidisciplinary Research University of Belgrade)

electrodeposit (neglecting oxygen), obtained by the linear EDS analysis, is presented in Fig. 7.27c, showing uniform distribution of both metals over the entire thickness of the electrodeposit [39].

7.2.3 Specific Structural Phenomena in Electrodeposition of Alloys

Concerning miscibility between the metal constituents of an alloy, all types of alloys could be obtained by electrodeposition: eutectic-type alloys, solid solution-type alloys, alloys with intermediate phases, and/or intermetallic compounds [1]. According to Krastev and Dobrovolska [40], self-organization phenomena during the electrodeposition of alloys, resulting in pattern and spatiotemporal structure formation on the surface of the coatings, are a consequence of the distribution of different phases in the coating, as well as on its surface, since the properties of heterogeneous alloy coatings are a mixture of the properties of the different phases present in the coatings.

In some cases, the electrodeposition of alloys is accompanied by electrochemical instabilities resulting in current or potential oscillations. This phenomenon was a subject of several review papers [41–43], being explained on the basis of autocatalytic processes, negative electric resistance of the system, etc. Usually, spontaneous formation of layered structures without applying external electrical pulses is a result of oscillatory processes. Such electrodeposits are observed in many systems [3, 44–47]. Multilayered structures could also be obtained by certain current regimes (see Sect. 7.3). In some cases, ordered distribution of the different phases of the heterogeneous alloy coating could be observed, being expressed as spatiotemporal structures.

7.2.3.1 Spatiotemporal Structures in Electrodeposited Alloys

The first spatiotemporal structure was observed in 1938 by Raub and Schall [48] in the system Ag–In. According to their observation, the transition from chaotically distributed phases of the alloy into well-defined periodic spiral structures was caused by the fact that the In-rich phase sometimes forms well-defined crystallization spirals on the surface of the Ag-rich alloy coating. About 50 years later, similar structure formation was observed during electrodeposition of Ag–Sb alloys on Cu substrates in non-agitated electrolytes [49]. At higher contents of Sb, the transition of the α-phase (solid solution of Sb in Ag) of the electrodeposit into a heterogeneous mixture of different textured phases of this alloy system is observed [50]. It was concluded that well-expressed spatiotemporal structures appear in the transition region of the α-phase into hexagonal ξ-phase. An example of the surface of the Ag–Sb coating with spatiotemporal structures is shown in Fig. 7.28a, while in Fig. 7.28b, an example of spatiotemporal structures for the Co–In alloy is shown [51].

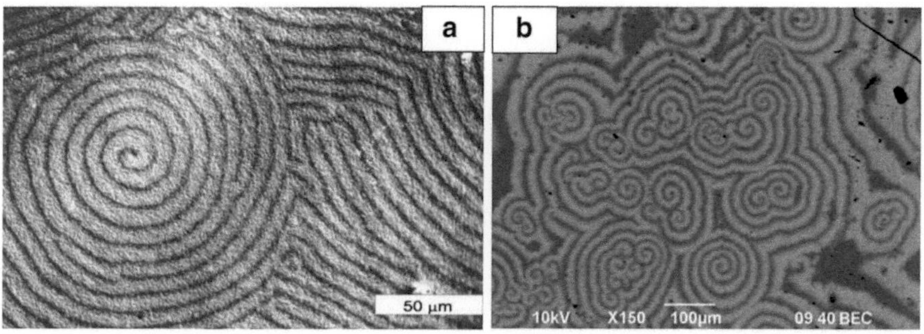

Fig. 7.28 Spatiotemporal structures obtained in the system Ag–Sb (**a**) and Co–In (**b**) (**a** – reprinted from Ref. [40] with kind permission from Springer; **b** – reprinted from Ref. [51] with kind permission from Springer)

The oscillations observed in the Ag–Sb electrodeposition system were found to correspond to the formation of coarse travelling waves consisting of phases with different Sb content and their propagation in front of the Haber–Luggin capillary, leading to the formation of layered structures visible in the cross sections of the electrodeposit [47]. At more negative potentials, the waves become finer, a certain number of them cover the cathode surface in front of the capillary, and further oscillations cannot be registered with this configuration of the capillary. In this case, the formation of smaller waves, spirals with different number of arms, and target patterns is microscopically observed. The electrodeposition of Sb on its own substrate is easier than the one on the Ag substrate [52], ensuring necessary autocatalytic step for the appearance of the self-organization phenomenon. It is believed that the movement of the coarse waves in vertical direction on the electrode surface takes place due to the upwards flow of the electrolyte as a consequence of the natural convection, which seems to be one of the crucial factors for the appearance of this instability and the periodical pattern formation. This hypothesis was confirmed by experiments in a strong magnetic field (5 T) [52] where the natural convection was suppressed by the magnetohydrodynamic flow in horizontal direction at an appropriate orientation of the cell to the magnetic field. The results showed that the waves are situated perpendicularly to the flow of the electrolyte and moves in the direction of the flow. The observed instability was called "flow-induced electrochemical instability" [52]. The appropriate combination of the different factors during electrodeposition and the resulting velocity of the wave propagation probably play a very important role in the formation of the periodic structures as well. The waves in the case of the Ag–Sb spatiotemporal structures move with a velocity of several micrometers per second, and the rotation speed of the spiral structures is of about 8° per second [40].

7.2.4 *Electrodeposition of Alloys with Periodically Changing Currents*

7.2.4.1 Electrodeposited Au–Sn Alloy

Au–Sn eutectic solders ("hard solder" with superior mechanical and thermal properties relative to "soft" solders, such as the Pb–Sn system) are commonly used in the optoelectronic and microelectronic industries for chip bonding to dies [53]. In order to avoid the use of cyanide solutions, chloride solutions were developed at the beginning of 1990. The most stable one was found to be solution containing [53] 200 g dm^{-3} $(NH_4)_3C_6H_5O_7$, 5 g dm^{-3} $KAuCl_4$, 60 g dm^{-3} Na_2SO_3, 15 g dm^{-3} L-ascorbic acid, and 5 g dm^{-3} $SnCl_2 \times 2H_2O$. This solution was used to investigate the morphology of the electrodeposits obtained by direct (DC) and pulsating (PC) current regimes. Taking into account that Au belongs to the intermediate metals, while Sn belongs to normal metals, it could be expected that PC regimes would produce less rough electrodeposits. Due to low concentrations of Au and Sn

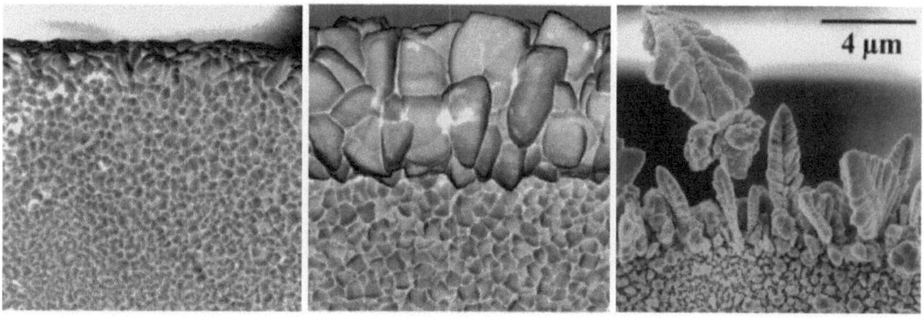

Fig. 7.29 SEMSE images showing edge effects in PC plating at different average current densities. The on-time and off-time for PC plating are 2 ms and 8 ms, respectively (Reprinted from Ref. [54] with kind permission from Springer)

ions, small values of average cathodic current densities (up to $-3.4\ \text{mA cm}^{-2}$) were used in DC and PC plating, varying on- and off-times, and average current density in PC plating. Electrodeposits obtained at low current densities (from -1.6 to $-2.0\ \text{mA cm}^{-2}$) were similar for both types of plating with practically no influence on their microstructure. At slightly higher current densities (from -2.0 to $-3.2\ \text{mA cm}^{-2}$), both DC and PC electrodeposits showed a tendency toward coarser microstructures, with this effect being more pronounced for DC electrodeposits. The cathodes were either InP or Si wafers, coated with Ti (25 nm)/Au (250 nm) blanket metallizations. Wafers were sectioned into smaller pieces, each having an exposed area ~1 cm^2 defined by stop-off lacquer. The increase of the average current density in PC regime caused coarser grain size all over the electrodeposit, while this increase was the most pronounced on the appearance of the edge of electrodeposit, as shown in Fig. 7.29. The actual current density at the edge is higher than the average current density, approaching the limiting current density and the electrodeposits at the edge exhibit microstructures characteristic of limiting current conditions (coarse grains and even dendritic growth, $i = -3.2\ \text{mA cm}^{-2}$) [54].

The on-time period showed the most significant effect on the microstructure characteristics of the Au–Sn electrodeposits. As can be seen on the SEM micrographs presented in Fig. 7.30, the sample electrodeposited with 1 ms on-time possessed much finer microstructure than the other coatings. This is caused by two factors: composition and grain size. High Au content or low Sn content (16.7 at.%) relative to the others (content of Sn higher than 33 at.%) and lower roughness, which was found to increase with increasing on-time. At longer on-times, the average current density increases leading to thicker and coarser electrodeposits. At 4 ms of on-time, the microstructure approaches that obtained from DC plating [54]. Cleaved cross sections of the electrodeposits in Fig. 7.30 are shown in Fig. 7.31. The 1 ms on-time sample exhibits ductile fracture, most probably due to its high Au content. Electrodeposits obtained at 2 and 3 ms of on-time are dense and uniform and adhere well to the substrate, with the microstructures similar to typical solidified cast

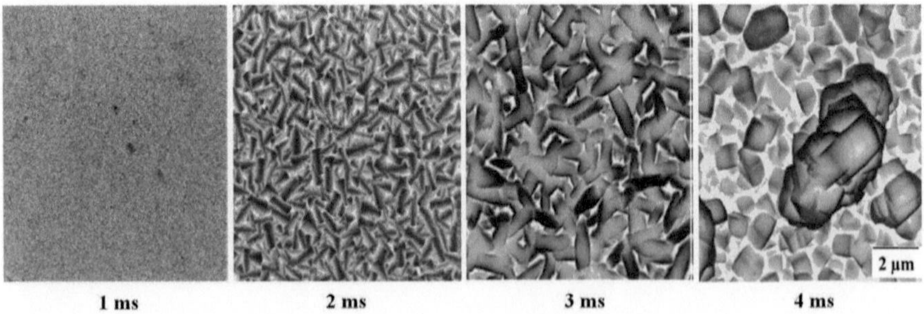

Fig. 7.30 SEMSE images of electrodeposits obtained at different on-times, with a constant peak current density of -10 mA cm^{-2} and an off-time of 8 ms (Reprinted from Ref. [54] with kind permission from Springer)

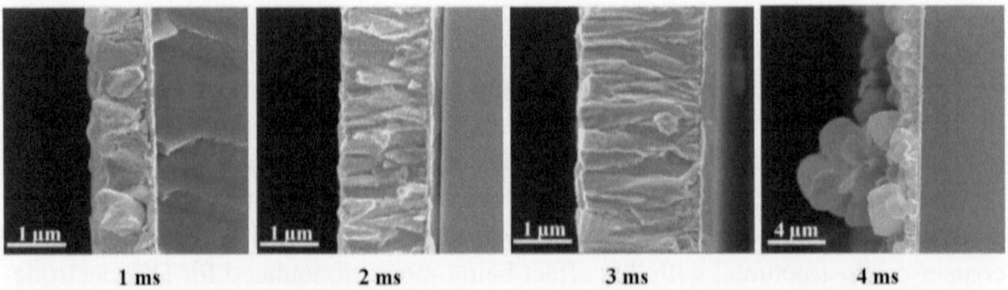

Fig. 7.31 SEM cleaved cross-sectional images of electrodeposits obtained at different on-times, with a constant peak current density of -10 mA cm^{-2} and an off-time of 8 ms (Reprinted from Ref. [54] with kind permission from Springer)

structures (the initial electrodeposit is fine grained and subsequent grains growing in a columnar manner). The 4 ms on-time sample is very rough – its thickness varies from 1.4 to 8.9 μm – which is due to the high average current density (-3.3 mA cm^{-2}) approaching the limiting current density. Generally speaking, on-times of 1–3 ms produce suitable microstructures and reasonable plating rates (0.7–1.7 μm h^{-1}). Higher plating rates could be achieved by increasing the Au and Sn contents in the plating bath (increasing the limiting current density).

7.2.4.2 Electrodeposited Co–Cu Alloy

This is one example of codeposition of one intermediate (Cu) and one inert (Co) metal. The electrolyte for electrodeposition [55] contained 0.7 M $CoSO_4 \cdot 7H_2O$, 0.025 M $CuSO_4 \cdot 5H_2O$, 0.18 M $Na_3C_6H_5O_7 \cdot 2H_2O$, and 1:50 parts of 0.04 % premixed FC99 fluorinated surfactant from 3 M® with the pH 4.3. The effect of pulse off-time on the phase distribution of Co–Cu alloys pulse-plated on the IrRDE (inverted recessed rotating disk electrodes) was studied in order to get additional information on the role of the displacement reaction.

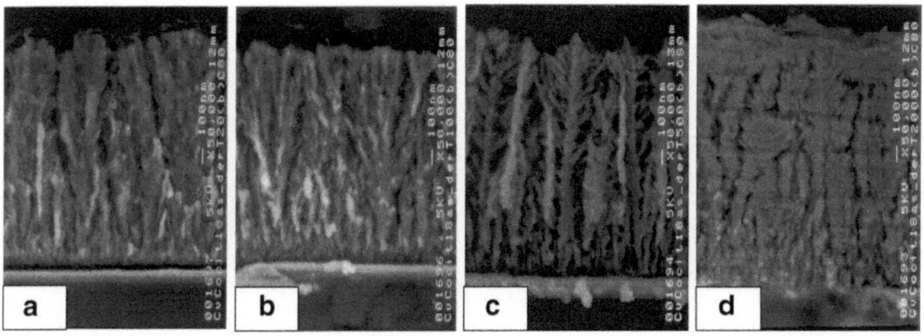

Fig. 7.32 SEM pictures of Co–Cu alloy electrodeposit cross sections after selective chemical etching of the Cu phase. Bright parts represent the Co-rich phase, while dark parts represent the etched Cu phase. Off-time (ms): 16 (**a**), 80 (**b**), 400 (**c**), and 1600 (**d**) (Reprinted from Ref. [55] with the permission of Elsevier)

The experimental conditions were pulse off-times 16, 80, 400, and 1600 s, duty cycle 0.2, pulse current density $i_{on} = -100$ mA cm^{-2}, current density during off-time $i_{off} = -0.2$ mA cm^{-2}, and RPM = 1400. Under such hydrodynamic conditions, a limiting current density for copper electrodeposition was $i_L(Cu) = -9.7$ mA cm^{-2}. The number of cycles (depending on the on-time and off-time) was 11022, 2204, 441, and 110, respectively, in order to obtain approximately 1.3 mm thick electrodeposits. The SEMs of etched cross sections presented in Fig. 7.32 illustrate the phase distribution obtained in the experiments. Except for the electrodeposit shown in Fig. 7.32d, which had the lowest Cu content, the electrodeposits exhibited a columnar structure. The Co alloy columns apparently go through the entire electrodeposit, indicating that atoms are added to existing growth sites in each phase [55].

7.3 Multilayered Structures

7.3.1 Introduction

According to the theory of composite systems [56], the existence of intermediate layer between two metals is inevitable in multilayered laminar metal structures. The thickness of the intermediate layer is usually of the order of couple atomic layers, and the parameters of the system (concentration of individual metals, crystal structure, Young's modulus, density, coefficient of thermal expansion, etc.) change from one layer to another. Usually such intermediate layer is composed of two layers of solid solutions and one layer of intermetallic compound, as it is schematically presented in Fig. 7.33.

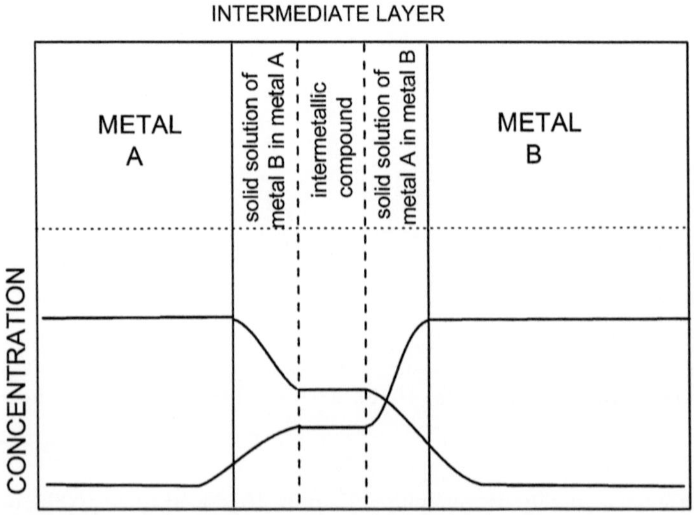

Fig. 7.33 Schematic representation of the intermediate layer in multilayered laminar metal structures (Reprinted from Ref. [5] with kind permission from Springer)

By using appropriate techniques for the formation of such structures (chemical or physical evaporation, electrodeposition, etc.), it is possible to achieve high value of total surface area of the intermediate layer amounting to 3000 cm^2 cm^{-3}. Accordingly, the larger the surface of the intermediate layer, the higher the difference in the properties of the multilayered laminar metal structure (higher values of Young's modulus, hardness, tensile strength, etc.) [56]. Taking into account that the electrodeposition of such structures is the cheapest technique, its advantage over the other techniques is obvious.

It should be emphasized here that multilayered laminar metal structures could be obtained by electrodeposition of individual metal layers from separate baths, which is a time-consuming process. On the other side, it is possible to obtain such electrodeposit from one bath containing ions of both metals by appropriate choice of metal ion concentration and applied current density. Hence, by appropriate choice of electrodeposition conditions, one can adjust the composition and the thickness of each layer and, accordingly, properties of such structures.

7.3.2 Electrodeposition of Laminar Metal Structures

As already stated above, multilayered laminar metal structures could be obtained by electrodeposition of individual metal layers from separate baths or from one bath containing ions of both metals. Only the second case is of interest for this chapter and such conditions will be discussed [1, 5].

7.3.2.1 Spontaneous Formation of Layered Deposits

In the investigations of electrodeposition of alloys, laminar metal structures, parallel to the cathode surface, were observed in quite a few instances [1, 5]. This phenomenon has been accompanied with the fluctuations of potential ranging from 0.1 mV to 0.1 V depending on the system and applied current density. Any fluctuations of periodic nature (oscillations) must be the consequence of superposition of two counteracting processes. Accordingly, this phenomenon could be explained in a following way: At the beginning of the process under galvanostatic conditions, the rate of electrodeposition of the more noble metal is sufficient for its nuclei to be formed and its grains to grow. However, after some time, due to depletion of the diffusion layer, the concentration of the ions of the more noble metal at the cathode surface becomes practically zero, and the discharge of the ions of the less noble metal must compensate for the decrease of the partial current density of the more noble metal. This requires a certain increase in cathodic overpotential and nucleation and growth of the less noble metal mainly. After some time, the concentration of the ions of the more noble metal at the cathode surface recovers to the point where its nucleation can start again, causing decrease of the cathodic overpotential and, accordingly, the nucleation and growth of the less noble metal subsides. Hence, the process starts all over again.

7.3.2.2 Formation of Laminar Deposits by Pulsating Current Regimes

Instead of letting laminar metal structures to be formed spontaneously during the alloy electrodeposition, there was significant motivation to obtain well-defined multilayered structures of controlled compositions and thicknesses of individual layers. Electrodeposition of such structures was first demonstrated by Brenner [3], while Cohen et al. [57] obtained an Ag–Pd alloy with periodically changing composition by alternating the current density and such structure showed improved wear performance of electrical contacts. In order to obtain laminar metal structure with the thickness of individual layers (with the second layer being Cu–Ni alloy with small amount of Ni) of 300 nm in the system Cu–Ni, pulsating regimes were used [58, 59]. It was shown that such structures possessed higher tensile strength [58] and microhardness [59].

In the application of pulsating regimes [57–59], three types of pulses, schematically presented in Fig. 7.34, were used.

The first attempt in theoretical treatment of the variation of the composition of the electrodeposit obtained by pulsating current from a bath containing two different metal ions, using complicated mathematical procedure, was presented in 1985 [60]. The quantitative theory of laminar metal electrodeposition based on fundamental concepts of electrode kinetics, using much simpler mathematical approach, was presented in a series of papers of Despić and co-workers [61–63].

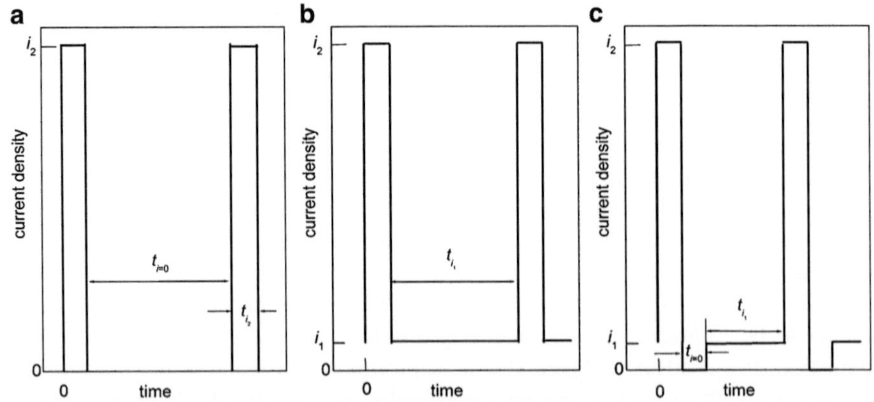

Fig. 7.34 Schematic presentation of pulsating regimes used for electrodeposition of laminar metal structures (Reprinted from Ref. [5] with kind permission from Springer)

Single Current Pulse Regime

For the electrodeposition of laminar metal coatings, two conditions must be fulfilled: (1) The reversible potentials for metals A and B must be sufficiently different so that at a given current density, the less noble one (B) virtually does not electrodeposit during the electrodeposition of the more noble one (A) until complete concentration polarization with respect to ions of metal A takes place; (2) within the duration of the current density pulse, Send's equation [15] for diffusional polarization is obeyed with respect to concentration change, resulting in transition from electrodeposition of metal A to electrodeposition of metal B after well-defined transition time.

Hence, the first layer contains pure metal A and its thickness is, according to Faraday's law, defined by equation

$$d(\mathrm{I}) = \left(\frac{M_{\mathrm{A}}}{z_{\mathrm{A}}\rho_{\mathrm{A}}F}\right) i\tau_{\mathrm{A}} \tag{7.24}$$

where M_A, ρ_A, and z_A are, respectively, atomic weight, density, and charge on the ions of metal A, and i is the current density in the pulse, while τ_A is the transition time with respect to the ions of metal A. Assuming that the convection and migration of ions of the electrolyte is negligible, τ_A is defined by Send's equation [15]:

$$\tau_{\mathrm{A}} = \frac{(z_{\mathrm{A}}F)^2 \pi D_{\mathrm{A}}}{4} \left(\frac{C_{\mathrm{A}}^{\mathrm{o}}}{i}\right)^2 \tag{7.25}$$

where $C_A{}^o$ and D_A are, respectively, the bulk concentration and diffusion coefficient of the ions of metal A. Introducing Eq. (7.25) into Eq. (7.24) and rearranging one obtains

$$\frac{\left(C_{\mathrm{A}}^{\mathrm{o}}\right)^2}{i} = \frac{4\rho_{\mathrm{A}}}{z_{\mathrm{A}}F\pi D_{\mathrm{A}}M_{\mathrm{A}}}\mathrm{d}(I) \tag{7.26}$$

Considering Eq. (7.26), it could be concluded that desired thickness of the layer of metal A could be achieved by proper choice of metal ion concentration and the current density of the pulse.

Taking into account that Sand's equation is valid only as long as the change of concentration occurs within a stagnant layer undisturbed by convection and introducing the Nernst diffusion layer boundary δ and hydrodynamic layer boundary Δh, the minimum current density that must be applied in the first pulse for electrodeposition of the second layer to take place is given by

$$i^{\min} = \frac{z_{\mathrm{A}}F\pi D_{\mathrm{A}}}{2\Delta h}C_{\mathrm{A}}^{\mathrm{o}} \tag{7.27}$$

and, accordingly, by introducing Eq. (7.26) into Eq. (7.27), the maximum achievable thickness of the first layer (metal A) is

$$\mathrm{d}(I)^{\max} = \frac{M_{\mathrm{A}}C_{\mathrm{A}}^{\mathrm{o}}\Delta h}{2\rho_{\mathrm{A}}} \tag{7.28}$$

Electrodeposition of the second layer starts after τ_{A} is reached and is accompanied by sudden change of potential to the reversible potential of metal B electrodeposition plus the overpotential for its electrodeposition [1, 5].

The shape of the potential versus time (*E* vs. *t*) response in such a case depends on the property of the alloy and its constituents in the investigated solution. Two situations could be considered: (*a*) replacement reaction takes place during the off-time ($i = 0$) period; (*b*) replacement reaction does not take place during the off-time ($i = 0$) period.

If the difference between the reversible potentials of metals A and B is sufficient, and the constituents of the alloy do not mix in the solid state (eutectic-type alloy), and/or electrodeposition/dissolution of metal B is reversible (case *a*), replacement reaction must take place during the off-time ($i = 0$). Such a case is schematically presented in Fig. 7.35. The current density change is presented in (a), while corresponding potential change is presented in (b). As can be seen, after reaching τ_{A}, metal A continues to electrodeposit but at a decreasing partial current density (actually by its diffusion limiting current density, $i_{\mathrm{d}}(\mathrm{A}) = i_{\mathrm{L}}(\mathrm{A})$), while the partial current density of the metal B, $i_{\mathrm{d}}(\mathrm{B})$, increases to make a constant current density of the pulse, i_{d}. If the concentration of ions of metal B is much larger than that of the metal A, the growth of the second layer could be virtually unlimited and the content of metal A in the second layer will continuously decrease with the thickness of the second layer (following the decrease of $i_{\mathrm{d}}(\mathrm{A})$). Corresponding potential response, schematically presented in (b), is characterized by a sudden increase of cathodic potential to the value higher than $E_{\mathrm{r}}(\mathrm{B})$ at the position of τ_{A}. At the position of the

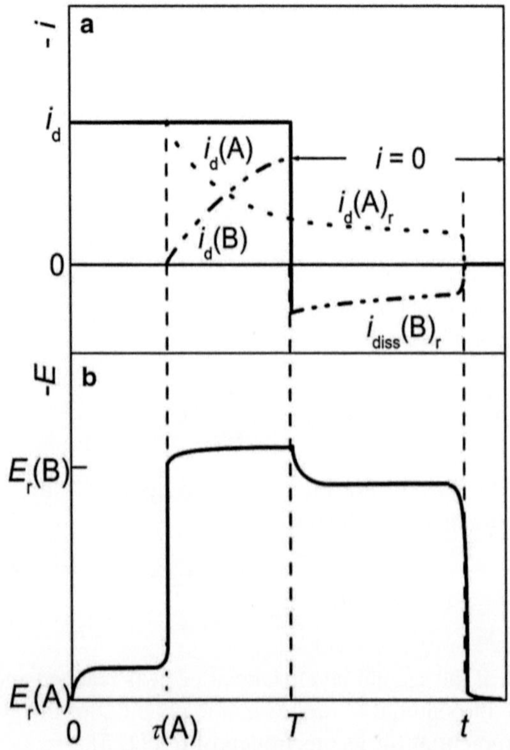

Fig. 7.35 Schematic representation of the partial current density changes (**a**) and corresponding potential response (**b**) during the electrodeposition of two-layer of metals A and A + B by constant current density pulse (i_d) up to time T and during the replacement reaction at $i = 0$. Partial current density for electrodeposition of metal A after reaching τ_A, $i_d(A)$; partial current density for electrodeposition of metal B after reaching τ_A, $i_d(B)$; partial current density for electrodeposition of metal A during the replacement reaction, $i_d(A)_r$; partial current density for dissolution of metal B during the replacement reaction, $i_{diss}(B)_r$ (Reprinted from Ref. [5] with kind permission from Springer)

end of the current pulse (T) and the beginning of the current density off period ($i = 0$), replacement reaction will take place. The electrodeposition of metal A will continue with $i_d(A) = i_L(A) = i_d(A)_r$, while, in order to keep total current density at zero, the less noble metal B will start to dissolve with the positive partial current density $i_{diss}(B)_r$ equal to that of metal A electrodeposition. Once this process is finished, either due to dissolution of the whole amount of metal B in the second layer or prevention of its dissolution for some reasons (e.g., at some places of a second layer atoms of metal B become covered with a monolayer of metal A), the current density $i_{diss}(B)_r$ will drop to zero and corresponding potential to the value of $E_r(A)$ [1, 5].

Although the replacement reactions have extensively been studied [64], the only example for such reaction in the layered electrodeposits was demonstrated for the system Cu–Pb [1, 5, 61]. A two-layer structure, composed of a first layer of pure Cu

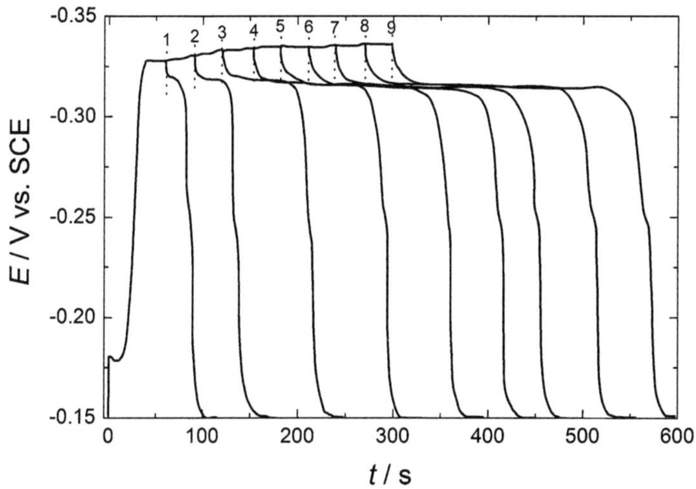

Fig. 7.36 Potential responses recorded during constant current density $i_d = -1$ mA cm^{-2} pulse trains on a stationary glassy carbon electrode from a solution containing 0.01 M Cu $(CH_3COO)_2 + 0.01$ M $Pb(CH_3COO)_2 + 1$ M HBF_4. After the electrodeposition, zero current density ($i = 0$) was applied (positions marked in the figure with 1–9) (Reprinted from Ref. [5] with kind permission from Springer)

and a second layer of Cu–Pb alloy, has been electrodeposited at a constant current density $i_d = -1$ mA cm^{-2} for different times (from 60 to 300 s) on a stationary glassy carbon electrode [61] from a solution containing 0.01 M Cu $(CH_3COO)_2 + 0.01$ M $Pb(CH_3COO)_2 + 1$ M HBF_4. After the electrodeposition, zero current density ($i = 0$) was applied (positions marked in the figure with 1, 2, 3, 4, 5, 6, 7, 8, and 9) and corresponding potential responses are presented in Fig. 7.36. As can be seen after applying zero current density (positions 1–9), the potential remains for some time at a level slightly below the reversible potential of Pb, after which it falls to the potential of Cu. That time is virtually equal to duration of the current density pulse, indicating that the dissolution of Pb from a second layer occurs all the time. When the dissolution process is finished, potential returned to the reversible potential of Cu. By the EDS analysis, it was confirmed that the whole amount of electrodeposited Pb has been dissolved during the off-time periods. Hence, all conditions (*a*) for the replacement reaction to take place were fulfilled: Cu–Pb is eutectic-type alloy and electrodeposition/dissolution of Pb is reversible [1, 5, 61].

If the difference between the reversible potentials of metals A and B is sufficient, and the constituents of the alloy mix in the solid state forming solid solution and a metal B passivates in the electrolyte used (case *b*), replacement reaction will not take place during the off-time ($i = 0$). Such a case is schematically presented in Fig. 7.37. The current density change is presented in (a), while corresponding potential change is presented in (b). During the current density pulse, everything is the same as in a previous case. The absence of replacement reaction is

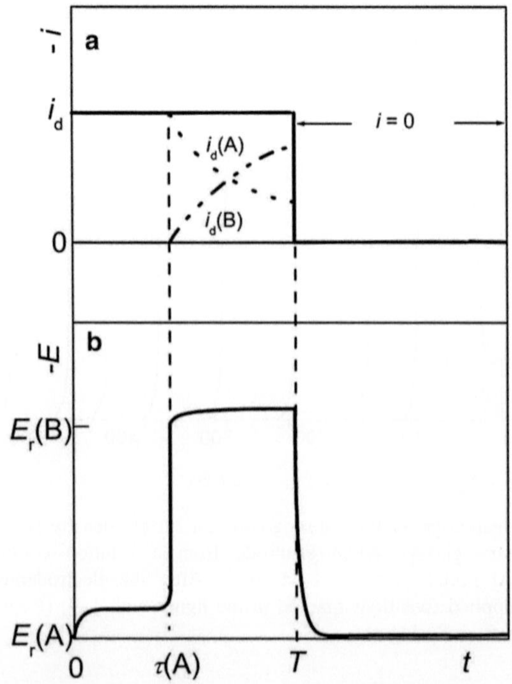

Fig. 7.37 Schematic representation of the partial current density changes (**a**) and corresponding potential response (**b**) during the electrodeposition of two layers of metals A and A + B by constant current density pulse (i_d) up to time T and after, in the absence of the replacement reaction at $i = 0$. Partial current density for electrodeposition of metal A after reaching $\tau_A - i_d(A)$, partial current density for electrodeposition of metal B after reaching $\tau_A - i_d(B)$ (Reprinted from Ref. [5] with kind permission from Springer)

characterized by sudden potential change from E_r(B) to E_r(A) at the position of the end of the current pulse (T) and the beginning of the current density off period ($i = 0$) [5, 58].

Typical example for such behavior is the system Cu–Ni [58, 59, 63, 65, 66]. This system belongs to the solid solution-type alloys, and the dissolution of Ni is prevented by its passivation in the electrolyte for electrodeposition.

Dual Current Pulse Regime

Single current pulse regime imposes serious limitations on both the thickness of the layers of the more noble metal and the content of that metal in the second layer. The application of a dual current pulse scheme was found more suitable for manipulation of the thickness of each layer, as well as the content of the more noble metal in the second layer. The scheme consists of current pulses in two different intensities each. A lower current density i(I) for a certain time period T(I) is followed by a higher current density $i(\mathrm{II})$ for a period $T(\mathrm{II})$ before the current is interrupted (or the sequence repeated). It can be shown that with such a scheme and proper choice of parameters, any desired thickness of both layers can be achieved and the content of metal A in the second layer can be reduced to a desired level. The process can be improved, in terms of shortening the electrodeposition time of the first layer while maintaining a desired content of metal A in the second layer, by synchronously modulating the hydrodynamic conditions. Such a process can have significant practical value for obtaining bilayers (or multilayers) [1, 5, 62].

In the dual current pulse regime, it is desirable to have

$$C_{\mathrm{B}}^{\mathrm{o}} >> C_{\mathrm{A}}^{\mathrm{o}} \tag{7.29}$$

the best choice being that which satisfies equality

$$\frac{C_{\mathrm{B}}^{\mathrm{o}}}{i(\mathrm{II})} = \frac{C_{\mathrm{A}}^{\mathrm{o}}}{i(\mathrm{I})} \tag{7.30}$$

An attempt was made to electrodeposit two-layer structure composed of pure (first) Cu layer of about 2 μm and the second layer of different Cu–Ni alloy compositions of a thickness of about 12 μm by applying dual current pulse regime presented in Fig. 7.38. The two layers were deposited onto Ag rotating disc electrode (RPM = 1000) from the solution 0.02 M $CuSO_4$ + 2 M $NiSO_4$ + 0.5 M $Na_3C_6H_5O_7$. The value of the current density in the first pulse, $i(\mathrm{I})$, was adjusted at 75 % of the value of the diffusion limiting current density for copper ions at the given concentration and rotation speed [62].

Relatively smooth electrodeposits, consisting of two well-defined layers, were obtained. Results of the EDS analysis of the composition of the second layer (Cu–Ni alloy) of the electrodeposits obtained by the current regimes shown in Fig. 7.38 are compared in Table 7.4 with the composition of the second layer calculated from the current density ratio (c.r.) in the second pulse assuming 100 % current efficiency for alloy electrodeposition, c.r. – $i(\mathrm{Ni})/i(\mathrm{Cu})_{\mathrm{L}}$. Taking into account precision of the EDS analysis, it is seen that a relatively good agreement is obtained.

A multilayer structure consisting of ten relatively thick layers shown in Fig. 7.39 was obtained by the current regime presented in Fig. 7.40.

Presented analysis confirmed that the thickness and the composition of layered electrodeposits could be controlled with sufficient precision [62].

7.3.2.3 Intermediate Layer Detection

Taking into consideration the fact that the intermediate layer in electrodeposited laminar metal structures plays an important role in obtaining improved mechanical and magnetic properties [56], from 1980 to 2000, particular attention has been given to the pulsating regimes which would provide as thin as possible layers (of the

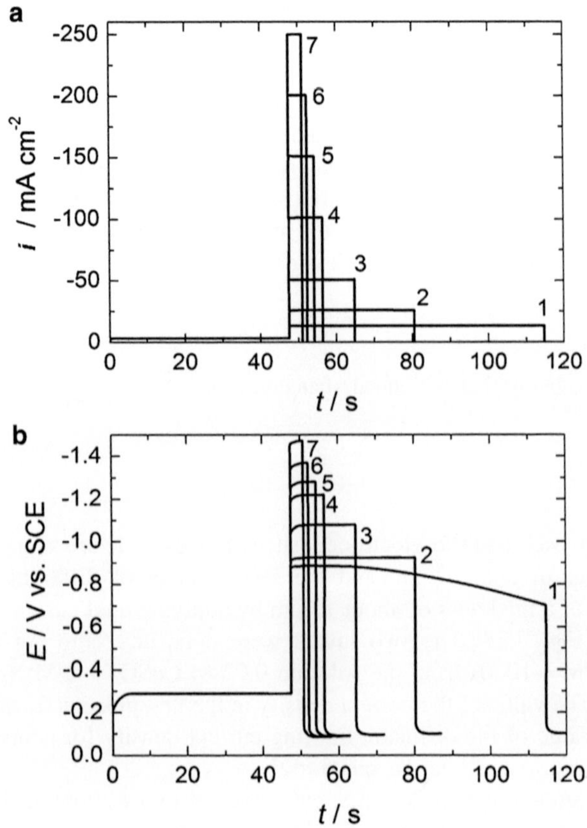

Fig. 7.38 A sequence of high current density pulses (**a**) and corresponding potential responses (**b**) (marked with numbers 1–7) during the formation of two-layer structure composed of a pure (first) Cu layer of about 2 μm and the second layer of different Cu–Ni alloy compositions of a thickness of about 12 μm. The current density ratio, c.r. $-i(Ni)/i(Cu)_L$, for pulse no. 1–2.0, no. 2–5.3, no. 3–13.3, no. 4–27.6, no. 5–36.5, no. 6–56.1, no. 7–61.5 (Reprinted from Ref. [5] with kind permission from Springer)

Table 7.4 The composition of the second layer of the electrodeposits as a function of the current density ratio (c.r.)

		Composition of the second layer			
		EDS analysis		Calculated from c.r.	
Pulse no.	c.r.	at. % Cu	at. % Ni	at. % Cu	at. % Ni
1	2.0	35.4	64.6	33.3	66.7
2	5.3	20.8	79.2	16.0	84.0
3	13.3	7.9	92.1	7.0	93.0
4	27.6	4.9	95.1	3.5	96.5
5	36.5	4.7	95.3	2.7	97.3
6	56.1	3.9	96.1	1.7	98.3
7	61.5	4.2	95.8	1.6	98.4

Reprinted from Ref. [5] with kind permission from Springer

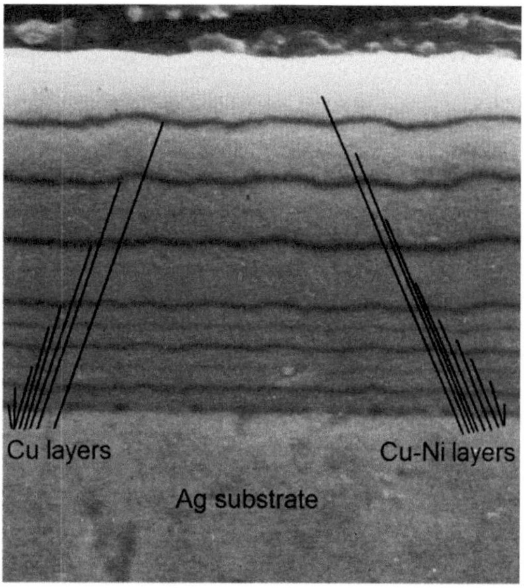

Fig. 7.39 Multilayer structure obtained by the current pulses presented in Fig. 7.40 (Reprinted from Ref. [5] with kind permission from Springer)

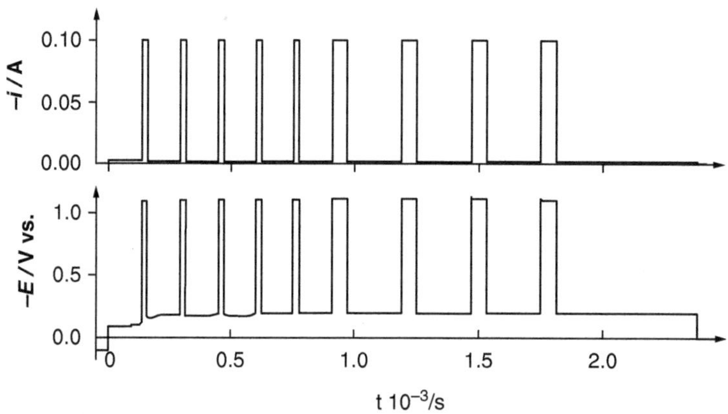

Fig. 7.40 Current density pulses and corresponding potential responses applied for the formation of multilayer structure presented in Fig. 7.39 (Reprinted from Ref. [5] with kind permission from Springer)

order of nanometers) [65–69] in multilayered metal structures, and, accordingly, much higher surface area of the intermediate layer.

In theoretical consideration of multilayer electrodeposition of Cu–Ni alloy under the conditions of convective diffusion [63], it was shown that a constant concentration of Cu in the second layer (Cu–Ni alloy) could be established at different thicknesses of the second layer, depending on the value of the current density of a second pulse and the rotation rate. As the value of the current density in the second pulse increases the thickness at which a constant concentration of Cu could be established also increases, indicating that the thickness of the intermediate layer also increases, reaching the maximum value of about 150 nm [63]. Unfortunately, this theoretical consideration was not experimentally confirmed. Most of the experimentally confirmed theoretical treatments of the pulsed electrodeposition of multilayered structure have been applied to the Cu–Ni system [70–73]. Equations for partial currents for Cu and Cu–Ni electrodeposition [71] (including the evolution of hydrogen) from a single bath in the case of galvanostatic [70] and potentiostatic [71] electrodeposition of multilayered Cu–Ni coatings under the conditions of convective diffusion were developed. The experimental results were in good agreement with the theoretical predictions. The subject of two papers [72, 73] was the process taking place during the off-time period in pulsed electrodeposition, showing that during the off-time period Cu electrodeposits by a displacement of Ni ("replacement reaction"). Their models were found to agree well with the experimental findings. Hence, the existence of an intermediate layer in these investigations has not been confirmed experimentally.

However, it was shown in the system Cd–Ni [74] that the intermediate layer could be detected by the application of the ALSV technique. Alloy samples used for the ALSV analysis were galvanostatically electrodeposited at the amount of charge of -1 C cm^{-2} in the solution containing 2 M $NiSO_4 + 0.2$ M $Na_3C_6H_5O_7 +$ 0.002 M $CdSO_4$. Pure Ni and pure Cd layers were also electrodeposited galvanostatically at $i_d = -10$ mA cm^{-2} ($Q_d = -1$ C cm^{-2}) from the solutions containing 2 M $Ni(NH_2SO_3)_2 + 0.5$ M H_3BO_3 and 0.5 M $CdSO_4 + 0.01$ M H_2SO_4, respectively. The samples composed of a layer of alloy and a layer of pure Cd on top of the alloy layer, used for the analysis of the intermediate layer existence, were galvanostatically deposited from two different baths: 2 M $NiSO_4 + 0.2$ M $Na_3C_6H_5O_7 +$ 0.002 M $CdSO_4$ and 2 M $Ni(NH_2SO_3)_2 + 0.5$ M $H_3BO_3 + 0.002$ M $CdSO_4$. All samples were electrodeposited onto Au disc electrode at RPM = 1000 and were dissolved in a solution of 1 M NaCl (pH 2) at RPM = 0.

(a) *Intermediate layer between Ni and Cd electrodeposited from a separate baths*: For this purpose, layers of pure Cd and pure Ni were electrodeposited from the appropriate separate baths onto gold disc electrode. To provide normal dissolution of these metals, Ni was first electrodeposited as a more noble metal and a layer of Cd was electrodeposited on top of it. From such electrodeposit, Cd dissolves at about −0.8 to −0.7 V versus SCE, while Ni starts to dissolve at about −0.33 V versus SCE with the peak appearing at about −0.1 V versus SCE.

Hence, to detect an intermediate layer between these two metals, ALSVs of pure Ni and pure Cd dissolution were first recorded. Then, after the electrodeposition of a layer of Cd on top of a layer of Ni, an ALSV of dissolution of such electrodeposit consisting of two metal layers was recorded. Since these two metals make two intermediate phases [8], in the case of interaction between the two layers, one would expect one or two small ALSV peaks, pertaining to the dissolution of either of these phases, between the peaks of dissolution of pure metals. There was no indication, however, of the existence of any additional ALSV peak even after prolonged thermal treatment (deposit was kept in an atmosphere of purified nitrogen at a temperature of 50 °C for 24 h) of such deposit. Knowing that the ALSV technique is very sensitive and that it is possible to detect clearly a monolayer of metals by this technique, it appeared that in the case of Ni and Cd layers, electrodeposited from a separate baths, no intermediate layer between them has been formed.

(b) *Intermediate layer between Cd–Ni alloy and pure Cd electrodeposited from a single bath*

The system Cd–Ni should be very suitable for the detection of intermediate layer since the ALSVs of Cd–Ni alloy dissolution were seen to be characterized by the existence of two ALSV peaks corresponding to the dissolution of phases β and γ of the formula Cd_5Ni and CdNi, respectively [8, 74]. Hence, if there exists an intermediate layer, it should be composed of these phases [1, 5, 74]. By comparing ALSV of dissolution of a single layer of the alloy with the ALSV of dissolution of the two-layer electrodeposit (Cd–Ni alloy layer with Cd layer on top of it), it was possible to detect an intermediate layer between these two layers. An example is shown in Fig. 7.41. Curve 1 represents the ALSV of the alloy dissolution without the Cd layer on top, curve 2 the ALSV of pure Cd dissolution, while curve 3 represents the ALSV of the two-layer electrodeposit dissolution. As can be seen, curves 1 and 3 are different, with the peaks of β and γ phases dissolution being somewhat higher on the curve 3, while the peak of pure Cd dissolution did not change. As the content of Ni in the alloy layer increased, the difference between these two ALSV curves became smaller.

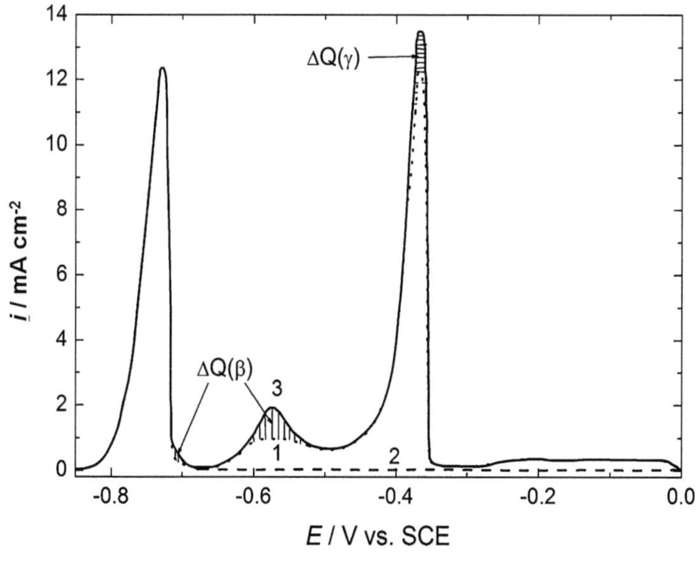

Fig. 7.41 The example of the ALSVs of Cd–Ni alloy dissolution (*dotted line*, *1*), pure Cd dissolution (*dashed line*, *2*), and dissolution of a two-layer electrodeposit (*solid line*, *3*) composed of a layer of Cd–Ni alloy and a layer of pure Cd electrodeposited on top of a layer of the alloy (Reprinted from Ref. [5] with kind permission from Springer)

Hence, the difference between curves 3 and 1 (shaded areas in Fig. 7.41) should represent the intermediate layer. By integration of the surface of shaded areas, it was possible to obtain the charges (amounts) corresponding to the increase of β and γ phases in the two-layer electrodeposit (marked in the figure as $\Delta Q(\beta)$ and $\Delta Q(\gamma)$), which should be a consequence of the formation of an intermediate layer between a layer of alloy and a layer of pure Cd. Thus, the intermediate layer is mainly composed of β and γ phases, which is in good agreement with the theory of an intermediate layer in laminar metal structures [56].

These investigations were performed with the two-layer electrodeposits obtained from two different baths (as stated above). The recorded charges $\Delta Q(\beta)$ and $\Delta Q(\gamma)$, obtained for both investigated baths for alloy electrodeposition, are shown in Fig. 7.42 as a function of the alloy layer composition. As can be seen in the figure, the intermediate layer composition is slightly different in the two cases. The amount of the β phase in the intermediate layer is seen to reach its maximum value at about the same alloy composition (30–40 at.% Ni) in both cases, while the contribution of the γ phase in the intermediate layer was different in two-layer electrodeposits obtained from different baths. Such a behavior clearly indicates that the composition of an intermediate layer is sensitive to the composition of the electroplating bath, since the conditions of electrodeposition were identical in both cases [5, 74].

Assuming that both phases are homogeneously distributed in the intermediate layer, the intermediate layer thickness can be calculated from the charges recorded on the ALSVs. From the formula of the β phase (Cd_5Ni), one can determine the charge corresponding to Ni as 0.167 and to Cd as 0.833 of the total charge $\Delta Q(\beta)$

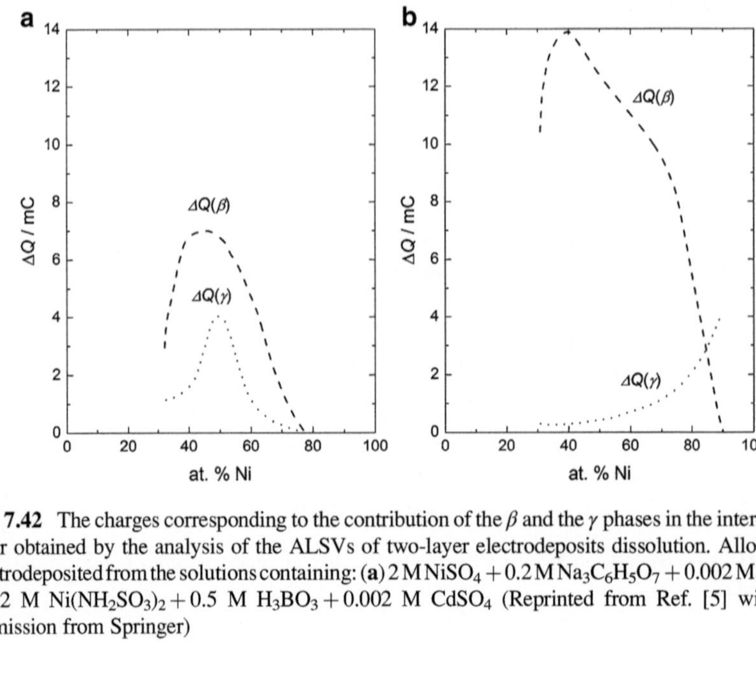

Fig. 7.42 The charges corresponding to the contribution of the β and the γ phases in the intermediate layer obtained by the analysis of the ALSVs of two-layer electrodeposits dissolution. Alloys were electrodeposited from the solutions containing: (**a**) 2 M $NiSO_4$ + 0.2 M $Na_3C_6H_5O_7$ + 0.002 M $CdSO_4$; (**b**) 2 M $Ni(NH_2SO_3)_2$ + 0.5 M H_3BO_3 + 0.002 M $CdSO_4$ (Reprinted from Ref. [5] with kind permission from Springer)

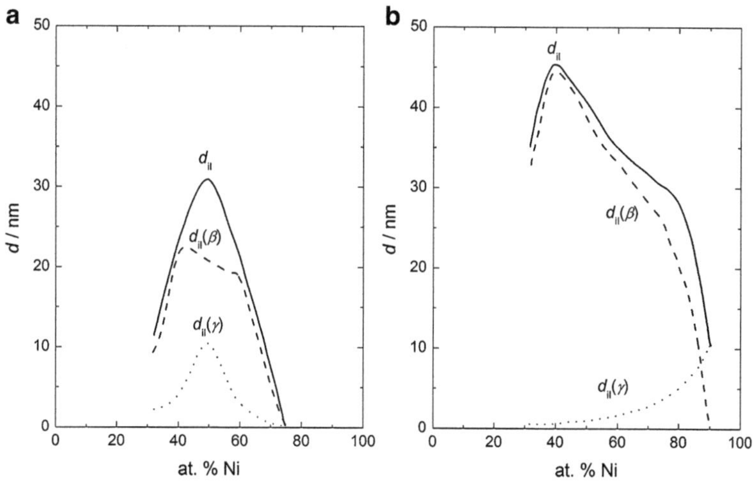

Fig. 7.43 The thickness of the individual phases, $d_{il}(\beta)$ and $d_{il}(\gamma)$, and the intermediate layer, d_{il}, obtained by the analysis of the results presented in Fig. 7.42. (**a**) Samples electrodeposited from the solution 2 M $NiSO_4$ + 0.2 M $Na_3C_6H_5O_7$ + 0.002 M $CdSO_4$; (**b**) Samples electrodeposited from the solution 2 M $Ni(NH_2SO_3)_2$+ 0.5 M H_3BO_3 + 0.002 M $CdSO_4$ (Reprinted from Ref. [5] with kind permission from Springer)

and in a similar way for the γ phase (Ni 0.5 and Cd 0.5 of $\Delta Q(\gamma)$). Using these charges and corresponding values of the density and molecular weights of the two metals, the thicknesses of β and γ phases in the intermediate layer can be calculated. Hence, the total thickness of the intermediate layer should represent the sum of the thickness of individual phases. Figure 7.43 shows the dependence of the thickness of β and γ phases and the total thickness of an intermediate layer (d_{il}) as a function of alloy composition, obtained by the analysis of the results presented in Fig. 7.42. As can be seen, the thickness of the intermediate layer depends on the solution for alloy electrodeposition and its maximum value is about 45 nm. Hence, in certain systems, it is possible to estimate the thickness and the surface of intermediate layer by the application of ALSV technique [5, 74].

7.3.2.4 Mechanical and Magnetic Properties of Multilayered Structures

Mechanical properties of multilayered structures were only investigated for electrodeposits containing Cu/Cu–Ni layers. Following mechanical properties were mainly investigated: Young's modulus [58], hardness [75], and tensile strength [67, 75–77]. It was shown that all investigated properties depend on the thickness of the individual layers and that in all cases multilayered structures showed better properties than that of pure metals and/or their alloys.

The results of hardness investigation [75] are presented in Fig. 7.44. As can be seen, maximum hardness for the electrodeposit containing Cu/Ni-100 layers was

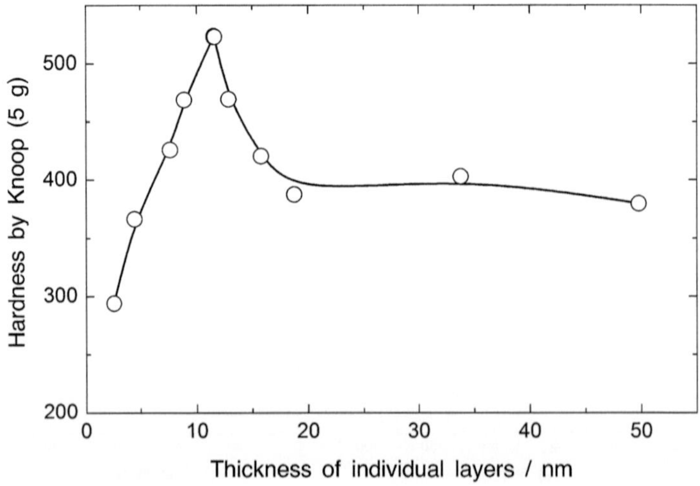

Fig. 7.44 The hardness of the electrodeposit containing Cu/Ni-100 layers as a function of the thickness of individual layers (Reprinted from Ref. [5] with kind permission from Springer)

achieved at the thickness of individual layers of 12 nm. These results are not in agreement with those detected for evaporated layers Cu/Ni-111 [76], where the maximum hardness was obtained for individual layer thickness of 1.6 nm, indicating that, most probably, the hardness depends on the thickness and the total surface area of the intermediate layer which is different for different techniques of multilayered structure formation.

In several investigations, the influence of the individual layer thickness on the tensile strength was reported [66, 75–77]. The results of these investigations are presented in Fig. 7.45. Significant difference in the obtained results is mainly the consequence of different parameters of the electrodeposition process (bath composition, pulse regimes, temperature, etc.). A common explanation for the decrease of tensile strength with the increase of individual layer thickness (after the maximum value) is the increase of the coherence of the intermediate layer. It is concluded that with the decrease of the number of dislocations in the intermediate layer, the probability of their transfer from it to previous or subsequent layer increases, resulting in the decrease of tensile strength.

The dependence of the tensile properties of multilayered Cu-(Ni–Cu) electrodeposits, with the nominal overall composition 90 at.% Ni–10 at.% Cu, was investigated in the work of Tench et al. [68] as a function of the Cu layer thickness (varying from 1 to 15 nm). Multilayers of the nominal thickness of about 50 μm were electrodeposited from a commercial sulfamate bath with the addition of 5 mM $CuSO_4$. Pure Cu layers were plated at −0.14 V versus SCE with cathode rotation at RPM = 750, while Ni layers (with 0.8 at.% Cu) were plated at $i = -105$ mA cm^{-2} from the stagnant bath. The results of the tensile property (ultimate tensile strength, UTS, yield strength, YS, modulus, and strain) measurements are presented in Table 7.5. Considering X-ray data, it was concluded that a decrease in the enhanced strength for such electrodeposits, producing peak at a Cu

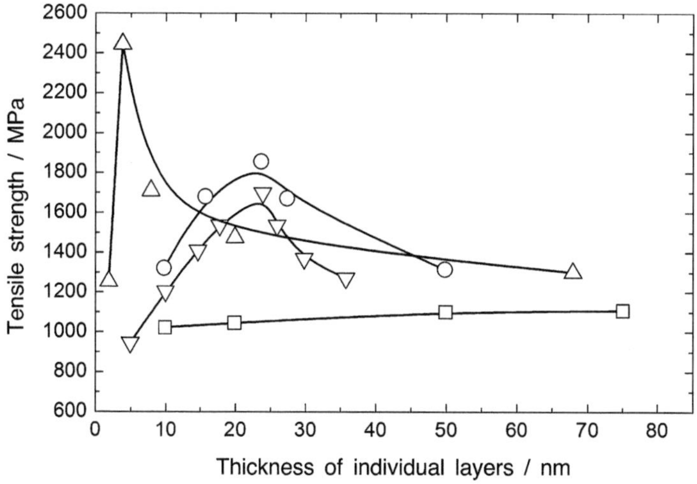

Fig. 7.45 The tensile strength of the multilayered electrodeposits as a function of the thickness of individual layers: ○ – Ref. [67], □ – Ref. [77], Δ – Ref. [78], ∇ – Ref. [68] (Reprinted from Ref. [5] with kind permission from Springer)

Table 7.5 Tensile test for electrodeposited 90 % Ni–10 % Cu multilayers versus Cu layer thickness

Cu layers (nm)	UTS (MPa)	YS (MPa)	Modulus (GPa)	Strain (%)
1.0	1021	795	160	2.4
2.0	1056	840	152	2.4
2.0	1029	881	141	1.4
5.0	1107	893	147	2.0
7.5	1116	844	154	2.1
7.5	1116	848	151	2.7
10.0	1069	807	145	2.8
12.5	978	789	154	1.5
15.0	863	724	130	1.2
		Averages	**148 (±18)**	**2.1 (±0.9)**

Reprinted from Ref. [5] with kind permission from Springer

layer thickness of about 7 nm, is associated with a decrease in the electrodeposit (100) texture (normally established for Ni layers) and an increase in the (110) orientation preferred for Cu deposits.

Concerning the magnetic properties, giant magnetoresistance (GMR), which is significant decrease of electric resistance in the presence of magnetic field, characteristic effect for multilayered structures, has been discovered in 1988 for the systems Fe/Cr/Fe and Fe/Cr by Grünberg and Fert (these scientists received Nobel Prize for physics in 2007). This phenomenon has also been recognized for electrodeposited multilayered structures [79–83]. The strongest effect is detected for multilayered Co–Cu/Cu [81] and Co–Ni–Cu/Cu [82, 83] structures, while less

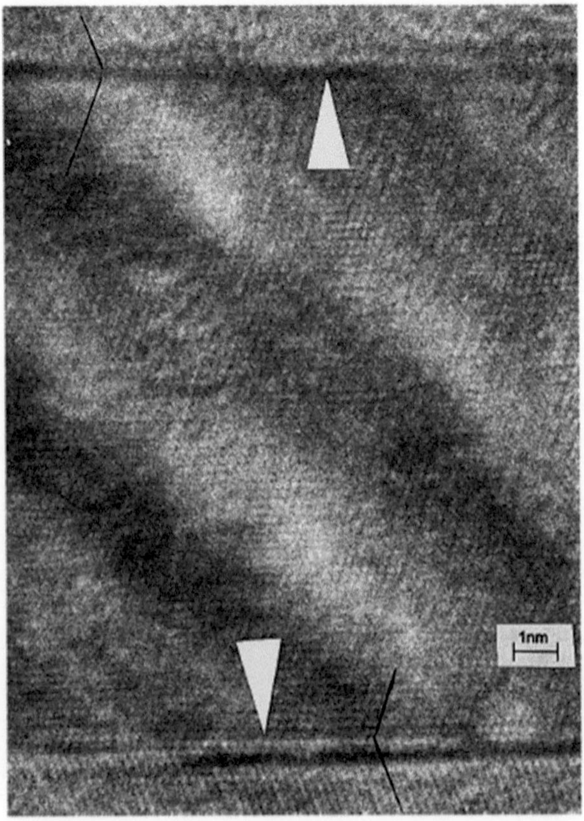

Fig. 7.46 HRTEM picture taken on the Ni–Cu(3.8 nm)/Cu(1.4 nm) sample which shows the atomic structure of a twin boundary (indicated by the *arrow*) and the magnetic/nonmagnetic layers (Reprinted from Ref. [5] with kind permission from Springer)

pronounced effect is detected for the multilayered Ni–Cu/Cu [81] structures. It is important to note that the presence of the intermediate layer of the thickness of 2–3 nm has been detected for the first time by high-resolution transmission electron microscopy (HRTEM) for the multilayered Ni–Cu/Cu structures [79], as shown in Fig. 7.46.

7.3.2.5 Multilayered Structures for Packaging Optoelectronic and Microelectronic Devices

Plating experiments for Au–Sn coatings were carried out from the same solution as in Ref. [53] on pieces sectioned from either metallized AlN (75×75 mm^2) or Si (100 mm diameter) substrates [84]. Examination of the Sn content versus current density revealed that there are two composition plateaus, one at current densities less than ~1.0 mA cm^{-2} and one at current densities greater than ~2.0 mA cm^{-2}. The former corresponds to a specific Au–Sn intermediate phase, disordered Au_5Sn (ζ), while the latter corresponds to the intermediate phase AuSn (δ) [85]. By utilization of two pulses, one at −0.8 mA cm^{-2} and one at −2.4 mA cm^{-2}, a multilayered structure presented in Fig. 7.47 has been obtained. The duty cycle for each pulse was maintained at 20 % (2 ms on and 8 ms off), but the duration for plating of each phase was varied to yield the desired composition. The plating times for layers shown in Fig. 7.47a were 20 min for −0.8 mA cm^{-2} and 5 min for −2.4 mA cm^{-2}.

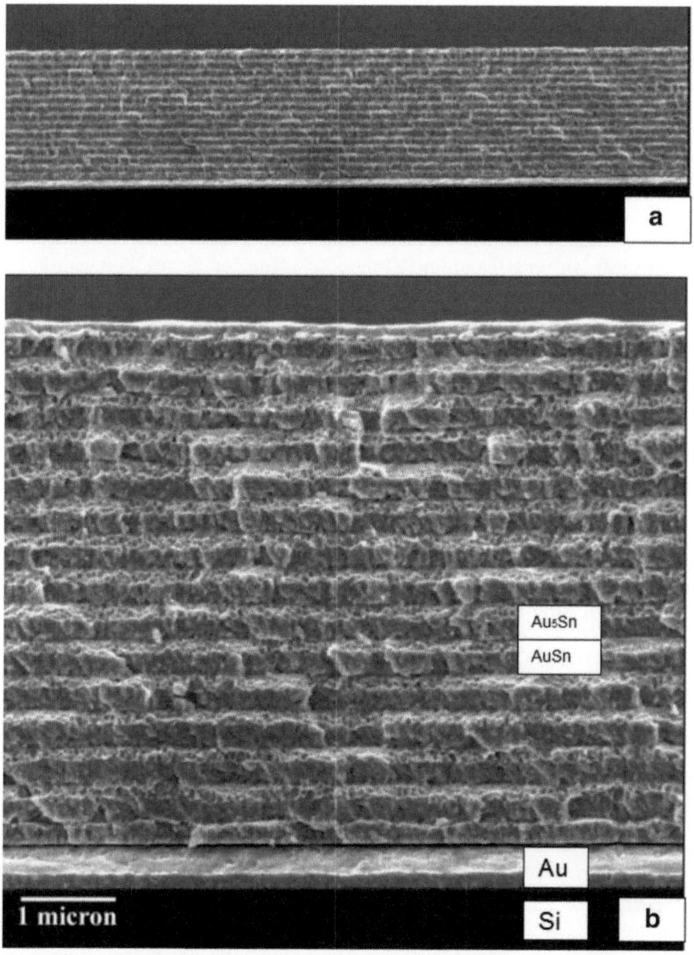

Fig. 7.47 Cross-sectional SEM (SE) images of multilayered electrodeposit at low (**a**) and higher magnification (**b**) (Reprinted from Ref. [85] with kind permission from Springer)

References

1. Despić AR, Jović VD (1995) Electrochemical deposition and dissolution of alloys and metal composites – fundamental aspects. In: White RE, Bockris JO'M, Conway BE (eds) Modern aspects of electrochemistry, vol 27. Plenum Press, New York, pp 143–232
2. Srivastava RD, Mukerjee RC (1976) Electrodeposition of binary alloys: an account of recent developments. J Appl Electrochem 6:321–331
3. Brenner A (1963) Electrodeposition of alloys: principle and practice. Academic, New York
4. Bondar VV, V. Grimina V, Pavlov VN (1980) Itogi nauki i tehniki, Elektrokhimiya, vol 16. Izd. Viniti, Moscow
5. Jović VD, Lačnjevac UČ, Jović BM (2014) Electrodeposition and characterization of alloys and composite materials. In: Djokić SS (ed) Modern aspects of electrochemistry, vol 57. Springer, New York/Heidelberg/Dodrecht/London, pp 1–84
6. Lačnjevac U, Jović BM, Jović VD (2012) Electrodeposition of Ni, Sn and Ni-Sn alloy coatings from pyrophosphate-glycine bath. J Electrochem Soc 159:D310–D318
7. Weast RC (ed) (1976–1977) Handbook of chemistry and physics, 57th edn. CRC Press Inc., Cleveland, pp D141–D146
8. Jović VD, Tošić N (1998) Qualitative and quantitative assessment of phases in electrodeposited Ni + Cd alloys by the ALSV technique. J Electroanal Chem 441:69–76
9. Daen JA (ed) (1999) Lange's handbook of chemistry, 15th edn. McGraw-Hill Inc., New York, pp 920–927
10. Duffield JR, Williams DR, Kron I (1991) Speciation studies of the solubility and aqueous solution chemistry of tin(II)- and tin(IV)-pyrophosphate complexes. Polyhedron 10:377–387
11. Han C, Liu Q, Ivey DG (2008) Kinetics of Sn electrodeposition from Sn(II)-citrate solutions. Electrochim Acta 53:8332–8340
12. Jović VD, Lačnjevac U, Jović BM, Karanović LJ, Krstajić NV (2012) Ni-Sn coatings as cathodes for hydrogen evolution in alkaline solution. Chemical composition, phase composition and morphology effects. Int J Hydrogen Energy 37:17882–17891
13. Dobrovolska TS, Krastev I, Jović BM, Jović VD, Beck G, Lačnjevac U, Zielonka A (2011) Phase identification in electrodeposited Ag-Cd alloys by anodic linear sweep voltammetry and X-ray diffraction techniques. Electrochim Acta 56:4344–4350
14. Dobrovolska TS, Krastev I, Zielonka A (2010) Pattern formation in electrodeposited Silver-Cadmium alloys. ECS Trans 25:1–9
15. Delahay P (1954) New instrumental methods in electrochemistry. Interscience, New York
16. Jović VD, Stojanović MV, Jović BM, Gajić-Krstajić LJ (1992) Electrochemical deposition of Ag, Pd and Ag-Pd alloy from chloride containing electrolyte. J Serb Chem Soc 57:951
17. Sturzenegger B, Puippe JC (1984) Electrodeposition of Palladium-Silver alloys from ammoniacal electrolytes. Plat Met Rev 3:117–124
18. Nakano H, Oue S, Uranka M, Masuda M, Fukushima H, Saka Y, Sawada S, Hattori Y (2010) Electrodeposition of Sn-Ag alloys and evaluation of connection reliability for automotive connectors. Mater Trans 51:712–719
19. Jović VD, Tošić N, Stojanović M (1997) Characterization of electrodeposited Co + Ni alloys by application of the ALSV technique. J Electroanal Chem 420:43–51
20. Dahms H, Croll J (1965) The anomalous codeposition of iron-nickel alloys. J Electrochem Soc 112:771–775
21. Higashi K, Fukushima H, Urakawa T, Adaniya T, Matsuko K (1981) Mechanism of the electrodeposition of zinc alloys containing a small amount of cobalt. J Electrochem Soc 128:2081–2085
22. Horkans T (1979) On the role of buffers and anions in NiFe electrodeposition. J Electrochem Soc 126:1861–1867
23. Horkans T (1981) Effect of plating parameters on electrodeposited NiFe. J Electrochem Soc 128:45–49

24. Wang L, Gao Y, Xue Q, Lui H, Xu T (2005) Microstructure and tribological properties of electrodeposited Ni-Co alloy deposits. Appl Surf Sci 242:326–332
25. Wu BYC, Ferreira PJ, Schuh CA (2005) Nanostructured Ni-Co alloys with tailorable grain size and twin density. Metall Mater Trans A 36A:1927–1936
26. Kanagasabapathy M, Jayakrishnan S (2011) Textural and morphological studies on zinc-iron alloy electrodeposits. J Chem Sci 123:357–364
27. Abdel-Karim R, Reda Y, Muhammed M, El-Raghy S, Shoeib M, Ahmed H (2011) Electrodeposition and characterization of nanocrystalline Ni-Fe alloys. J Nanomater 2011:1–8
28. Podlaha EJ, Landolt D (1996) Induced codeposition I. An experimental investigation of Ni-Mo alloys. J Electrochem Soc 143:885–892
29. Podlaha EJ, Landolt D (1996) Induced codeposition II. A mathematical model describing the electrodeposition of Ni-Mo alloys. J Electrochem Soc 143:893–899
30. Podlaha EJ, Landolt D (1997) Induced codeposition III. Molybdenum alloys with nickel, cobalt, and iron. J Electrochem Soc 144:1672–1680
31. Marlot A, Kern P, Landolt D (2002) Pulse plating of Ni-Mo alloys from Ni-rich electrolytes. Electrochim Acta 48:29–36
32. Zeng Y, Li Z, Ma M, Zhou S (2000) In situ surface Raman study of the induced codeposition mechanism of Ni-Mo alloys. Electrochem Comm 2:36–38
33. Sanches LS, Domingues SH, Carubelli A, Mascaro LH (2003) Electrodeposition of Ni-Mo and Fe-Mo alloys from sulfate-citrate acid solutions. J Braz Chem Soc 14:556–563
34. Jović VD, Jović BM, Lačnjevac U, Branković G, Bernik S, Rečnik A (2010) An attempt to predict the mechanism of Mo-Ni-O powders electrodeposition from the results of their TEM analysis. Electrochim Acta 55:4188–4193
35. Sanches L, Domingues SH, Marino CEB, Mascaro LH (2004) Characterization of electrochemically deposited Ni-Mo alloy coatings. Electrochem Comm 6:543–548
36. Krstajić NV, Jović VD, Gajić-Krstajić LJ, Jović BM, Antozzi AL, Martelli GN (2008) Electrodeposition of Ni-Mo alloy coatings and their characterization as cathodes for hydrogen evolution in sodium hydroxide solution. Int J Hydrogen Energy 33:3676–3687
37. Grgur BN, Krstajić NV, Elezović N, Jović VD (2005) Electrodeposition and characterization of Fe-Mo alloys as cathodes for hydrogen evolution in the process of chlorate production. J Serb Chrem Soc 70:879–889
38. Elezović NR, Jović VD, Krstajić NV (2005) Kinetics of the hydrogen evolution reaction on Fe-Mo film deposited on mild steel support in alkaline solution. Electrochim Acta 50:5594–5601
39. Jović V, Elezović N (2009) Electrochemical deposition and characterization of alloys. Institute for Multidisciplinary Research University of Belgrade, Belgrade, pp 1–101
40. Krastev I, Dobrovolska TS (2013) Pattern formation during electrodeposition of alloys. J Solid State Electrochem 17:481–488
41. Wojtowicz J (1972) Oscillatory behavior in electrochemical systems. In: Bockris JO'M, Conway B (eds) Modern aspects of electrochemistry, vol 8. Butterworths, London, pp 47–115
42. Krischer K (2001) Spontaneous formation of spatiotemporal patterns at the electrode|electrolyte interface. J Electroanal Chem 501:1–21
43. Koper MTM (1996) Oscillations and complex dynamical bifurcations in electrochemical systems. In: Prigogine I, Rice SA (eds) Advances in chemical physics, vol 92. Wiley, Hoboken, pp 161–298
44. Raub E (1953) Galvanische Legierungsniederschläge. Metalloberfläche 7:A17–A27
45. Krastev I, Baumgärtner ME, Raub CJ (1992) Stromoszillationen bei der galvanischen Abscheidung Untersuchungen zur Silber-Antimon-Legierungsabscheidung, Teil 2. Metalloberfläche 46:115–120
46. Hrussanova A, Krastev I, Beck G, Zielonka A (2010) Properties of silver-tin alloys obtained from pyrophosphate-cyanide electrolytes containing EDTA salts. J Appl Electrochem 40:2145–2151

47. Krastev I, Baumgärtner ME, Raub CJ (1992) Stromoszillationen bei der galvanischen Abscheidung. Untersuchungen zur Silber-Antimon-Legierungsabscheidung. Teil 1. Metalloberfläche 46:63–66
48. Raub E, Schall A (1938) Silber-Indium-Legierungen. Ein beitrag zur frage der anlaufbeständigen Silber legierungen. Zeitschr Metallkunde 30:149–151
49. Kristev I, Nikolova M (1986) Structural effects during the electrodeposition of silver-antimony alloys from ferrocyanidethiocyanate electrolytes. J Appl Electrochem 16:875–878
50. Kristev I, Nikolova M (1986) Phase composition and structure of silver–antimony alloy deposits plated from ferrocyanidethiocyanate electrolytes. J Appl Electrochem 16:867–874
51. Krastev I, Dobrovolska T, Lacnjevac U, Nineva S (2012) Pattern formation during electrodeposition of indium-cobalt alloys. J Solid State Electrochem 16:3449–3456
52. Nakabayashi S, Krastev I, Aogaki R, Inokuma K (1998) Electrochemical instability of Ag/Sb co-deposition coupled with a magnetohydrodynamic flow. Chem Phys Lett 294:204–208
53. Sun W, Ivey DG (1999) Development of an electroplating solution for codepositing Au-Sn alloys. Mater Sci Eng B 65:111–122
54. Sun W, Ivey DG (2001) Microstructural study of co-electroplated Au/Sn alloys. J Mater Sci 36:757–766
55. Bradley PE, Landolt D (1999) Pulse-plating of copper-cobalt alloys. Electrochim Acta 45:1077–1087
56. Metcalfe AG (1974) Interfaces in metal composites. Academic, New York, pp 65–102
57. Cohen U, Koch FB, Sard R (1983) Electroplating of Cyclic Multilayered Alloy (CMA) coatings. J Electrochem Soc 130:1987–1995
58. Tench D, White J (1984) Enhanced tensile strength for electrodeposited nickel-copper multilayer composites. Metall Trans 15A:2039–2040
59. Ogden C (1986) High-strength, composite Copper-Nickel electrodeposits. Plat Surf Finish 73:130–134
60. Verbruge MW, Tobias CW (1985) A mathematical model for the periodic electrodeposition of multicomponent alloys. J Electrochem Soc 132:1298–1307
61. Despić AR, Jović VD (1987) Electrochemical formation of laminar deposits of controlled structure and composition: I. Single current pulse galvanostatic technique. J Electrochem Soc 134:3004–3011
62. Despić AR, Jović VD, Spaić S (1989) Electrochemical formation of laminar deposits of controlled structure and composition: II. Dual current pulse galvanostatic technique. J Electrochem Soc 136:1651–1657
63. Despić AR, Trišović T (1993) Transition layer thickness in microlaminar deposits. J Appl Electrochem 23:662–668
64. Power CP, Ritchie IM (1975). In: Conway BE, Bockris JO'M (eds) Modern aspects of electrochemistry, Metal displacement reactions, vol. 11, Plenum Press, New York, pp 199–250
65. Yahalom J, Zadok O (1987) Formation of composition-modulated alloys by electrodeposition. J Mater Sci 22:499–503
66. Lashmore DS, Dariel MP (1988) Electrodeposited Cu-Ni textured superlattices. J Electrochem Soc 135:1218–1221
67. Tench DM, White JD (1990) Considerations in electrodeposition of compositionally modulated alloys. J Electrochem Soc 137:3061–3066
68. Tench M, White J (1991) Tensile properties of nanostructured Ni-Cu multilayered materials prepared by electrodeposition. J Electrochem Soc 138:3757–3758
69. Ebrahimi F, Liscano AJ (2001) Microstructure/mechanical properties relationship in electrodeposited Ni/Cu nanolaminates. Mater Sci Eng A301:23–34
70. Yang C-C, Cheh HY (1995) Pulsed electrodeposition of copper/nickel multilayers on a rotating disk electrode: I. Galvanostatic deposition. J Electrochem Soc 142:3034–3040
71. Yang C-C, Cheh HY (1995) Pulsed electrodeposition of copper/nickel multilayers on a rotating disk electrode: II. Potentiostatic deposition. J Electrochem Soc 142:3040–3043

72. Roy S, Landolt D (1995) Effect of off-time on the composition of pulse-plated Cu-Ni alloys. J Electrochem Soc 142:3021–3027
73. Bradley PE, Landolt D (1997) A surface coverage model for pulse-plating of binary alloys exhibiting a displacement reaction. Electrochim Acta 42:993–1003
74. Despić AR, Jović VD, Tošić N (1998) Intermediate layers in electrodeposited CMA coatings. Surf Coat Techn 105:206–212
75. Oberle RR, Cammarata RC (1995) Dependence of hardness on modulation amplitude in electrodeposited Cu-Ni compositionally modulated thin films. Scr Metall Et Mater 32:583–588
76. Cammarata RC, Schleisinger TE, Kim C, Qadri SB, Edelstein AS (1990) Nanoindentation study of the mechanical properties of copper-nickel multilayered thin films. Appl Phys Lett 56:1862–1864
77. Meneyes S, Anderson DP (1990) Wavelength-property correlation in electrodeposited ultrastructured Cu-Ni multilayers. J Electrochem Soc 137:440–444
78. Simunovich D, Schleisinger M, Snyder DD (1994) Electrochemically layered Copper-Nickel nanocomposites with enhanced hardness. J Electrochem Soc 141:L10–L11
79. Cziraki A, Pierron-Bohnes V, Ulhaq-Bouillet C, Toth-Kadar E, Bakonyi I (1998) A cross-section high-resolution transmission electron microscopy study of electrodeposited Ni-Cu/Cu multilayers. Thin Solid Films 318:239–242
80. Bakonyi I, Toth-Kadar E, Becsei T, Toth J, Tarnoczi T, Cziraki A, Gerӧcs I, Nabiyouni G, Schwarzacher W (1996) Giant magnetoresistance in self-supporting electrodeposited Ni-Cu/Cu multilayers. J Magn Magn Mater 156:347–349
81. Bird KD, Schleisinger M (1995) Giant magnetoresistance in electrodeposited Ni/Cu and Co/Cu multilayers. J Electrochem Soc 142:L65–L66
82. Hart R, Alper M, Attenborough K, Schwarzacher W (1994). In: Romankiw LT, Herman DAJr (eds) Proc. 3rd int. symp. on magnetic mater. processes and devices, electrodeposition division of the electrochem. soc, vol. 94–96, Pennington, p 215
83. Hua SZ, Lashmore DS, Salamanca-Riba L, Schwarzacher W, Swartzendruber LJ, McMichael RD, Bennett LH, Hart R (1994) Giant magnetoresistance peaks in CoNiCu/Cu multilayers grown by electrodeposition. J Appl Phys 76:6519–6521
84. He A, Djurfos B, Akhlaghi S, Ivey DG (2002) Pulse plating of gold-tin alloys for microelectronic and optoelectronic devices. Plat Surf Finish 89:48–53
85. Djurfos B, Ivey DG (2001) Pulsed electrodeposition of the eutectic Au/Sn solder for optoelectronic packaging. J Electron Mater 30:1249–1254

Chapter 8
Electrodeposited Alloy Powders

8.1 Introduction

The alloy powders of the iron-group metals (anomalous codeposition) are of great interest for many industrial applications [1]. The Co-Ni powders showed significant promise for future development of hard magnetic materials, commercial batteries, catalysts, catalyzing electroplates, hydrogen-absorbing alloy anodes, and magneto-resistive sensors, being made by several different techniques [1]. Unfortunately, limited number of papers concerning Co-Ni powder electrodeposition exists in the literature [2–7]. As shown in some of these papers [4–7], the morphology and composition of electrodeposited powders were found to be sensitive to the solution composition (to the ratio of Ni^{2+}/Co^{2+} ion concentration). Fe-Ni-based alloy powders are known as promising soft magnetic materials with low coercivity and high permeability [1]. Electrodeposition of Fe-Ni alloy powders was the subject of only few papers [8–12]. The influence of the composition of electrolyte (Ni^{2+}/Fe^{2+} ions ratio) on the powder morphology was investigated in Ref. [8, 9]. Zhelibo et al. [10, 11] suggested a method for producing very fine Fe-Ni alloy powder by the electrolysis in a two-layer electrolytic bath, using a hydrocarbon solvent from an oil-refining fraction as an upper organic layer with evaporation at 180 °C and subsequent reduction annealing in a hydrogen atmosphere. The influence of the reduction annealing temperature [10] and the electrolysis temperature [11] on the formation, chemical and phase composition, structure, and magnetic properties of highly dispersed Fe-Ni alloy powders was investigated, and the optimal thermal conditions for the production of powders with micro-sized particles were determined [10, 11]. The effect of complexing agents (citric and oxalic acid) on the process of Fe-Ni alloy powder electrodeposition was also investigated [12]. It was shown that complexing agents influence the kinetics of powder electrodeposition as well as the morphology of the Fe-Ni powders. Finer powders were produced in the presence of citric acid in comparison with those obtained in the presence of oxalic acid [1, 12].

K.I. Popov et al., *Morphology of Electrochemically and Chemically Deposited Metals*, DOI 10.1007/978-3-319-26073-0_8

The Mo-Ni alloys (induced codeposition) possess several useful properties, with the most important one being exceptional corrosion and wearing resistance [1]. The Mo-Ni alloys can be produced by several methods, from which metallurgical ones are not convenient because of easy oxidation and high melting temperature of molybdenum. The powder metallurgy and mechanical alloying, spark plasma sintering, and laser cladding are mostly used for the Mo-Ni alloy preparation [1].

All abovementioned methods are expensive in comparison with the electrodeposition of Mo-Ni alloy coatings. Although molybdenum cannot be separately electrodeposited from aqueous solutions, it can be codeposited with the iron-group metals (Fe, Co, Ni) in the presence of appropriate complexing agents, by the type of alloy electrodeposition defined by Brenner [13] as induced codeposition.

It should be emphasized here that only two papers concerning electrodeposition of Mo-Ni alloy powders (actually powders of the system Mo-Ni-O) and their characterization are published so far [14–16].

8.2 Anomalous Codeposition of Alloy Powders

8.2.1 *Electrodeposited Co-Ni Powders*

Electrodeposition of Co-Ni powders was performed in two types of ammonium hydroxide containing electrolytes: 1 M $(NH_4)_2SO_4 + 0.7$ M NH_4OH and 1 M $NH_4Cl + 0.7$ M NH_4OH. In sulfate-containing electrolytes, cobalt and nickel sulfates were used, while in chloride-containing electrolytes, cobalt and nickel chlorides were used as a source of Co^{2+} and Ni^{2+} ions [3–6]. The Co-Ni alloy powders were also electrodeposited from a solution containing 0.4 M $H_3BO_3 + 0.2$ M Na_2SO_4 and Ni and Co sulfate salts [7]. Three different Ni^{2+}/Co^{2+} concentration ratios were used: (1) $Ni^{2+}/Co^{2+} = 1.00$ (0.01 M $NiSO_4$/0.01 M $CoSO_4$); $Ni^{2+}/Co^{2+} = 0.50$ (0.01 M $NiSO_4$/0.02 M $CoSO_4$); $Ni^{2+}/Co^{2+} = 0.33$ (0.01 M $NiSO_4$/0.03 M $CoSO_4$). From these solutions, Co-Ni alloy powders were electrodeposited at a constant current density of approximately 70 mA cm^{-2}, corresponding to the cell voltage of about 5.0 V [7].

8.2.1.1 Polarization Curves for Co-Ni Alloy Powder Electrodeposition from the Sulfate-Containing Supporting Electrolyte

In Fig. 8.1a are shown polarization curves corrected for IR drop for the processes of Co, Ni, and Co-Ni alloy powder electrodeposition from 1 M $(NH_4)_2SO_4 + 0.7$ M NH_4OH containing supporting electrolyte. Their shape of all polarization curves is identical, characterized with two inflection points, A and B. For Co electrodeposition, sharp increase of current occurs at about -1.19 V versus Ag/AgCl, while for Ni electrodeposition, this phenomenon is moved to more negative potentials

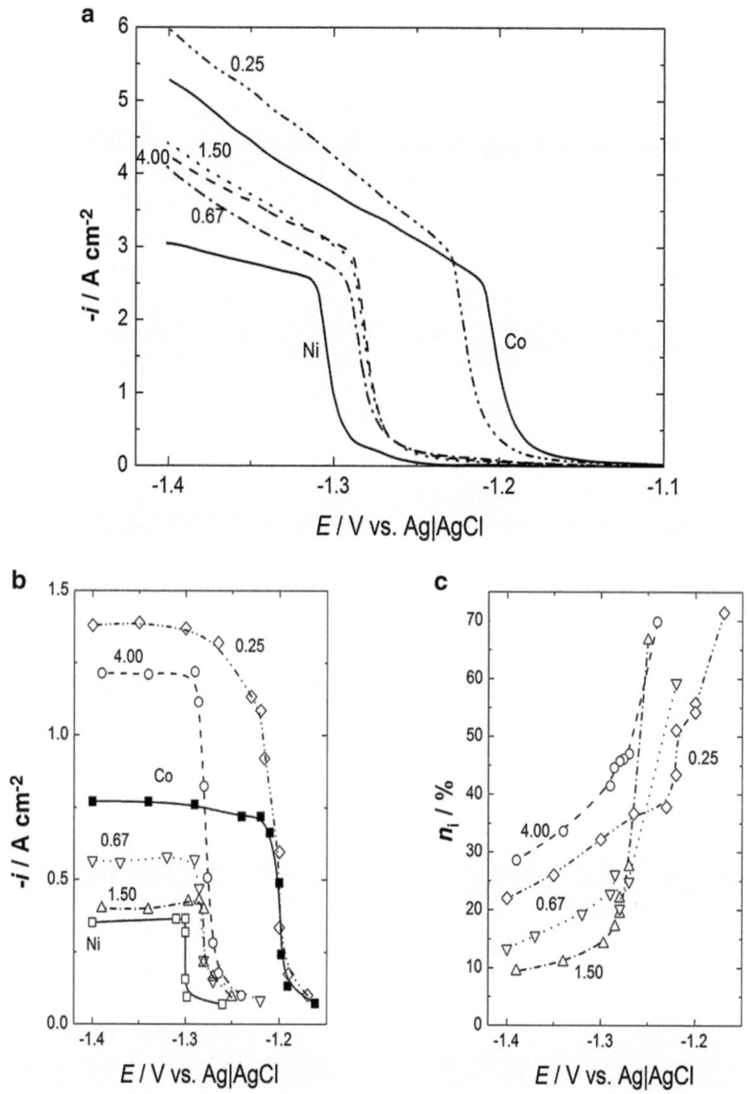

Fig. 8.1 (**a**) Polarization curves for the electrodeposition of cobalt (Co), nickel (Ni), and Co-Ni alloy powders after IR drop correction recorded for different Ni^{2+}/Co^{2+} ions ratios: 4.00, 1.50, 0.67, and 0.25 (marked in the figure). (**b**) The same polarization curves after hydrogen evolution current density subtraction. (**c**) Corresponding η_i vs. E curves for (**b**) (Reprinted from Ref. [1] with kind permission from Springer)

(at about −1.30 V versus Ag/AgCl), indicating that the overpotential for Ni electrodeposition in this solution is for about 0.1 V higher than that for Co electrodeposition, as it is the case for hydrogen evolution from pure supporting electrolytes [1]. In the case of alloy electrodeposition, sharp increase of current density on the polarization curves recorded for Ni^{2+}/Co^{2+} ratios 4.00, 1.50, and 0.67 takes place at almost identical potentials, while the one for $Ni^{2+}/Co^{2+} = 0.25$ is moved to more positive potential, close to the polarization curve for pure Co.

Polarization curves corresponding only to the processes of electrodeposition of pure metals and alloy powders (after subtracting hydrogen evolution current density) are shown in Fig. 8.1b. Corresponding η_i vs. E curves for alloy powder electrodeposition are presented in Fig. 8.1c. It is interesting to note that the current density plateaus (diffusion-limiting current densities) for the electrodeposition of Co-Ni alloys at the Ni^{2+}/Co^{2+} ratios 4.00 and 0.25 are higher than that for pure Co electrodeposition (Fig. 8.1b). The reason for such behavior is the fact that the total concentration of cations in these two solutions was not 0.1 M, since the compositions of these solutions were 0.1 M $NiSO_4 + 0.025$ M $CoSO_4 + 1$ M $(NH_4)_2SO_4 + 0.7$ M NH_4OH and 0.025 M $NiSO_4 + 0.1$ M $CoSO_4 + 1$ M $(NH_4)_2SO_4 + 0.7$ M NH_4OH (total concentration of cations 0.125 M). At the same time, the current efficiency for Co-Ni powder electrodeposition is the highest in these two solutions. Such behavior is most probably the consequence of the independence of the current density for hydrogen evolution on the total concentration of cations, so that the same curve for hydrogen evolution is subtracted from the higher values of total current densities for alloy powder electrodeposition [1].

8.2.1.2 The Morphology of the Co-Ni Alloy Powders Electrodeposited from the Sulfate-Containing Supporting Electrolyte

The morphology of the Co-Ni alloy powders electrodeposited at the current density $i = -0.5$ A cm^{-2} from the electrolytes containing different Ni^{2+}/Co^{2+} ions ratio is shown in Figs. 8.2, 8.3, 8.4, and 8.5. It is obvious that the morphology of Co-Ni alloy powder is sensitive to the Ni^{2+}/Co^{2+} ions ratio, i.e., to the composition of the alloy powder.

In the powder containing the highest amount of Ni, of about 64 at.% Ni, electrodeposited at the $Ni^{2+}/Co^{2+} = 4.00$, more than 90 % of agglomerates were typical spongy agglomerates varying in the size from about 50 μm to about 200 μm, shown in Fig. 8.2. Very few cauliflower-like agglomerates of slightly different shape than those typical for pure Ni powder were also detected in the alloy powder. One such particle is marked with the arrow in the upper right corner of Fig. 8.2. It is important to note that this particle has a flat surface on one side, indicating that this type of growth belongs to the beginning of the powder formation with this particle being detached from the electrode surface before the bigger, typical spongy agglomerate is formed. Spongy agglomerates are more similar to the ones detected in pure Co powder, with the presence of large number of cavities, but contrary to the situation in the case of pure Co powder [1], here practically none of the cavities possess fern-like dendrites on their bottom, indicating that most of the agglomerates has been detached from the electrode surface before the hydrogen bubble in the cavity was liberated. It is interesting that practically no dendrites are found in this powder, although about 26 wt.% of Co are present in the powder.

Fig. 8.2 Typical powder particle detected in the powder electrodeposited at the $Ni^{2+}/Co^{2+} = 4.00$ (Reprinted from Ref. [1] with kind permission from Springer)

Fig. 8.3 Typical powder agglomerates detected in the powder electrodeposited at the $Ni^{2+}/Co^{2+} = 1.50$. (**a**) Few bigger aglomerates (about 200 μm). Number of cavities significantly reduced in comparison with the one presented in Fig. 8.2. (**b**) Fern-like dendrites covering all agglomerates (Reprinted from Ref. [1] with kind permission from Springer)

As the Ni^{2+}/Co^{2+} ions ratio in the electrolyte decreases from 4.00 to 1.50, the composition of the Co-Ni alloy powder changes from about 78 at.% Ni – 22 at.% Co to about 56 at.% Ni – 44 at.% Co, and this change influences the morphology of the powder. All powder agglomerates electrodeposited at $Ni^{2+}/Co^{2+} = 1.50$ are covered with the fern-like dendrites (Fig. 8.3b). Only few of agglomerates are bigger (about 200 μm), and the number of cavities observed on them is significantly

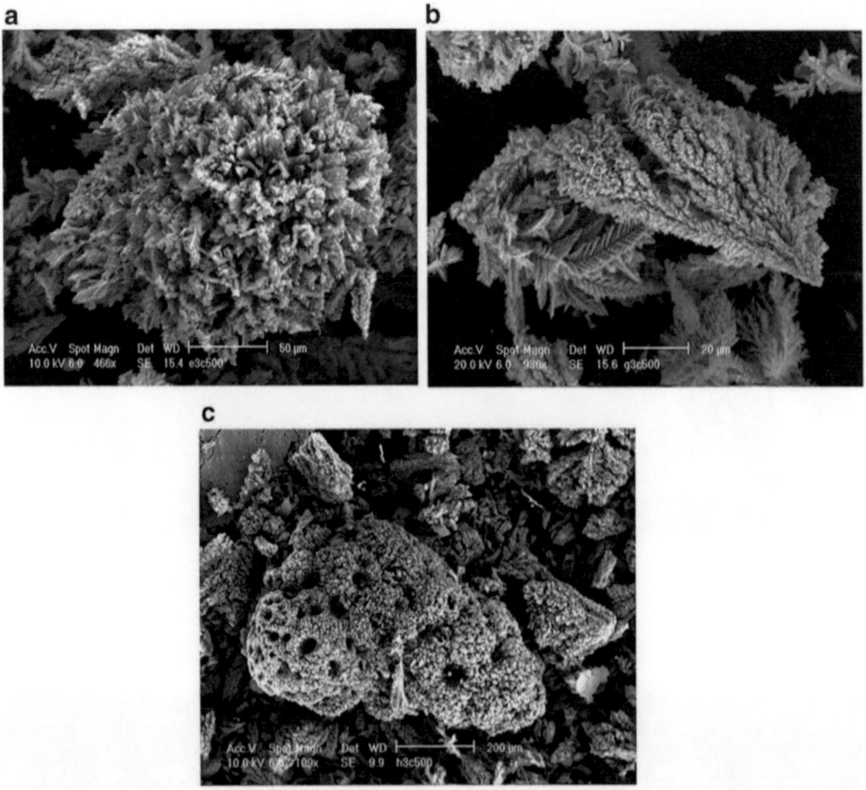

Fig. 8.4 Typical powder agglomerates detected in the powder electrodeposited at the $Ni^{2+}/Co^{2+} = 0.67$. (**a**) Agglomerate covered with fern-like dendrites, (**b**) Higher amount of dendritic particles around the agglomerates, (**c**) Compact agglomerate covered with small fern-like dendrites and the presence of cavities (Reprinted from Ref. [1] with kind permission from Springer)

Fig. 8.5 Typical powder agglomerates detected in the powder electrodeposited at the $Ni^{2+}/Co^{2+} = 0.25$. (**a**) Spongy-like agglomerates of the size of about 500 μm covered with the fern-like dendrites, (**b**) Agglomerates composed of closely packed fern-like dendrites about 200 μm long (Reprinted from Ref. [1] with kind permission from Springer)

reduced (Fig. 8.3a) in comparison with the one obtained at the ratio $Ni^{2+}/Co^{2+} = 4.00$. As can be seen in Fig. 8.3a, the significant number of dendritic agglomerates is detected in this powder deposit (around the agglomerate). Very few spongy-like agglomerates of the size of about 500 μm are also present in the powder, and they are also covered with the fern-like dendrites, indicating that the presence of higher amount of Co in the powder, dendritic growth prevails. With further decrease of the Ni^{2+}/Co^{2+} ions ratio (0.67) in the electrolyte, the composition of the Co-Ni alloy powder changes toward lower content of Ni, and the morphology of powder agglomerates is slightly different. In this case, the number of agglomerates increases in comparison with the ratio $Ni^{2+}/Co^{2+} = 1.50$, with the shape of agglomerates being different, as can be seen in Fig. 8.4a–c. A common characteristic of all agglomerates obtained at $Ni^{2+}/Co^{2+} = 0.67$ is that they are covered with the fern-like dendrites. At the same time, higher amount of dendritic particles (around the agglomerate shown in Fig. 8.4b) is formed during Co-Ni powder electrodeposition. The size of dendrite particles varies from about 5 μm to about 50 μm.

In the Co-Ni alloy powder electrodeposition with the highest percentage of Co (85 at.%), Fig. 8.5, different types of agglomerates could be detected: (*i*) spongy-like agglomerates of the size of about 500 μm covered with the fern-like dendrites, also found in the powder obtained at $Ni^{2+}/Co^{2+} = 0.67$ (Fig. 8.4c); (*ii*) compact agglomerates, characteristic for pure Co powder (Fig. 8.5a); and (*iii*) agglomerates composed of closely packed fern-like dendrites about 200 μm long (Fig. 8.5b). By the EDS analysis of the compact agglomerate, shown in Fig. 8.5a, it is found that it does not represent pure Co particle, but the Co-Ni alloy of almost identical composition as the rest of the powder, indicating that the formation of alloy agglomerates follows a growth mechanism for Co powder formation [1, 3].

Considering all detected types of agglomerates in electrodeposited Co-Ni alloy powders of different compositions, one can conclude that only one powder (with 78 at.% of Ni) contains very small amount of cauliflower-like agglomerates typical for pure Ni powder, but slightly different, while the rest of them contain some of the agglomerates detected in pure Co powder. The difference between the agglomerates detected in pure Co powder and in Co-Ni alloy powder (except for the agglomerates obtained from the electrolyte with the highest Ni^{2+}/Co^{2+} ions ratio, Fig. 8.1) is the presence of fern-like dendrites on the surface of these agglomerates. Such a behavior indicates that for these three alloy compositions the presence of Ni^{2+} and Co^{2+} ions promotes dendrite formation on the surface of all agglomerates detected in Co-Ni alloy powders. Hence, it appears that although some similarities between the type of agglomerates detected in pure Co and Co-Ni alloy powders exist, it is practically not possible to strictly correlate composition of alloy powders with their morphology [1].

In order to explain the difference in the morphology as a function of alloy composition (Ni^{2+}/Co^{2+} ions ratio), one should refer to the polarization curves presented in Fig. 8.1b. As can be seen in this figure, at the applied current density of -0.5 A cm^{-2}, only one alloy, obtained from the solution containing $Ni^{2+}/Co^{2+} = 1.50$, electrodeposits at its diffusion limiting current density. The one obtained at the $Ni^{2+}/Co^{2+} = 0.67$ electrodeposits at a slightly lower current density than its diffusion limiting current density, while for the alloys electrodeposited at $Ni^{2+}/Co^{2+} = 4.00$ and 0.25, applied current density lies in the region of mixed control of electrodeposition, being much less than the diffusion limiting current density for those electrolyte compositions. Bearing in mind that the currents presented in this figure represent only 20 % of the total current and that massive hydrogen evolution takes place together with the process of alloy electrodeposition, it is quite difficult to find good explanation for the phenomena observed. Different theories concerning powder formation during metal electrodeposition exist in the literature [1] with one of them claiming that powder formation takes place at the diffusion limiting current at potentials of hydrogen evolution, which is necessary process for powder formation. In the case of Co-Ni alloy powder electrodeposition, hydrogen evolution commences already at the beginning of the electrodeposition process, being catalyzed by the electrodeposition. Hence, any comparison between these two processes is not realistic, but it is obvious that hydrogen evolution plays an important role in the process of powder electrodeposition.

8.2.1.3 Polarization Curves for Co-Ni Alloy Powder Electrodeposition from the Chloride-Containing Supporting Electrolyte

The polarization curves corrected for IR drop for the processes of pure cobalt (Co), pure nickel (Ni), and Co-Ni alloy (powders) electrodeposition from ammonium chloride–ammonium hydroxide (1 M $NH_4Cl + 0.7$ M NH_4OH) containing supporting electrolyte are presented in Fig. 8.6a (in this case, total concentration of cations in all investigated solutions was 0.1 M). As can be seen, cobalt electrodeposition (Co) commences at about −1.1 V versus Ag/AgCl, while sharp increase of current density (massive Co electrodeposition and hydrogen evolution) occurs at about −1.19 V versus Ag/AgCl. In the case of nickel electrodeposition (Ni), sharp increase of current density is moved to more negative potentials (at about −1.30 V versus Ag/AgCl), indicating that the overvoltage for massive Ni electrodeposition (and hydrogen evolution) in this solution is for about 0.1 V higher than that for Co electrodeposition, as it is the case for hydrogen evolution from pure supporting electrolytes [1]. In the case of Co-Ni alloy powder electrodeposition, polarization curves recorded for all Ni^{2+}/Co^{2+} ratios are placed between the polarization curves for pure Co and pure Ni.

Polarization curves corresponding only to the processes of electrodeposition of pure metals and alloy powders (after subtracting hydrogen evolution current density) are shown in Fig. 8.6b. Corresponding η_i vs. E curves for pure metals and alloy powder electrodeposition are presented in Fig. 8.6c [1].

As in the case of sulfate-supporting electrolyte (Fig. 8.1b), "diffusion limiting current density" (in this case not represented with the plateau) for Co is higher than that for Ni (Fig. 8.6b). At the same time, all polarization curves for Co-Ni alloy powder electrodeposition are characterized with higher values of current densities,

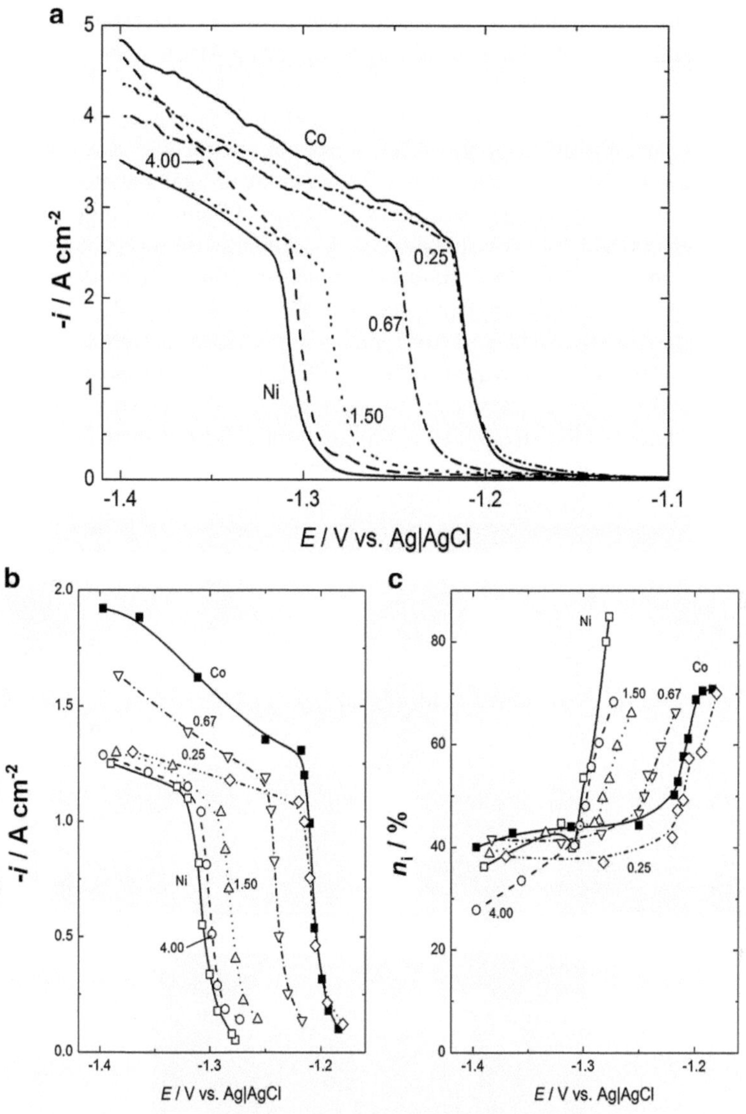

Fig. 8.6 (**a**) Polarization curves for the electrodeposition of cobalt (Co), nickel (Ni), and Co-Ni alloy powders after IR drop correction recorded for different Ni^{2+}/Co^{2+} ions ratios: 4.00, 1.50, 0.67, and 0.25 (marked in the figure). (**b**) The same polarization curves after hydrogen evolution current density subtraction. (**c**) Corresponding η_i vs. E curves for (**b**) (Reprinted from Ref. [1] with kind permission from Springer)

indicating more efficient process of powder formation in the chloride-supporting electrolyte. This statement is supported with the η_i vs. E curves presented in Fig. 8.6c, where the current efficiency for powder electrodeposition is much higher than that in the sulfate-supporting electrolyte (Fig. 8.1c).

8.2.1.4 The Morphology of the Co-Ni Alloy Powders Electrodeposited from the Chloride-Containing Supporting Electrolyte

In Figs. 8.7, 8.8, 8.9, and 8.10 are shown SEMs of Co-Ni alloy powders obtained from a solution containing different Ni^{2+}/Co^{2+} ions ratios. At the Ni^{2+}/Co^{2+} ions ratio of 4.00 (22 at.% Co in the powder), Fig. 8.7, two types of agglomerates are detected: cauliflower ones (a), varying in size between 60 and 100 μm, and bigger (varying in size between 200 and 400 μm), spongy ones (b), similar to the morphology of pure Ni powder. In the case of alloy, these two types of agglomerates are not equally distributed, but most of the agglomerates (over 80 %) are spongy-like agglomerates.

With increasing the content of Co^{2+} ions in the solution, as well as Co content in the powder (Ni^{2+}/Co^{2+} ions ratio of 1.50, 42 at.% Co in the powder), very few cauliflower-like agglomerates can be detected in the powder electrodeposit (Fig. 8.8a), being of the size of about 100 μm, while most of them are spongy-like

Fig. 8.7 Typical agglomerates detected in the powder electrodeposited at the $Ni^{2+}/Co^{2+} = 4.00$. (**a**) Cauliflower agglomerate, (**b**) Spongy agglomerate, (**c**) Top view of spongy agglomerates at higher magnification (Reprinted from Ref. [1] with kind permission from Springer)

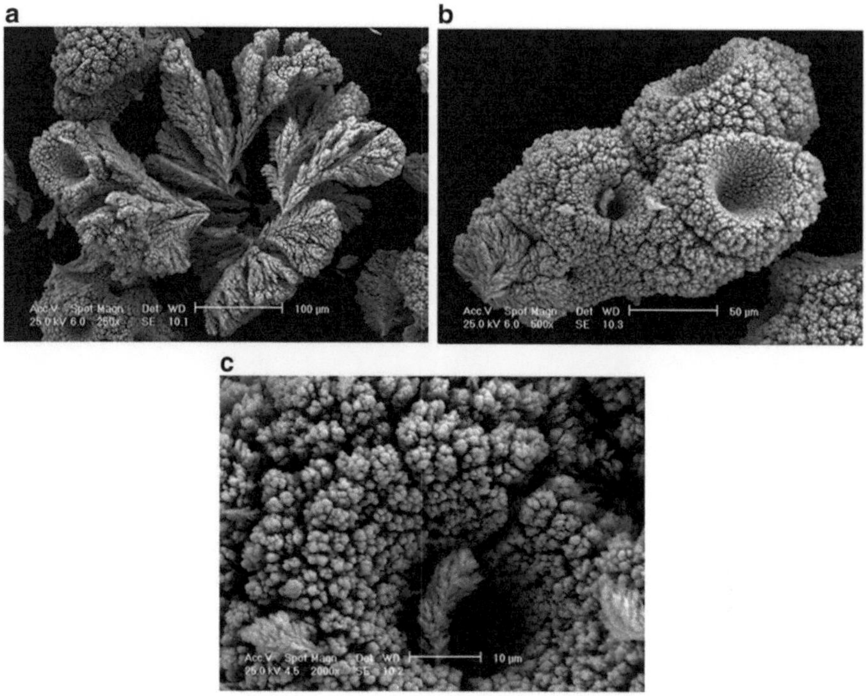

Fig. 8.8 Typical powder agglomerates detected in the powder electrodeposited at the $Ni^{2+}/Co^{2+} = 1.50$. (**a**) Few cauliflower-like agglomerates of about 100 µm, (**b**) Spongy-like agglomerates of about 200 µm, (**c**) Top view of spongy agglomerates at higher magnification (Reprinted from Ref. [1] with kind permission from Springer)

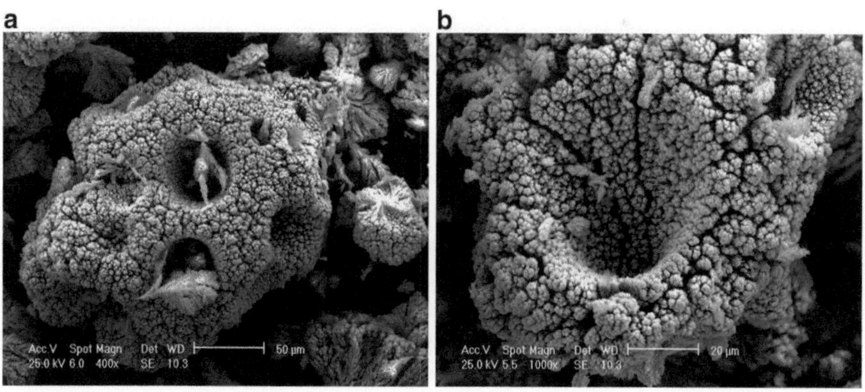

Fig. 8.9 Typical powder agglomerates detected in the powder electrodeposited at the $Ni^{2+}/Co^{2+} = 0.67$. (**a**) Narrow (cylindrical) cavities with fern-like dendrites on their bottom, (**b**) Cone-shaped cavities without fern-like dendrites on their bottom (Reprinted from Ref. [1] with kind permission from Springer)

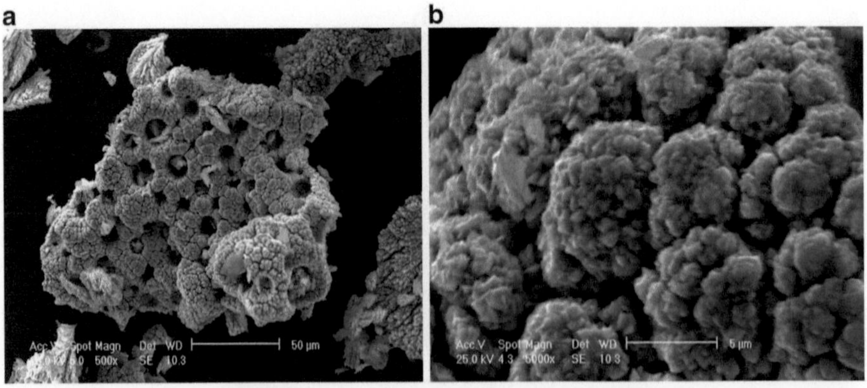

Fig. 8.10 Typical powder agglomerates detected in the powder electrodeposited at the $Ni^{2+}/Co^{2+} = 0.25$. (**a**) Typical spongy agglomerate, (**b**) Agglomerate with the presence of well-defined crystals on the cauliflower endings (Reprinted from Ref. [1] with kind permission from Springer)

ones of a bigger size of about 200 μm (Fig. 8.8b). It is characteristic for spongy agglomerates that two types of cavities could be detected on all powder agglomerates: narrow (cylindrical) cavities with fern-like dendrites formed at their bottom (Fig. 8.8c), indicating that before agglomerates detached from the electrode surface conditions of planar diffusion in the cavity were established causing formation of fern-like dendrites, and wider, cone-shaped cavities without fern-like dendrites on their bottom (Fig. 8.8a, b). It is most likely that the absence of fern-like dendrites at the bottom of cone-shaped cavities is the consequence of the fact that because of such a shape, with wide opening, no condition for planar diffusion in these cavities was established.

With further decrease of Ni^{2+}/Co^{2+} ions ratio to 0.67 (65 at.% Co in the powder), no cauliflower-like agglomerates are detected. All agglomerates are spongy-like, again with two types of cavities: narrow (cylindrical) ones with fern-like dendrites on their bottom (Fig. 8.9a) and cone-shaped ones (Fig. 8.9b) without fern-like dendrites on their bottom. The top surface of all agglomerates obtained from these three solutions is practically identical, being characterized by cauliflower nature, as shown in Fig. 8.7c.

At the lowest Ni^{2+}/Co^{2+} ions ratio of 0.25 (82 at.% Co in the powder), the same types of spongy agglomerates as ones obtained from the solution containing Ni^{2+}/Co^{2+} ions ratio of 0.67 are obtained (Fig. 8.10a). In this case, two types of spongy agglomerates, concerning the top surface of the agglomerates, are detected: agglomerates with flat noodle-like endings (Fig. 8.7c) and agglomerates with the presence of well-defined crystals on the cauliflower endings, as shown in Fig. 8.10b. It is most likely that the detachment of agglomerates shown in Fig. 8.10b occurred later than the detachment of agglomerates shown in Fig. 8.7c, allowing formation of a second layer of their growth to take place [1, 3]. Hence, in comparison with pure metal powders, agglomerates of alloy powders are characterized by the presence of two types of cavities: cylindrical ones with the fern-like dendrites on their bottom, indicating that hydrogen bubbles are liberated before the powder agglomerate detachment, and cone-shaped ones without fern-like dendrites on their bottom (because of the cavity shape, planar diffusion cannot be established in the cavity), detected at smaller powder agglomerates. The appearance of such cavity is most probably due to different (higher) rates of hydrogen bubble formation in this solution. Comparing results obtained in chloride-containing electrolyte with the results obtained in sulfate-containing electrolyte, the following remarks could be made: (1) polarization curves for hydrogen evolution in supporting electrolytes are different, with the one recorded in the chloride-containing supporting electrolyte being characterized with higher current densities, indicating faster hydrogen evolution in the presence of smaller anions (see Fig. 2.24, Ref. [17]); (2) correct polarization curves for pure metals and alloy powder electrodeposition are also characterized with higher current densities, as well as higher current efficiencies for powder electrodeposition, indicating faster and easier powder formation in chloride-containing electrolyte; (3) relatively similar morphologies of powder agglomerates are detected in both solutions, with the distribution of agglomerates being different in all cases. Characteristic feature of powder agglomerates obtained from chloride-containing electrolytes is the presence of cone-shaped cavities.

8.2.1.5 The Morphology of the Co-Ni Alloy Powders Electrodeposited from the Borate-Sulfate Containing Supporting Electrolyte

At the highest Ni^{2+}/Co^{2+} ions ratio in the solution ($Ni^{2+}/Co^{2+} = 1.00$, 55 at.% Co in the powder), all powder agglomerates possess 2D fern-like shape [7], varying in the size from about 100 μm to about 200 μm, as can be seen in Fig. 8.11 [1]. They are all practically two dimensional (Fig. 8.11a), although at higher magnification

Fig. 8.11 Typical powder agglomerates detected in the powder electrodeposited at the $Ni^{2+}/Co^{2+} = 1.00$. (**a**) Agglomerates with 2D fern-like shape, (**b**) 2D fern-like agglomerate at higher magnification (Reprinted from Ref. [1] with kind permission from Springer)

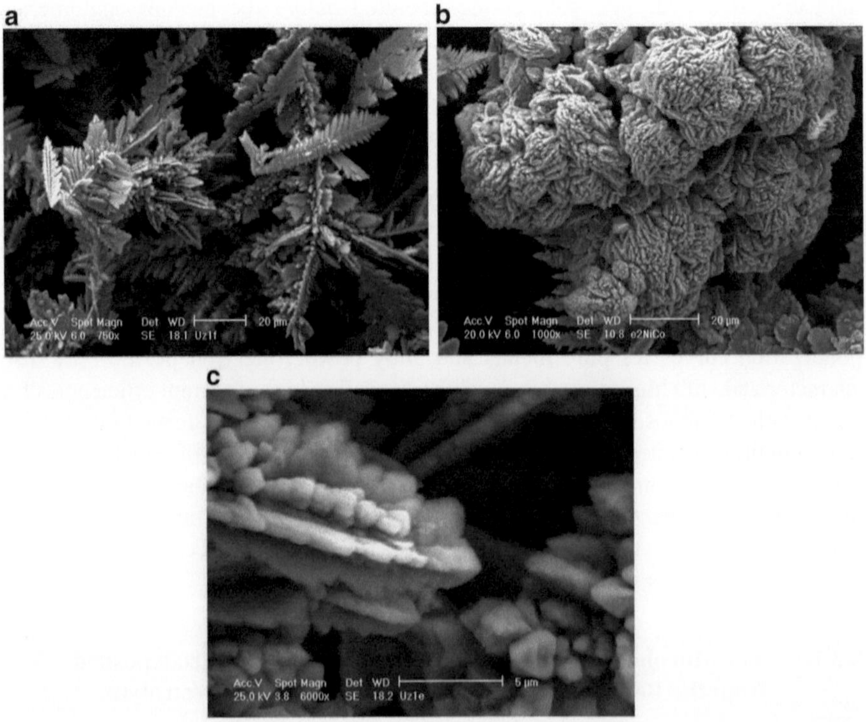

Fig. 8.12 Typical powder agglomerates detected in the powder electrodeposited at the $Ni^{2+}/Co^{2+} = 0.50$. (**a**) 2D fern-like agglomerates, (**b**) Agglomerates of the size of about 100 μm composed of a densely packed dendritic agglomerates, (**c**) Top view of dendritic agglomerates at higher magnification (Reprinted from Ref. [1] with kind permission from Springer)

(Fig. 8.11b) it can be seen that a third dimension of their growth also exists, but it seems that this is a practically two-dimensional layer-by-layer growth following original dendrite shape. It can also be seen that on the main branch of such dendrite, small dendritic agglomerates, growing in the direction normal to the layer-by-layer growth, could be detected (Fig. 8.11b).

With the decrease of the Ni^{2+}/Co^{2+} ions ratio in the solution (0.50), i.e., the increase of the Co content in the alloy powder (80 at.% Co in the powder), among 2D fern-like agglomerates of a smaller size than in a previous case (varying in the size between 50 and 100 μm – Fig. 8.12a), agglomerates of the size of about 100 μm composed of a densely packed dendritic agglomerates were detected, as can be seen in Fig. 8.12b. It is interesting to note that dendrites covering the surface of such agglomerates are not 2D fern-like type. As can be seen in Fig. 8.12c, they are branching in four directions, but they are more compact and branches are much smaller, as are the dendrites (about 10 μm long and about 2 μm wide).

With further decrease of the Ni^{2+}/Co^{2+} ions ratio in the solution (0.33, 83 at.% Co in the powder), compact agglomerates, typical for pure Co powder, could be clearly detected in the powder electrodeposit, as can be seen in Fig. 8.13a, together

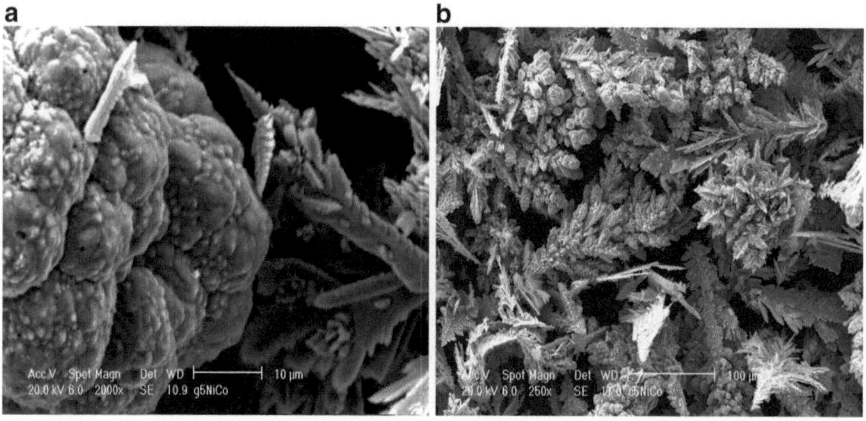

Fig. 8.13 Typical powder agglomerates detected in the powder electrodeposited at the $Ni^{2+}/Co^{2+} = 0.33$. (**a**) Compact agglomerates together with the 2D fern-like dendrites, (**b**) Densely packed 3D dendritic agglomerates (Reprinted from Ref. [1] with kind permission from Springer)

with the 2D fern-like dendrites. By the EDS analysis of compact agglomerates, it was confirmed that they do not represent pure Co particle, but the Co-Ni alloy, indicating that similar shapes of agglomerates characterize pure Co and Co-Ni alloy powder deposit. The presence of densely packed 3D dendritic agglomerates can also be seen in this powder, as shown in Fig. 8.13b. Again, these dendrites are more compact and smaller than those recorded on powder agglomerates electrodeposited at the Ni^{2+}/Co^{2+} ions ratio 0.50.

It is noteworthy that with the increase of the h.c.p. α-Co (100) phase in the powder electrodeposit and the decrease of the f.c.c. β-Ni (111) phase, the shape of dendrite agglomerates changes from typical 2D fern-like dendrites to 3D dendrites. This is in accordance with the statement [1] that in the presence of the (111) orientation (f.c.c. β-Ni (111) phase) 2D dendrite growth prevails (the highest intensity for f.c.c. β-Ni (111) phase is detected in sample electrodeposited at Ni^{2+}/Co^{2+} ions ratio 1.00). Such dendrites are denoted in the literature [1] as 2D{100} 60°, with the angle of 60° between the main tree of the dendrite and the branches. In order to explain the influence of other orientations (h.c.p. α-Co (100) etc.) on the growth of dendrites in the investigated powder and appearance of 3D dendrites in the powder deposit, additional experiments and more detailed analysis are needed.

8.2.2 *Electrodeposited Fe-Ni Alloy Powders*

Electrodeposition of Fe-Ni powders was performed in ammonium chloride–sodium citrate containing electrolyte of the composition 1 M NH_4Cl + 0.2 M $Na_3C_6H_5O_7$. Fe(III) and Fe(II) chloride and nickel chloride were used as sources of Fe^{2+} or Fe^{3+} and Ni^{2+} ions. Total concentration of cations was kept at 0.1 M. Four different

Ni/Fe concentration ratios were investigated: Ni/Fe = 9/1; Ni/Fe = 3/1; Ni/Fe = 1/1; and Ni/Fe = 1/3. In the case of Fe(III) ions, pH of the solution was adjusted to 4.5, while in the case of Fe(II) ions, pH of the solution was adjusted to 4.0. All solutions were made from analytical grade purity chemicals and distilled water by the following procedure: $Na_3C_6H_5O_7$ was first dissolved and the pH was adjusted to slightly higher value than desired by HCl; in the next step, metal (Fe(III), Fe(II), or Ni) salts were dissolved; finally, ammonium chloride was added and pH adjusted to the exact value. Concerning stability of solutions, it is well known that during the investigation, some Fe(II) become oxidized into Fe(III). This should be particularly pronounced during the electrodeposition of powders for 1 or 2 h. Taking into account that in all cases Fe(II) or Fe(III) made very stable complexes with citrate anions, no problems have been experienced in the case of polarization measurements (polarization curves were practically the same after 3–4 measurements), but for any case before each experiment, fresh solution was made and used for investigation as well as for powder electrodeposition [1, 8].

It should be mentioned here that an attempt was made to deposit Fe-Ni powders from the electrolytes of the same composition but of low pH 2. The powder agglomerates were successfully produced on the cathode, but immediately after detaching from the cathode surface, they started dissolving with gas evolution in all investigated electrolytes. Only in the case of very short time of electrolysis and removal of remained powder (which had not yet been dissolved) it was possible to obtain small amount of powder for further analysis. That was the reason why all experiments were performed in the solutions of pH 4.0 or 4.5, since in this solution powders were stable after electrodeposition.

From all solutions, Fe-Ni alloy powders were electrodeposited at a constant current density corresponding to the slightly lower value (i_{pd}) than the inflection point B (marked with (●)) on the polarization curves (see Fig. 8.14) [1, 8].

8.2.2.1 Polarization Curves for Fe-Ni Alloy Powder Electrodeposition from Chloride: Citrate-Supporting Electrolyte

The polarization curves corrected for IR drop for the processes of Fe, Ni, and Fe-Ni alloy powder electrodeposition from ammonium chloride–sodium citrate containing supporting electrolyte in the presence of Fe(II) and Ni(II) species are shown in Fig. 8.14. In the case of Fe(II) salts, polarization curve for iron electrodeposition (Fe) was placed at more positive potentials than that for nickel (Ni) as it is expected from the values of their reversible potentials. The polarization curves for Fe-Ni alloy powder electrodeposition are placed in between, and all of them were placed at more positive potentials than expected from the Ni/Fe ratio, indicating anomalous codeposition.

It was found that for the Fe(III) salt electrolytes, the current efficiency was very low, 1–2 % (polarization curves for powder electrodeposition (i_{tot}) and for hydrogen evolution (i_H) practically overlapped), and it was necessary to electrodeposit powders at least for 2 h in order to obtain amount of powder that could be used for

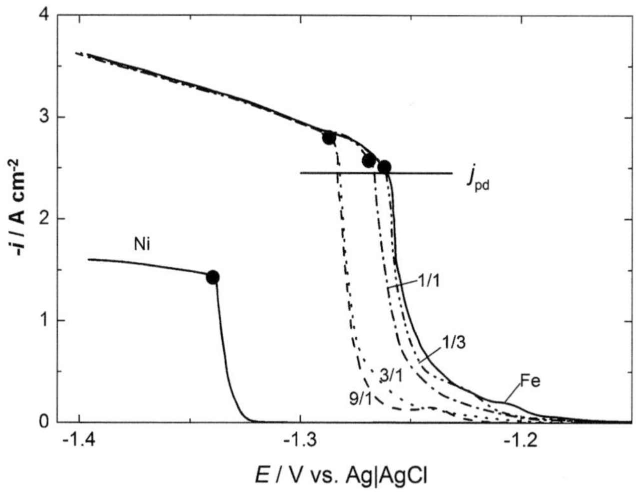

Fig. 8.14 Polarization curves for the electrodeposition of iron (Fe), nickel (Ni), and Fe-Ni alloy powders after IR drop correction recorded for different Ni/Fe ions ratios (9/1, 3/1, 1/1, and 1/3 (marked in the figure)), in the solution of Fe(II) and Ni(II) species (Reprinted from Ref. [1] with kind permission from Springer)

the morphology and composition analysis (SEM, EDS). In the case of Fe(II) salt electrolytes, current efficiency at the potentials more negative than the second inflection point (● in Fig. 8.14) varied between 8 % and 15 % depending on the Ni/Fe ratio, as shown in Fig. 8.15. The average values for the diffusion-limiting current densities for alloy powder electrodeposition were $i_{\text{Fe-Ni}} = -0.26$ A cm^{-2} for the ratio 1/3 and $i_{\text{Fe-Ni}} = -0.49$ A cm^{-2} for the ratio 9/1 [1, 8].

With the increase of iron concentration (as well as with the increase of the amount of iron in the powder – decrease of Ni/Fe ratio), the current efficiency for powder electrodeposition decreased, which is in accordance with the data obtained for compact Fe-Ni alloy electrodeposits [1, 18].

The polarization curves were recorded starting from the potential of −0.6 V vs. Ag|AgCl. By comparing polarization curves for Fe powder electrodeposition from Fe (III) (curve **a**) and Fe(II) (curve **b**) salts in ammonium chloride–sodium citrate electrolyte, shown in Fig. 8.16, certain cathodic current density (~ −2.5 mA cm^{-2}) has been detected already at the starting potential for curve **a** (inset of Fig. 8.16). This cathodic current density remained constant down to the potential of about −1.0 V versus Ag|AgCl, and it started rising at the same potential value at which the current density recorded in the presence of Fe(II) salt started to rise.

A low current efficiency for the Fe-Ni alloy powder electrodeposition from the solution containing Fe(III) salt is the consequence of the first step in the overall reaction being reduction of Fe(III) species into Fe(II) species, taking place at all potentials more negative than −0.2 V versus Ag|AgCl. Nevertheless, some experiments were performed in the solution of Fe(III) ions and the morphology of electrodeposited Fe-Ni alloy powders was investigated.

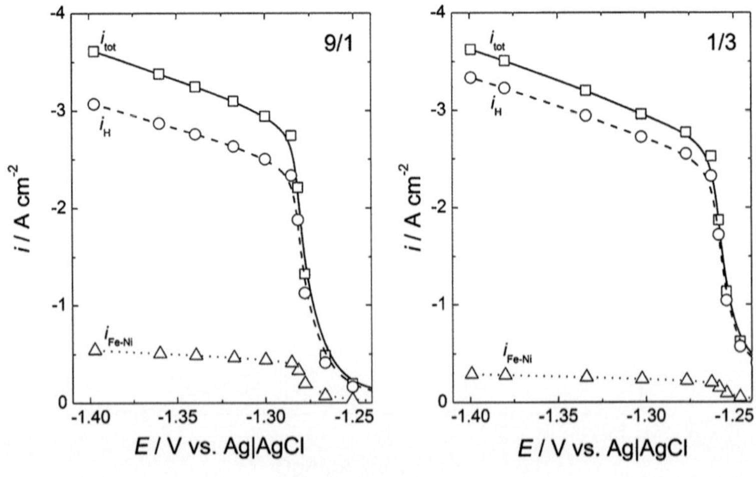

Fig. 8.15 Polarization curve for the Fe-Ni alloy powder electrodeposition (i_{tot}) (□), polarization curve for hydrogen evolution (i_H) (○), and polarization curve for Fe-Ni powder electrodeposition after subtraction of hydrogen evolution current ($i_{Fe\text{-}Ni}$) (Δ). The values of Ni/Fe ions ratios are marked in the figure (Reprinted from Ref. [1] with kind permission from Springer)

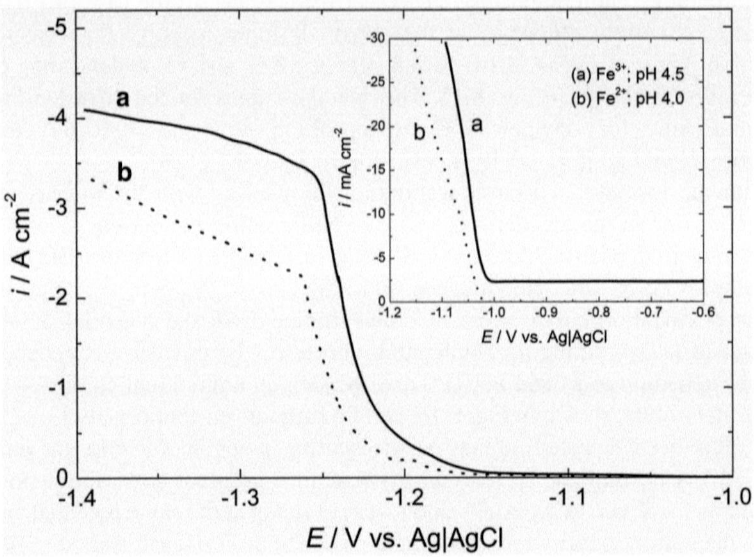

Fig. 8.16 The polarization curves for the electrodeposition of Fe powders from the electrolyte containing 0.1 M $FeCl_3 + 0.2$ M $Na_3C_6H_5O_7 + 1$ M NH_4Cl (**a**) and 0.1 M $FeCl_2 + 0.2$ M $Na_3C_6H_5O_7 + 1$ M NH_4Cl (**b**). Inset – the same polarization curves recorded at potentials more positive than −1.0 V versus Ag|AgCl (Reprinted from Ref. [1] with kind permission from Springer)

8.2.2.2 Morphology of the Fe-Ni Powders Electrodeposited from the Solutions Containing Fe(II) Species

The morphology of the Fe-Ni alloy powders electrodeposited at different Ni/Fe ratios is presented in Figs. 8.17, 8.18, 8.19, and 8.20. A common characteristic of all investigated Fe-Ni powders is the formation of big agglomerates (up to 500 μm).

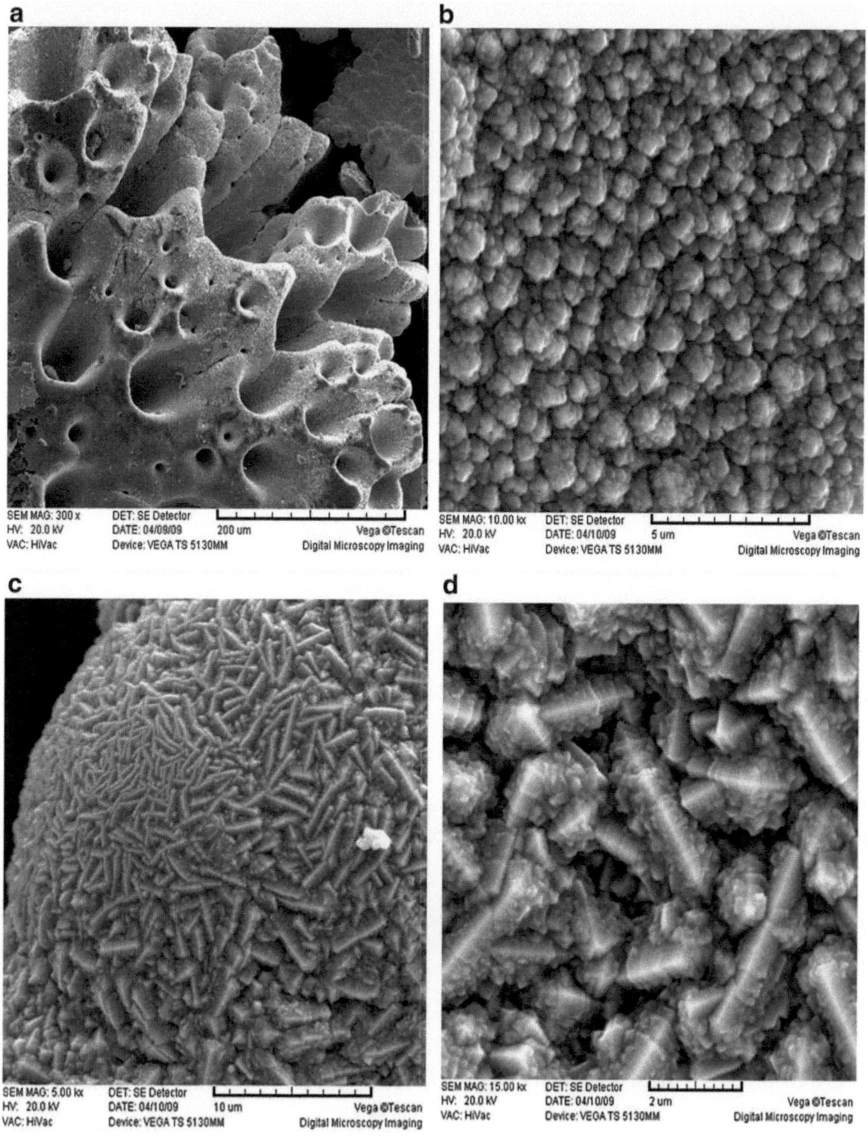

Fig. 8.17 Morphology of the Fe-Ni alloy powder electrodeposited at the Ni/Fe = 9/1. (**a**) Typical agglomerates with the large cone-shaped cavities and much smaller cylindrical cavities, (**b**) Pagoda-like crystals on nodular surfaces, (**c**) Crystals of the shape of elongated prism, (**d**) (**c**) at higher magnification (Reprinted from Ref. [1] with kind permission from Springer)

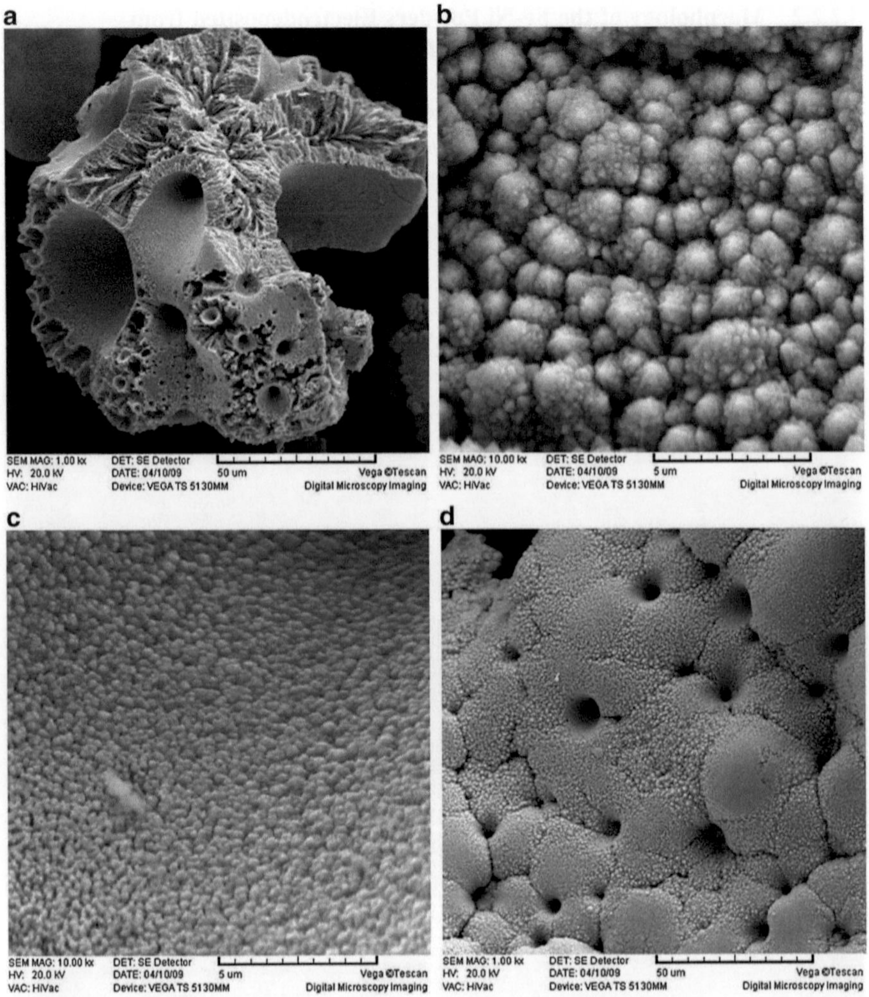

Fig. 8.18 Morphology of the Fe-Ni alloy powder electrodeposited at the Ni/Fe = 3/1 (Reprinted from Ref. [1] with kind permission from Springer)

Their surfaces are covered with large cone-shaped cavities, as well as with much smaller cylindrical cavities.

In Fig. 8.17 are presented typical agglomerates for Fe-Ni alloy powders electrodeposited at the Ni/Fe ratio 9/1. Large cone-shaped cavities and much smaller cylindrical cavities, corresponding to the places were hydrogen bubbles were formed, are clearly seen in Fig. 8.17a. Nodular surfaces are covered with two types of crystals: pagoda-like crystals (b) and crystals of the shape of elongated prism (c, d), indicating possible presence of different phases.

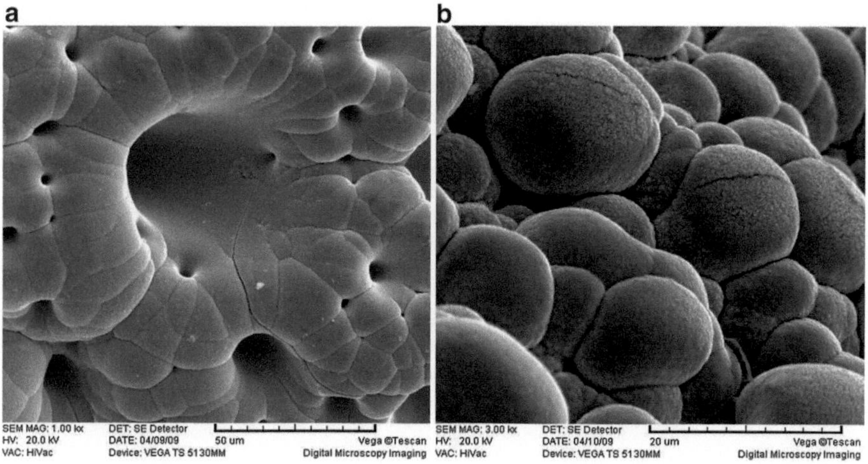

Fig. 8.19 Morphology of the Fe-Ni alloy powder electrodeposited at the Ni/Fe = 1/1. (**a**) Agglomerates with smooth surface of both types of cavities, (**b**) Top view of nodular endings (Reprinted from Ref. [1] with kind permission from Springer)

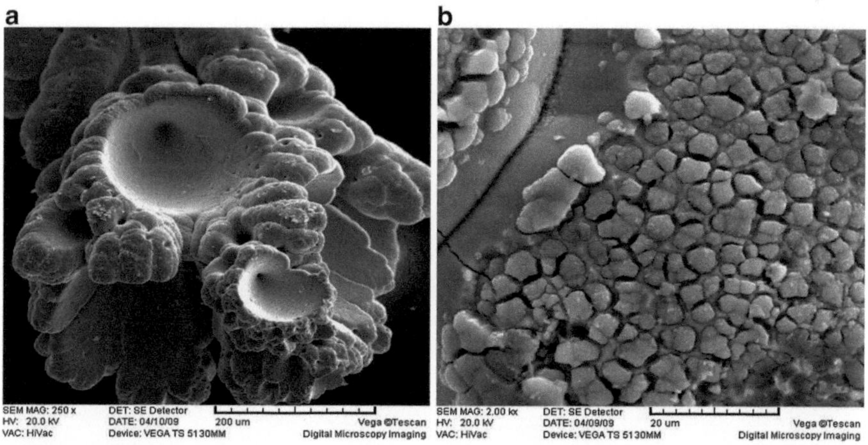

Fig. 8.20 Morphology of the Fe-Ni alloy powder electrodeposited at the Ni/Fe = 1/3. (**a**) Agglomerates with smooth surface of cavities, (**b**) Top view of nodular endings (Reprinted from Ref. [1] with kind permission from Springer)

Large cone-shaped cavities and much smaller cylindrical cavities are also present in the powder electrodeposited at Ni/Fe = 3/1 (Fig. 8.18a). Nodular surfaces (d) as well as the inside surface of cone-shaped cavities (c) are covered with spherical crystals (b).

Much smoother surface of both types of cavities and the nodular endings were obtained from the solution with Ni/Fe = 1/1 (Fig. 8.19).

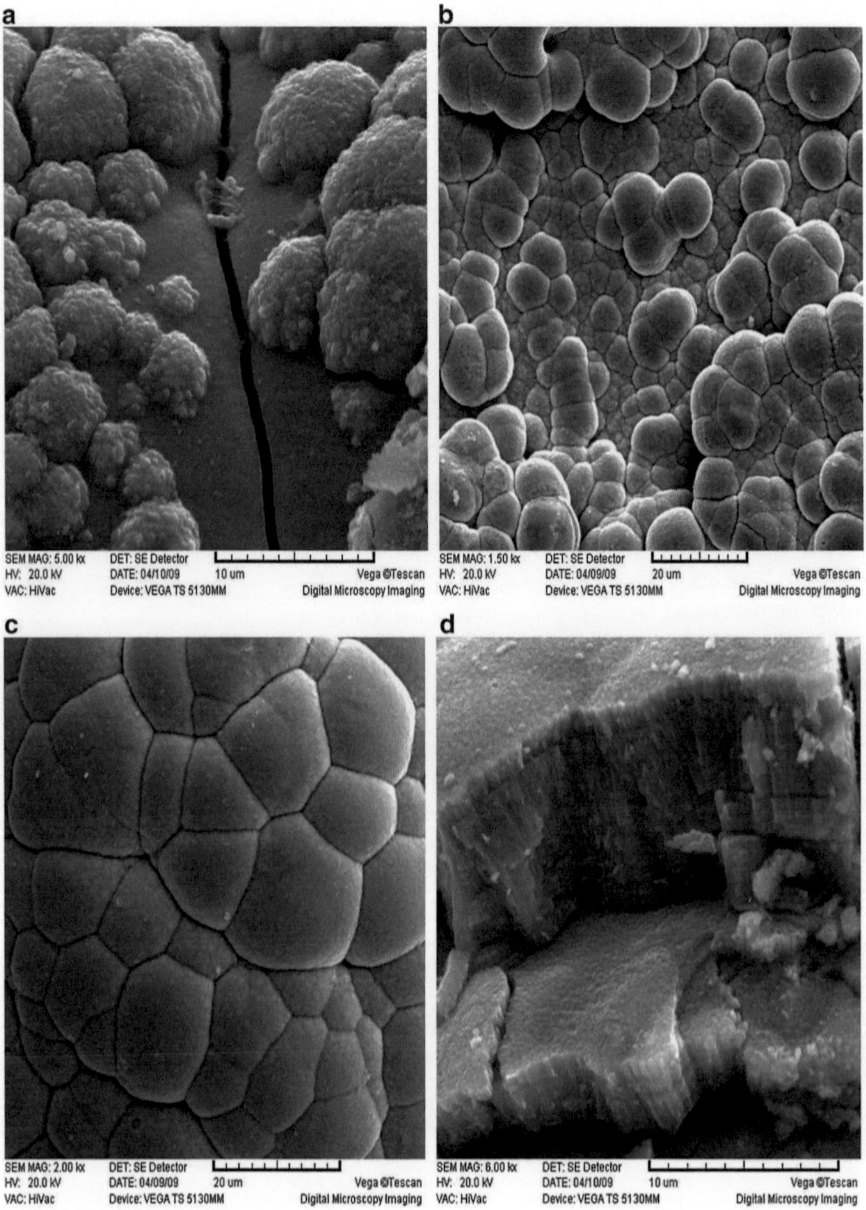

Fig. 8.21 Typical phases of the Fe-Ni powder growth. (**a**) Second generation of spherical grains, (**b**) Second generation of spherical grains on the compact surface, (**c**) Overlapping of spherical grains, (**d**) Layer-by-layer sticklike densely packed crystals of submicron dimensions (Reprinted from Ref. [1] with kind permission from Springer)

In the case of the highest content of iron in the powder (Ni/Fe = 1/3), a similar shape of cavities (Fig. 8.20a) was obtained as in Fig. 8.19a, while the surface of nodular endings was not smooth. On some parts of the agglomerate surface, new crystals were separated with the cracks, as shown in Fig. 8.20b.

Considering Figs. 8.17, 8.18, 8.19, and 8.20, it could be seen that on the compact surface of all agglomerates, smaller spherical (rounded) grains were formed (Figs. 8.21b–d and 8.19b). The formation of such spherical grains is typical for the electrodeposition of copper from acidic sulfate electrolytes under the conditions of activation–diffusion-controlled process [19]. Such mechanism has been explained for the electrodeposition of pure cobalt [17]. The development of spherical grains in the case of Fe-Ni electrodeposition is presented in Fig. 8.21. As can be seen in Fig. 8.21a, the second generation of spherical grains grow in all directions. On the compact surface, spherical grains are well defined (Fig. 8.21b), and since they are close to each other, they start to overlap. During the further growth, the overlapping of the diffusion zones formed over, and each spherical grain becomes more pronounced, producing practically flat surface of the agglomerates (Fig. 8.21c). It is interesting to note that the growth of compact agglomerates of spherical grains occurs through the formation of layer-by-layer sticklike densely packed crystals of submicron dimensions, as shown in Fig. 8.21d. Presented figures are in good agreement with already-given explanation for the electrodeposition of powders.

8.2.2.3 Morphology of the Fe-Ni Powder Electrodeposited from the Solutions Containing Fe(III) Species

The main characteristic of these powders is the presence of high number of both types of cavities on the surfaces of big agglomerates (up to 500 μm), as shown in Fig. 8.22. Taking into account that the current efficiency for the electrodeposition of these powders is only 1–2 %, massive hydrogen evolution must produce larger number of cavities than in previous cases. Also, one of the characteristics of these powders is the presence of several different types of crystals on the surface of agglomerates around the cavities, which are presented in Figs. 8.23 and 8.24. On the agglomerates, surfaces electrodeposited from the solutions with Ni/Fe = 9/1 and 3/1 mostly spherical crystals could be detected, as shown in Fig. 8.23a. At the same time, pagoda-like crystals (b) have been detected on a certain, not negligible, number of agglomerates.

Particularly interesting are crystals of the shape of pagoda. Such crystals were detected in the Fe-Ni powder synthesized in high yield by a simple and facile hydrothermal method without the presence of surfactants [1, 19]. According to this investigation, $FeNi_3$ {111} single crystals were formed during the described procedure. The products obtained at 120 °C were mixture of $FeNi_3$ and Fe-Ni hydroxides composed of monodispersed microspheres (probably spheres presented in Fig. 8.23a). With the increase of temperature to 140 °C, these microspheres became micropagodas very similar to those shown in Fig. 8.23b. At higher temperatures

Fig. 8.22 Typical distribution of cavities on the surface of the Fe-Ni alloy powder agglomerates electrodeposited from the solution containing Fe(III) species (Reprinted from Ref. [1] with kind permission from Springer)

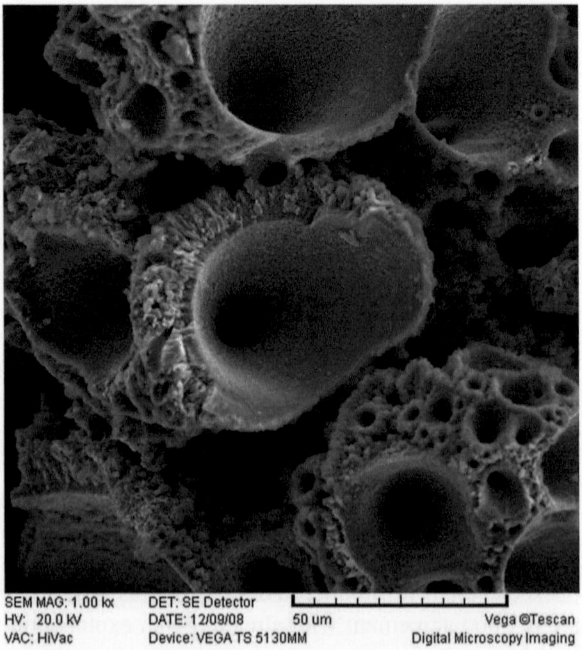

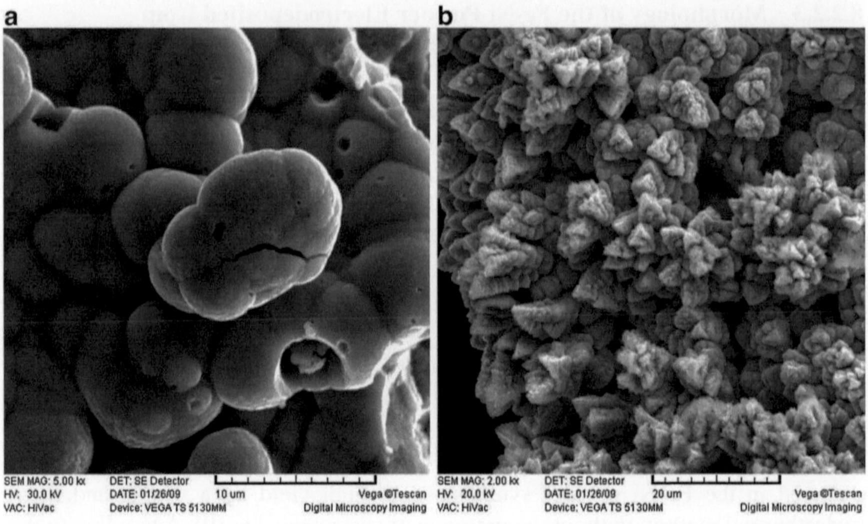

Fig. 8.23 Crystals formed on the surface of the Fe-Ni alloy powder agglomerates electrodeposited from the solution with Ni/Fe = 9/1 (Reprinted from Ref. [1] with kind permission from Springer)

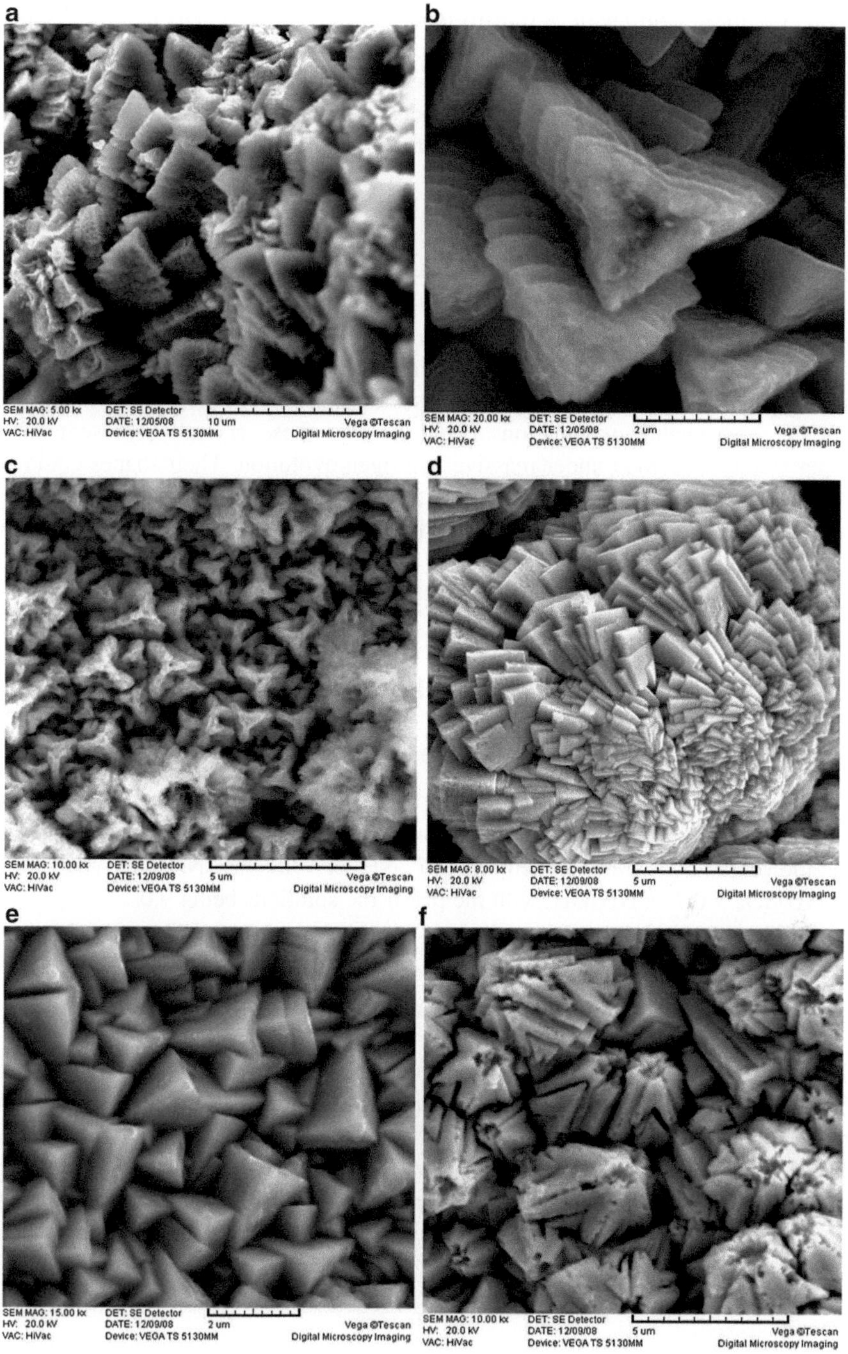

Fig. 8.24 Crystals formed on the surface of the Fe-Ni alloy powder agglomerates electrodeposited from the solution with Ni/Fe < 1 (Reprinted from Ref. [1] with kind permission from Springer)

(180 °C), these agglomerates transformed into perfect 3D $FeNi_3$ dendritic superstructures in certain directions [1, 19]. Hence, comparing the crystals presented in Fig. 8.23b with those obtained by hydrothermal method [1, 19], it seems reasonable to ascribe them to the $FeNi_3$ {111} single crystals (there is also an indication for the existence of $FeNi_3$ phase in electrodeposited powder – see XRD analysis presented in Fig. 28 of Ref. [1]).

As the amount of iron increases in the powders (above 50 at.%), the number of different crystals detected on the agglomerate surfaces also increases. In the powders electrodeposited from the solutions with Ni/Fe = 1/3, a very small amount of agglomerates is covered with spherical crystals, while most of them are characterized with the presence of several different shapes of crystals. These crystals are presented in Fig. 8.24a–f. It is quite difficult to explain the reasons for their appearance, since so many different phases do not exist in the Fe-Ni system and the only reasonable explanations could be that these crystals represent some superstructures formed under massive hydrogen evolution. Their shapes could be defined as follows: (a) Christmas tree-like crystals, (b) triangle-like crystals growing layer by layer, (c) propeller-like crystals, (d) plate-like crystals, (e) tetrahedral crystals, and (f) traversed polyhedron.

8.3 Induced Codeposition of Alloy Powders

8.3.1 Electrodeposited Mo-Ni-O Powders

The Mo-Ni-O powders were electrodeposited from two supporting electrolytes, as in the case of Co-Ni system: 1 M NH_4Cl + 0.7 M NH_4OH and 1 M $(NH_4)_2SO_4$ + 0.7 M NH_4OH, with the pH of the solutions being 9.0.

8.3.1.1 Polarization Curves in Chloride-Containing Electrolytes

Three different electrolytes were used for the investigations: 0.1 M $NiCl_2$ + 1 M NH_4Cl + 0.7 M NH_4OH + x M Na_2MoO_4, with x being 0.05 M, 0.1 M, and 0.3 M, respectively. In such a way, the Ni/Mo ion concentration ratio was 1/0.5, 1/1, and 1/3. All powders for microstructure, composition, and phase composition analysis were electrodeposited at the limiting current density (position of the inflection point B on the polarization curves; see Fig. 8.25). In all cases, small amount of the rough compact electrodeposit (necessary for powder formation) remained on the glassy carbon surface, and only powder agglomerates detached from the cathode surface were analyzed.

The polarization curves are presented in Fig. 8.25. As can be seen, the polarization curves characterized by two inflection points (Fig. 8.25a), as in all previous cases, were obtained. It is important to note that the potential of the beginning of

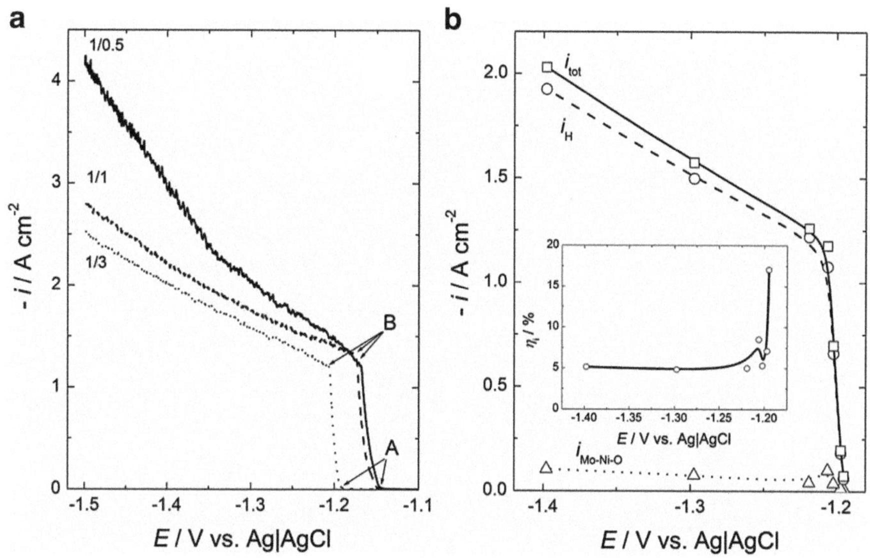

Fig. 8.25 (**a**) Polarization curves in chloride-containing electrolytes for different Ni/Mo ratios (marked in the figure). (**b**) Polarization curve for powder electrodeposition and hydrogen evolution (i_{tot}), polarization curve for hydrogen evolution (i_H), and polarization curve for powder electrodeposition after subtraction of the current density for hydrogen evolution ($i_{Mo\text{-}Ni\text{-}O}$). Inset: current efficiency for Mo-Ni-O powder electrodeposition (η_i) as a function of potential (Reprinted from Ref. [1] with kind permission from Springer)

alloy electrodeposition (A) becomes more negative with the increase of molybdate ion concentration (with the decrease of Ni/Mo ratio), as it could be expected, since the potential of the Mo electrodeposition is much more negative than that of Ni [13]. At the same time, electrodeposition of Mo can only take place in the presence of Ni (induced codeposition [13]). Taking into account that the concentration of Ni $^{2+}$ ions was constant, it is quite reasonable that the value of current density of the inflection point B did not change with changing Ni/Mo ion concentration ratio (being about -1.2 A cm^{-2}).

It should also be mentioned that the current efficiency for alloy electrodeposition in all cases was very low, about 5 %. This could be seen in the inset of Fig. 8.25b for the Ni/Mo ratio 1/3, where the η_i vs. E changes from 17 to 5 % in the region of sharp increase of current density, while at the potentials more negative than the second inflection point (B), the value of η_i is constant, being about 5 % [1, 14].

8.3.1.2 Polarization Curves in Sulfate-Containing Electrolytes

Three different electrolytes were used for the investigations: 0.1 M $NiSO_4 + 1$ M $(NH_4)_2SO_4 + 0.7$ M $NH_4OH + x$ M Na_2MoO_4, with x being 0.03 M, 0.1 M, and

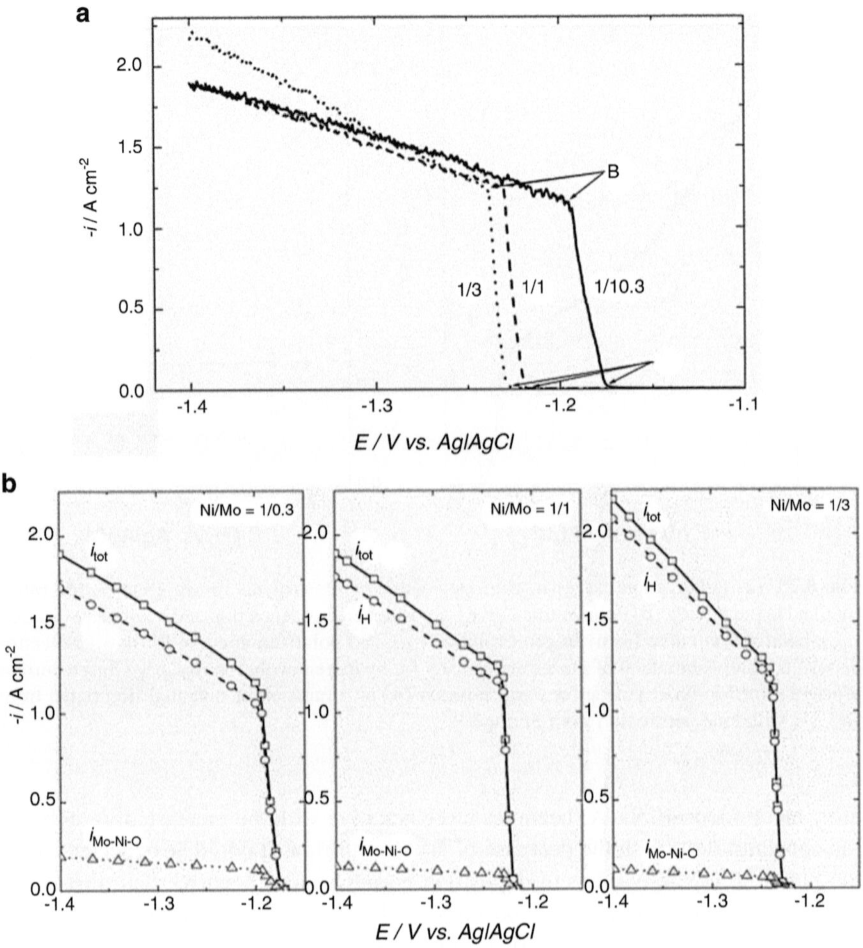

Fig. 8.26 (**a**) Polarization curves in sulfate-containing electrolytes for different Ni/Mo ratios (marked in the figure). (**b**) Polarization curves for powder electrodeposition and hydrogen evolution (i_{tot}), polarization curves for hydrogen evolution (i_H), and polarization curves for powder electrodeposition after subtraction of the current density for hydrogen evolution ($i_{Mo\text{-}Ni\text{-}O}$) (Reprinted from Ref. [1] with kind permission from Springer)

0.3 M, respectively. As in the case of chloride-containing supporting electrolytes, the Ni/Mo ion concentration ratio was 1/0.3, 1/1, and 1/3. The polarization curves recorded in the solution containing sulfate ions are presented in Fig. 8.26. In this case, all polarization curves are moved to slightly more cathodic potentials, indicating higher overvoltage for the process of Mo-Ni-O alloy electrodeposition in the presence of sulfate ions (Fig. 8.26a). At the same time, the total current densities in the range of potentials more negative than the inflection point (B) are smaller and the slopes of their changes are smaller in comparison with those recorded for chloride electrolyte (Fig. 8.25a). As well as in the case of chloride electrolyte, the potential of the beginning of alloy deposition (A) becomes more negative with the increase of molybdate ion concentration (with the decrease of Ni/Mo ratio). As shown in Fig. 8.26b, the current efficiency for alloy deposition is sensitive to the Ni/Mo ions ratio: for Ni/Mo = 1/0.3, $\eta_i = 10$ %, for Ni/Mo = 1/1, $\eta_i = 7$ %, and for Ni/Mo = 1/3, $\eta_i = 5$ % (all η_i values correspond to the potentials more negative than the second inflection point (B) [1, 15].

8.3.1.3 Morphology of Mo-Ni-O Powders Electrodeposited from Chloride-Containing Electrolytes

The morphology of powder agglomerates electrodeposited onto glassy carbon electrode from electrolytes with different Ni/Mo ions concentration ratios (1/0.5, 1/1, and 1/3, samples 1, 2, and 3, respectively) are shown in Figs. 8.27 and 8.28, respectively. For the highest Ni/Mo ratio of 1/0.5 (Fig. 8.27), size of the agglomerates varies in the range of about 2–50 μm. At the same time, sharp edges could be detected on top of the surfaces of all agglomerates, indicating crystallinity of electrodeposited powder (Fig. 8.27b) [1, 14].

With the decrease of the Ni/Mo ratio (1/1), cauliflower-type agglomerates, Fig. 8.28a, b, characterized with spherical edges (Fig. 8.28c) and the presence of cracks, were obtained. Their size is much higher than that of the powder particles electrodeposited at the 1/0.5 ratio, varying in the range of about 50–500 μm. In the case of Ni/Mo ratio of 1/3 (Fig. 8.28d), flat and thin parts of the powder,

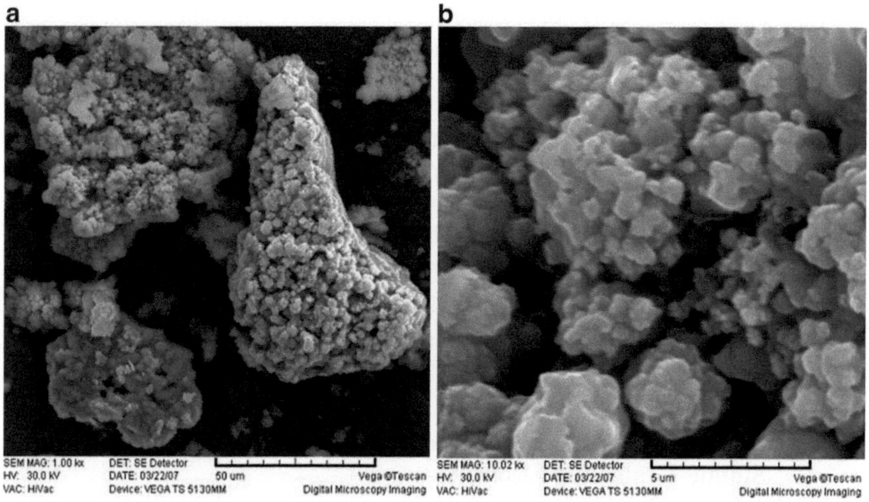

Fig. 8.27 Typical agglomerates for the Mo-Ni-O powders electrodeposited at the Ni/Mo = 1/0.5. (**a**) Agglomerates of the size in the range of about 2–50 μm, (**b**) Top view of sharp edges at the surfaces of agglomerates (Reprinted from Ref. [1] with kind permission from Springer)

Fig. 8.28 (**a**–**c**) Typical particles for the Mo-Ni-O powders electrodeposited at the Ni/Mo = 1/1. (**d**) Typical particles for the Mo-Ni-O powders electrodeposited at the Ni/Mo = 1/3 (Reprinted from Ref. [1] with kind permission from Springer)

corresponding most likely to the compact electrodeposit, were covered with spherical agglomerates, while the cracks are more pronounced. The appearance of a large number of cracks is the result of high tensile stresses present in the powder and the hydrogen evolution [20]. According to the presented results, it is obvious that the crystallinity of the particles becomes less pronounced with the increase of molybdate ions in the electrolyte, influencing at the same time the morphology of electrodeposited powder particles.

8.3.1.4 Morphology of Mo-Ni-O Powders Electrodeposited from Sulfate-Containing Electrolytes

The morphology of powder particles electrodeposited from electrolytes with different Ni/Mo ion concentration ratios (1/0.3, 1/1, and 1/3) is shown in Figs. 8.29, 8.30, and 8.31. For the Ni/Mo ratio, 1/0.3 typical spongy particles were detected

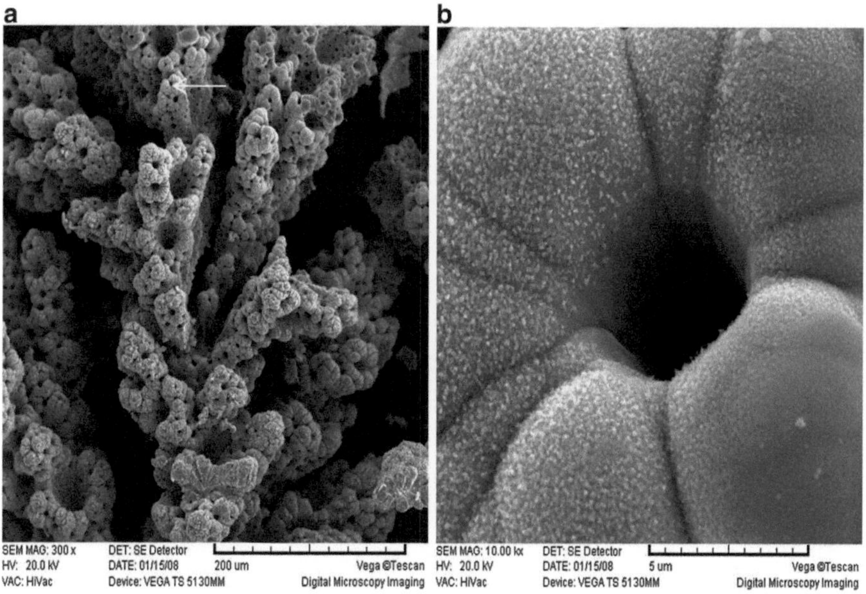

Fig. 8.29 (**a**) Typical spongy particles electrodeposited at the Ni/Mo = 1/0.3. (**b**) Cavity marked with *arrow* in (**a**) at higher magnification (Reprinted from Ref. [1] with kind permission from Springer)

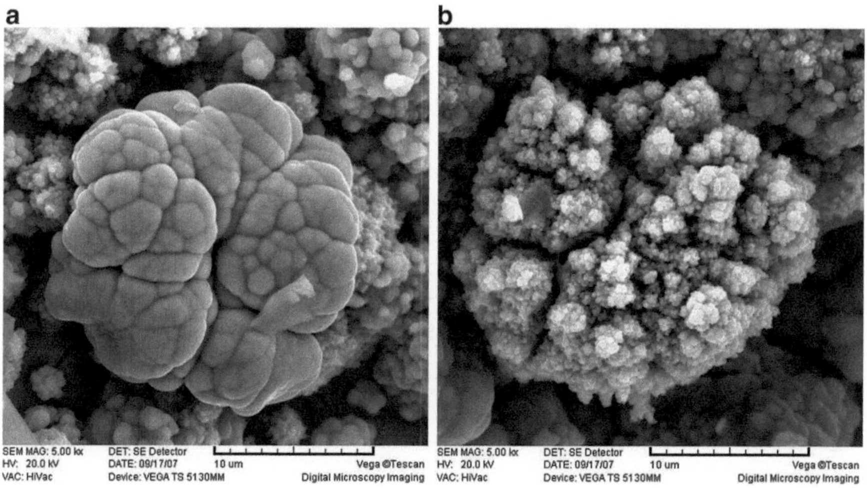

Fig. 8.30 Typical spongy particles electrodeposited at the Ni/Mo = 1/1 (Reprinted from Ref. [1] with kind permission from Springer)

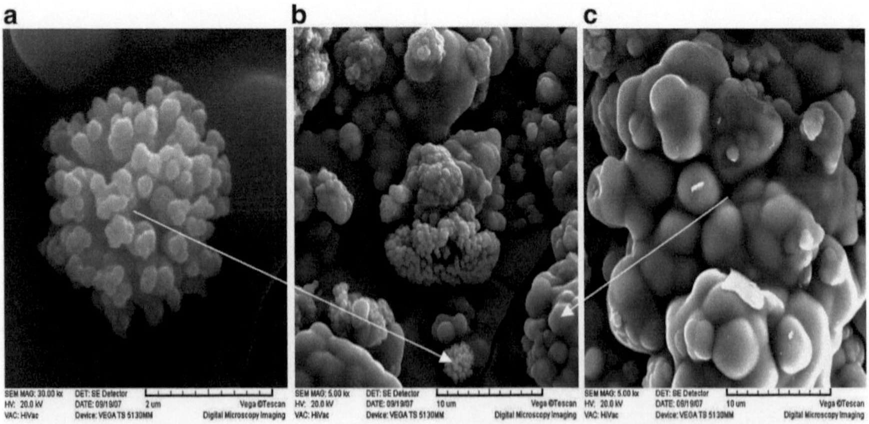

Fig. 8.31 (**b**) Typical powder electrodeposited at the Ni/Mo = 1/3. (**a**) Agglomerate with dendritic surface marked with the arrow in (**b**). (**c**) Agglomerate with flat surface marked with the arrow in (**b**) (Reprinted from Ref. [1] with kind permission from Springer)

(Fig. 8.29a). Characteristic of these particles is very flat surface around the holes, as shown in Fig. 8.29b.

With the decrease of the Ni/Mo ratio (1/1), two types of particles were obtained, as shown in Fig. 8.30. Both are spherical (rounded), one with flat edges (a) and another one with rough endings on top of the surface (b). Taking into account our previous investigations of alloy powder morphology [1, 3–6], this is not unexpected. Since the growth of powder particles is characterized by the presence of two or more growth zones, it appears that spherical particles with flat edges have been detached from the electrode surface before the beginning of a growth of new zone, while the particles with rough endings have been detached after the beginning of the growth of new crystals. It should be noted (not shown in this figure) that some particles were characterized by the presence of cracks.

In the case of Ni/Mo ratio of 1/3 (powder with the highest amount of molybdenum), a completely different morphology was detected, as shown in Fig. 8.31. The powder is composed of flat and thin parts with pronounced cracks, corresponding most likely to the compact electrodeposit, and spherical agglomerates growing on top of the flat electrodeposit (b). The surfaces of these agglomerates are either dendritic (a) or flat (c).

8.3.1.5 The Results of TEM Analysis of Mo-Ni-O Powders Electrodeposited from Sulfate Electrolyte

The TEM analysis was performed on two Mo-Ni-O powders: Ni-rich (Ni/Mo = 1/0.3) and Mo-rich (Ni/Mo = 1/3) powders. The common characteristic of both powders is the presence of amorphous and crystalline particles [1, 16].

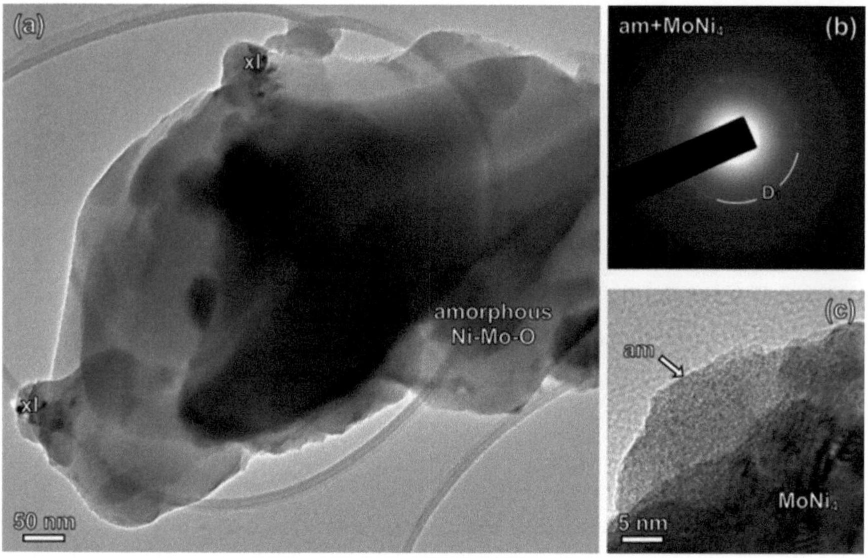

Fig. 8.32 (**a**) TEM image of the amorphous Mo-Ni-O particle detected in the powder electrodeposited at the Ni/Mo ratio 1/0.3. (**b**) SAED pattern recorded from this area (**a**) shows predominantly amorphous material (am) with weak reflections corresponding to $MoNi_4$ nanocrystals (x1). (**c**) High magnification of amorphous Mo-Ni-O particle with $MoNi_4$ nanocrystals (Reprinted from Ref. [1] with kind permission from Springer)

The TEM Analysis of Electrodeposited Ni-Rich Mo-Ni-O Powders

The specimen is composed of large amorphous clusters containing nanosized crystalline particles. Fig. 8.32 shows a TEM image of an amorphous particle (a) and the corresponding SAED pattern (b). Diffuse diffraction rings clearly indicate amorphous character of this powder particle. Weak spots in the diffraction pattern correspond to the {211} reflections of fine $MoNi_4$ nanoparticles (JCPDF # 65–1533) [21], shown in the close-up in Fig. 8.32c. In the crystalline parts of the Ni-rich powder, the following phases were detected: NiO, MoO_3, and $MoNi_4$, in form of idiomorphic crystallites measuring up to 70 nm in diameter. Figure 8.33 shows a cluster of well-crystallized nanoparticles. The SAED pattern (shown in Fig. 8.33b) from these clusters best corresponds to the monoclinic MoO_3 phase (JCPDF # 89–1554) [22], or one of its structural variants, depending on the position in the sample. Typical morphology of euhedral MoO_3 crystals is shown in Fig. 8.33a and on the close-up in Fig. 8.33c. Most of the crystals are tabular on {200}. The strongest SAED diffraction rings in the pattern correspond to $D_1 = \{200\}$, $D_2 = \{211\}$, $D_3 = \{220\}$, $D_4 = \{022\}$. In the Ni-rich areas of the sample, we observe crystalline $MoNi_4$ and NiO. $MoNi_4$ appears in form of irregularly intergrown grains measuring up to 10 nm in average.

An area of prevailing $MoNi_4$ composition is shown in Fig. 8.34a, with corresponding SAED pattern in Fig. 8.34b. The strongest reflections were indexed

Fig. 8.33 (**a**) TEM image of euhedral MoO_3 nanocrystals in the sample with the Ni/Mo ratio 1/0.3. (**b**) Electron diffraction pattern from this area indicates the presence of monoclinic MoO_3 phase. The strongest SAED diffraction rings in the pattern correspond to $D_1 = \{200\}$, $D_2 = \{211\}$, $D_3 = \{220\}$, $D_4 = \{022\}$. (**c**) MoO_3 crystals are terminated with well-defined {200} dominating pinacoidal faces (Reprinted from Ref. [1] with kind permission from Springer)

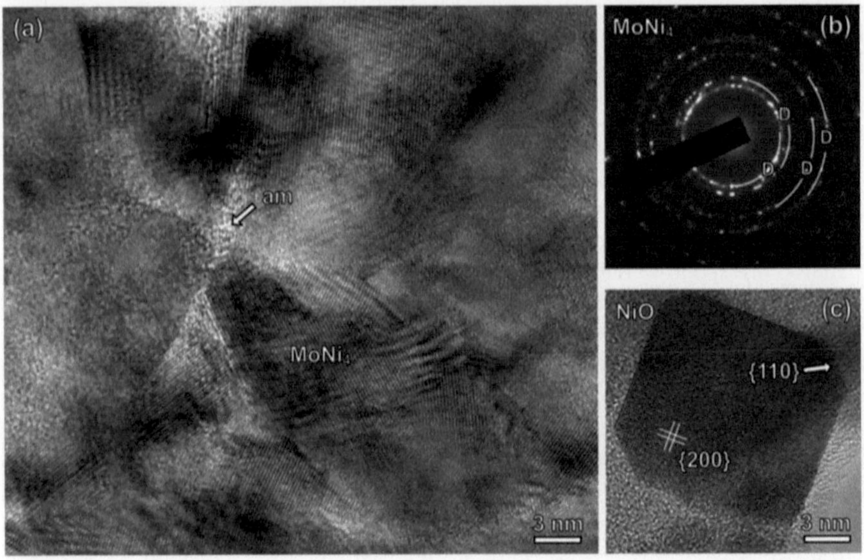

Fig. 8.34 (**a**) High-resolution TEM image of intergrown $MoNi_4$ nanocrystals. In the pockets of $MoNi_4$ grains, there are remnants of uncrystallized amorphous phase (am). (**b**) SAED pattern shows rings of high-intensity spots at reciprocal distances corresponding to $MoNi_4$ phase. The strongest reflections were indexed to $D_1 = \{211\}$, $D_2 = \{002\}$, $D_3 = \{132\}$, $D_4 = \{501\}$. (**c**) Idiomorphic NiO particle (rock-salt structure) with well-resolved {200} lattice planes. The crystal is faceted with cube {001} and dodecahedral {110} faces (Reprinted from Ref. [1] with kind permission from Springer)

to $D_1 = \{211\}$, $D_2 = \{002\}$, $D_3 = \{132\}$, $D_4 = \{501\}$ [21]. In some areas, we also find isolated crystals of NiO with a predominating cubic morphology and the edges truncated by rhombic dodecahedral faces. A close-up of a typical NiO nanocrystal is shown in Fig. 8.34c, showing well-resolved {200} lattice planes. The diffraction rings of the NiO phase cannot be seen most probably due to much stronger diffraction lines for the $MoNi_4$ and MoO_3 phases. Accordingly, its presence is determined by the measurement of *D* values, as shown in Fig. 8.34c [16].

On both SAED patterns, as well as on other SAED patterns, additional diffraction lines (more or less pronounced) are present, indicating that at the position of TEM analysis, other phases are also present (actually at each position of TEM analysis, a mixture of different phases exists), but the best defined are the ones determined by the analysis of the SAED patterns. Hence, it could be concluded that the $MoNi_4$ and MoO_3 phases prevail in the powder with high Ni content (electrodeposited at the Ni/Mo ratio 1/0.3), while some amount of NiO phase has also been detected.

The TEM Analysis of Electrodeposited Mo-Rich Mo-Ni-O Powders

A general appearance of Mo-rich sample does not differ much from the Ni-rich. The TEM of typical amorphous cluster is shown in Fig. 8.35a. In the small crystalline areas (x1), traces of $MoNi_4$ and MoO_3 phases were detected, and their diffraction

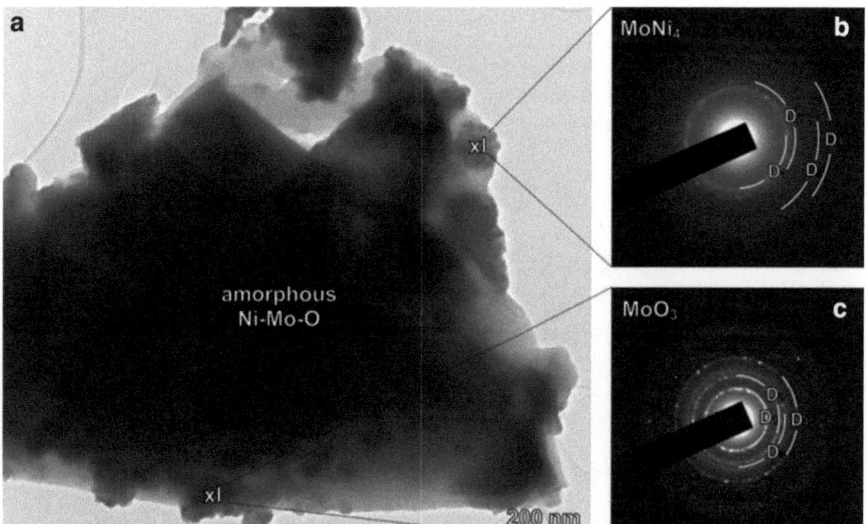

Fig. 8.35 (**a**) TEM image of the amorphous Mo-Ni-O cluster containing some crystalline areas in the powder electrodeposited at the Ni/Mo ratio 1/3. (**b**) SAED pattern from the area rich in $MoNi_4$ nanocrystals, similar to those in the Ni-rich sample, shown in Fig. 8.34. (**c**) SAED pattern of the area containing MoO_3 nanocrystals, similar to those in the Ni-rich sample, shown in Fig. 8.33 (Reprinted from Ref. [1] with kind permission from Springer)

patterns are presented in Fig. 8.35b, c, respectively. The diffraction lines in Fig. 8.35b are barely visible, indicating amorphous character of the $MoNi_4$ phase, while much stronger diffraction lines for the MoO_3 phase (Fig. 8.35c) indicate its more pronounced crystallinity. In this sample, the presence of NiO phase has not been detected.

It appears that in both samples, $MoNi_4$ phase is more strongly bound to the amorphous phase than the oxide phases. This might be the reason why in our previous investigation [1, 14, 15] this phase has clearly been detected by XRD in the recrystallized samples only after annealing at 600 °C, while at all other temperatures a broad peak around $2\theta = 44°$ was only an indication of its existence.

It should be emphasized here that most of the phases detected in the recrystallized electrodeposited Mo-Ni-O powders [14, 15] ($MoNi_4$ and MoO_3) were also found in as-electrodeposited Mo-Ni-O powders, indicating their formation during the process of alloy electrodeposition.

Taking into account that NiO, MoO_3, and $MoNi_4$ phases were detected in electrodeposited samples by TEM, a new mechanism of Mo-Ni-O alloy electrodeposition (see Eqs. (7.20) and (7.23)) has been proposed.

References

1. Jović VD, Lačnjevac UČ, Jović BM (2012) Morphology, chemical, and phase composition of electrodeposited Co-Ni, Fe-Ni, and Mo-Ni-O powders. In: Djokić S (ed) Modern aspects of electrochemistry, vol 54. Springer Science + Business Media, New York, pp 251–343
2. Abd El-Halim AM, Khalil RM (1986) Some characteristics of Ni-Co alloy powders electrodeposited from dilute sulphate baths. Surf Coat Technol 27:103–115
3. Jović VD, Maksimović V, Pavlović MG, Popov KI (2006) Morphology, internal structure and growth mechanism of electrodeposited Ni and Co powders. J Solid State Electrochem 10:373–379
4. Jović VD, Jović BM, Pavlović MG, Maksimović V (2006) Morphology and composition of Ni-Co alloy powders electrodeposited from ammoniacal electrolyte. J Solid State Electrochem 10:959–966
5. Jović VD, Jović BM, Maksimović V, Pavlović MG (2007) Electrodeposition and morphology of Ni, Co and Ni-Co alloy powders part II. Ammonium chloride supporting electrolyte. Electrochim Acta 52:4254–4263
6. Jović VD, Jović BM, Pavlović MG (2006) Electrodeposition of Ni, Co and Ni-Co alloy powders. Electrochim Acta 51:5468–5477
7. Jović VD, Maksimović V, Pavlović MG, Jović BM (2006) Morphology and phase composition of Ni-Co alloy powders electrodeposited from ammonium sulfate-boric acid electrolyte. Mater Sci Forum 518:307–312
8. Lačnjevac U, Jović BM, Jović VD (2009) Morphology and composition of the Fe-Ni powders electrodeposited from citrate containing electrolytes. Electrochim Acta 55:535–543
9. Lačnjevac U, Jović BM, Maksimović VM, Jović VD (2010) An attempt to produce $NiFe_2O_4$ powder from electrodeposited Fe-Ni alloy powders by subsequent recrystallization in air. J Appl Electrochem 40:701–708
10. Zhelibo EP, Kravets NN, Gamarkin MY, Remez SV (1995) Theory, preparation technology, and properties of powders and fibers. Effect of heat treatment on the properties, structure, and

magnetic properties of Iron-Nickel alloys with micron particle size. Powd Metall Metal Ceram 34:113–116

11. Zhelibo EP, Kravets NN (1997) Theory, preparation technology, and properties of powders and fibers. Influence of the electrolysis temperature on the formation, composition, and magnetic properties of highly dispersed Iron and Iron-Nickel alloy powders. Powd Metall Metal Ceram 36:264–268
12. Chu CM (2003) The effect of complexing agents on the electrodeposition of Fe-Ni powders. J Chin Inst Engrs 34:689–695
13. Brenner A (1963) Electrodeposition of alloys, vol II, Priniciples and practice. Academic, New York
14. Jović BM, Jović VD, Maksimović VM, Pavlović MG (2008) Characterization of electrodeposited powders of the system Mo-Ni-O. Electrochim Acta 53:4796–4804
15. Lačnjevac U, Jović BM, Baščarević Z, Maksimović VM, Jović VD (2009) Morphology and phase composition of as-deposited and recrystallized Ni-Mo-O powders. Electrochim Acta 54:3115–3123
16. Jović VD, Jović BM, Lačnjevac U, Branković G, Bernik S, Rečnik A (2010) An attempt to predict the mechanism of Mo-Ni-O powders electrodeposition from the results of their TEM analysis. Electrochim Acta 55:4188–4193
17. Jović VD, Nikolić ND, Lačnjevac UČ, Jović BM, Popov KI (2012) Morphology of different electrodeposited pure metal powders. In: Djokić SS (ed) Modern aspects of electrochemistry, vol 54. Springer Science + Business Media, New York, pp 63–123
18. Kieling VC (1997) Parameters influencing the electrodeposition of Ni-Fe alloys. Surf Coat Technol 96:135–139
19. Zhou XM, Wei XW (2009) Single crystalline FeNi dendrites: large scale synthesis, formation mechanism, and magnetic properties. Cryst Growth Des 9:7–12
20. Popov KI, Djokić SS, Grgur BN (2002) Fundamental aspects of electrometallurgy. Kluwer, New York
21. Harker D (1944) The crystal structure of Ni_4Mo. J Chem Phys 12:315–320
22. Parise JB, McCarron EM III, Von Dreele R, Goldstone JA (1991) β-MoO_3 produced from a novel freeze drying route. J Solid State Chem 93:193–201

Chapter 9
Chemical Deposition of Metals and Alloys from Aqueous Solutions

9.1 Introduction

The term *chemical deposition* of metals and/or alloys from aqueous solutions is usually used to refer to the production of metallic coatings or powders of various surface morphology and properties without an application of the external current source. As explained in previous chapters in the electrochemical deposition, electrons used for the reduction of metal ions are provided by an external current source. For the chemical deposition, electrons used for the reduction of metal ions are released under specific conditions from an appropriate reducing agent. These appropriate reducing agent may include compounds such as hypophosphite (NaH_2PO_2), borohydride ($NaBH_4$), formaldehyde (HCOH), ascorbic acid ($C_6H_8O_6$), etc., or metals which are less noble than the metal aimed to be deposited. These concepts of chemical deposition will be examined in details in the following text.

It is obvious from the present literature that many aspects of the chemical deposition of metals and/or alloys are ambiguous. Frequently, the chemical deposition is called as *electroless deposition*. As such, the word *electroless* is used in order to stress out that there is no application of the external electrical current to the system where the deposition takes place. In general terms, this is correct; however, clear differences among dissimilar systems are not precisely described. For example, heterogeneous processes involving a metal substrate and homogeneous aqueous solution containing ions of the more noble metal without a presence of the reducing agent, e.g., cementation of Cu on Fe, Ag on Cu, Zn on Al, and similar, are called electroless deposition. Similarly, deposition of metals from an aqueous solution containing a reducing agent for which a presence of a solid substrate (metallic or nonmetallic) is needed for smooth coatings but not absolutely required for powders is, again, called electroless deposition. Although these two processes are fundamentally different, in the present literature, they are discussed or grouped together as electroless processes. In order to clarify for the reader, in this book, all

K.I. Popov et al., *Morphology of Electrochemically and Chemically Deposited Metals*, DOI 10.1007/978-3-319-26073-0_9

processes of deposition of metals and/or alloys without an application of the external current will be considered as *electroless* or *chemical* processes. To further clarify, all electroless processes are divided into the two following groups: *galvanic displacement deposition* (heterogeneous processes, exclusively) and *autocatalytic deposition* (combined homogenous and heterogeneous processes).

9.2 Types of Chemical Deposition of Metals from Aqueous Solutions

As explained in the introduction section of this chapter, there are two main types of chemical deposition of metals from aqueous solutions, and they can be distinguished as:

(a) Galvanic displacement deposition
(b) Autocatalytic deposition

Follows the detailed description of these two processes.

9.2.1 *Galvanic Displacement Deposition*

By the definition, the galvanic displacement deposition is a heterogeneous process in which the noble metal is deposited at the surface of an active metal [1]. The consequence is that the less noble (or active) metal is oxidized or dissolved in the appropriate solution. As a result, the ions of a more noble metal present in the solution are reduced leading to the deposition of the more noble metal. This situation can be described using the electrochemical half reactions in the following way.

(a) Oxidation and consequent dissolution of the less noble metal, M_1:

$$M_1^{\mathrm{o}} \rightarrow M_1^{z_1^+} + z_1 \mathrm{e}^- \tag{9.1}$$

(b) Reduction and consequent deposition of the more noble metal, M_2:

$$M_2^{z_2^+} + z_2 \mathrm{e}^- \rightarrow M_2^{\mathrm{o}} \tag{9.2}$$

In the galvanic displacement deposition, the electrons produced due to oxidation or dissolution of the metal M_1, as shown by the reaction (9.1), are further used for the reduction of the metallic ions of the metal M_2 (reaction (9.2)). Consequently, a combination of the reactions (9.1) and (9.2) leads to:

$$M_2^{z_2^+} + \frac{z_2}{z_1} M_1^{\mathrm{o}} \rightarrow \frac{z_2}{z_1} M_1^{z_1^+} + M_2^{\mathrm{o}} \tag{9.3}$$

The reaction (9.3) represents a generic description of the galvanic displacement deposition. This relatively simple process does not require any complicated equipment. Practically speaking, only a beaker (or a reactor) for a solution containing ions of the more noble metal and a less noble metallic substrate immersed into the solution is needed for this process to proceed. Although not very descriptive, other names for the galvanic displacement reaction used in practice include *cementation* or *immersion plating*.

Even not recognized as such, the galvanic displacement deposition of noble metals such as Au or Ag onto Fe, Zn, Cu, or similar substrates is known since the times of early Mediterranean cultures and, possibly, before. In the sixteenth century, the recovery of copper from copper mine waters by contacting dilute process streams with iron scrap was successfully achieved [2]. Since that time, many different galvanic displacement deposition processes have been developed. Examples used on industrial scale include application of aluminum, iron, or zinc powders for the removal of copper, silver, gold, or other noble metals from waste solutions. Similar approaches are used for the solution purification in hydrometallurgical plants, electronics, electrochemical experiments, etc.

9.2.1.1 Kinetics and Mechanism of the Galvanic Displacement Reaction

The kinetics of the galvanic displacement reactions is usually studied using the following generic equation:

$$\frac{\mathrm{d}C}{\mathrm{d}t} = -kC^n \tag{9.4}$$

where $\mathrm{d}C/\mathrm{d}t$ is the rate of deposition, C is the concentration of ions of the more noble metal, t is time, and n is the reaction order. Most of the galvanic displacement reactions follow the first order and, consequently, $n = 1$. k is the rate constant, and it can be expressed as:

$$k = \frac{DA}{\delta V} \tag{9.5}$$

where D is diffusion coefficient, A is the surface area on which the deposition takes place, δ is the boundary layer thickness, and V is the volume of electrolyte.

Practical application of Eqs. (9.4) and (9.5) can be illustrated on the example of aluminum metal immersed into an alkaline solution containing Cu(II) ions [3]. Deposition of copper onto aluminum from the complexed alkaline solution containing Cu(II) ions is a galvanic process resulting from the dissolution of aluminum. The dissolution of aluminum in strong alkaline solutions can be expressed with the following reaction:

$$2Al + 6H_2O + 2OH^- \rightarrow 2Al(OH)_4^- + 3H_2 \tag{9.6}$$

The reaction (9.6) has $\Delta G^o = -857.946$ kJ/mol < 0, confirming that the process is spontaneous, and it represents a net reaction combined with a cathodic reaction:

$$2H_2O + 2e^- \rightarrow H_2 + 2OH^- \tag{9.7}$$

and an anodic reaction:

$$Al + 4OH^- \rightarrow Al(OH)_4^- + 3e^- \tag{9.8}$$

The deposition of copper on the surface of aluminum metal, as a consequence of its dissolution, is a cathodic process that can simply be presented by the reaction:

$$Cu^{2+} + 2e^- \rightarrow Cu \tag{9.9}$$

The whole process of copper deposition onto the aluminum metal surface can simply be described by the following equation, which is obtained by a combination of the reactions (9.7, 9.8, and 9.9):

$$4Al + 10\,OH^- + 6\,H_2O + 3\,Cu^{2+} \rightarrow 3\,H_2 + 4\,Al(OH)_4^- + 3Cu \tag{9.10}$$

The Gibbs energy for the reaction (9.10), $\Delta G^o = -2195.2$ kJ/mol < 0 is negative, evidencing that the deposition of copper onto aluminum from an alkaline solution is the spontaneous process under the standard physicochemical conditions. According to the reaction (9.10), besides copper deposition, hydrogen evolution takes place as well in this process. The simultaneous hydrogen evolution during the galvanic displacement deposition of copper onto aluminum in the alkaline solutions can further significantly influence the morphology of the deposits as it will be shown in the later section.

The kinetics of the deposition of copper onto aluminum from alkaline solutions, therefore, can be expressed with the equation:

$$\frac{d[Cu(II)]}{dt} = -k'A\frac{[Cu(II)]}{V} \tag{9.11}$$

where [Cu(II)] is the concentration of Cu(II) ions, t is time, A is the surface area, V is the volume of the electrolyte, and k' is the rate constant defined as:

$$k' = \frac{D}{\delta} \tag{9.12}$$

In Eq. (9.12), D is the diffusion coefficient and δ is the diffusion layer thickness.

The results of the deposition of copper onto aluminum can then be studied by integrating Eq. (9.11), which leads to:

$$\log \frac{[\mathrm{Cu(II)}]_t}{[\mathrm{Cu(II)}]_o} = -\frac{k'A}{2.303V}t \tag{9.13}$$

where $[\mathrm{Cu(II)}]_o$ is the initial Cu(II) concentration and $[\mathrm{Cu(II)}]_t$ is the concentration of copper ions at a specific time.

It is obvious from Eq. (9.13) that the dependence of $\log\{[\mathrm{Cu(II)}]_t/[\mathrm{Cu(II)}]_o\}$ on time should be linear. Indeed, the experimental results on the deposition of copper from an alkaline solution onto aluminum powder via the galvanic displacement reaction show that there is a linear dependence between $\log\{[\mathrm{Cu(II)}]_t/[\mathrm{Cu(II)}]_o\}$ and time as illustrated in Fig. 9.1 [3]. As can be seen from Fig. 9.1, the linear relationship was confirmed for all the investigated surface areas of the aluminum substrate. Furthermore, these results clearly show that an increase in the surface area of the substrate leads to the increase in the rate of copper deposition, which is a consequence of Eq. (9.13).

Considering that during the galvanic displacement deposition of copper the aluminum substrate is dissolved, alternatively, the rate of this process can be studied in the following way. Due to dissolution of aluminum, the concentration of Al(III) ions in the solution should increase with time. Eventually, complete aluminum substrate could be dissolved under proper conditions, during the deposition of copper. By applying Faraday's law to Eq. (9.8), the concentration of Al(III)

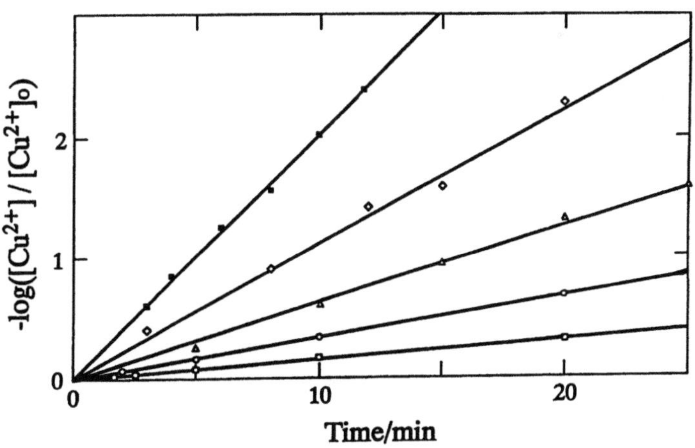

Fig. 9.1 Linear plots of – log ($[\mathrm{Cu}^{2+}/[\mathrm{Cu}^{2+}]_o$) vs. time for various masses of Al powder: (□) 0.1116 g, (○) 0.234 g, (Δ) 0.5617 g, (◇) 1.055 g, and (■) 1.562 g (initial Cu^{2+} concentration 3.56 g/dm^3, pH 13.5) (Reproduced from Ref. [3] with permission from The Electrochemical Society)

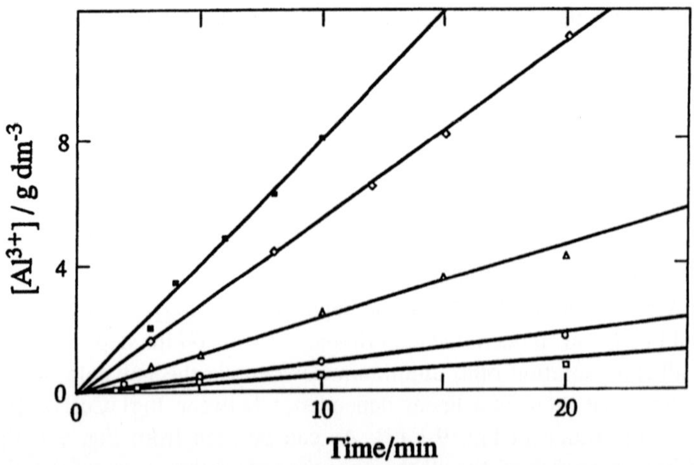

Fig. 9.2 Dependence of $[Al^{3+}]$ in solution on time for various initial amounts of Al powder: (□) 0.1116 g, (○) 0.234 g, (◇) 1.055 g, and (■) 1.562 g. (Initial Cu^{2+} concentration 3.56 g/dm3, pH 13.5) (Reproduced from Ref. [3] with permission from The Electrochemical Society)

in the solution, as a function of time, can be calculated according to the following reaction:

$$\left[\mathrm{Al}^{3+}\right] = \frac{A\, i_{\mathrm{Al}}}{F\, V\, n_{\mathrm{Al}}} t \tag{9.14}$$

where $[Al^{3+}]$ is the concentration of the Al(III) in solution, i_{Al} is the current density of the aluminum dissolution, n_{Al} is the number of electrons, F is the Faraday's constant, and V is the volume of the electrolyte. Obviously, the concentration of aluminum ions, $[Al^{3+}]$, should linearly increase with time, according to Eq. (9.14). This is experimentally confirmed, as illustrated in Fig. 9.2. Figure 9.2 shows that there is a linear increase in $[Al^{3+}]$ with an increase in time. As in the former case, an increase in the surface area leads to an increase in the rate of dissolution, which is in the agreement with Eq. (9.14). In this way, the rate of the process described by Eq. (9.14), or similarly developed expressions for other systems, can successfully be used for studying the kinetics of the galvanic displacement reactions.

The rate of the galvanic displacement reactions depends also on pH and temperature. In general terms, when dealing with the amphoteric metals, e.g., aluminum, zinc, or similar, that act as substrates on which the deposition takes place in the alkaline solutions (above pH 8 or so), an increase in pH leads to an increase of deposition of the more noble metal. This is due to increase in the rate of oxidation or dissolution of the substrate.

For the presently analyzed Cu/Al system in the alkaline solution, the dependence of the rate of deposition on pH is presented in Fig. 9.3. The rate of deposition of Cu

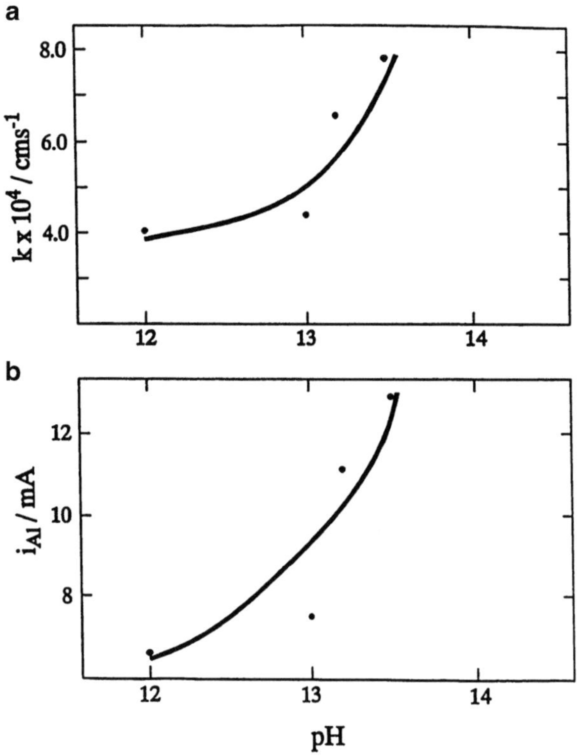

Fig. 9.3 (**a**) dependence of the reaction rate constant on pH (initial Cu^{2+} concentration 3.56 g/dm^3, initial amount of Al powder 1.5 g). (**b**) Dependence of calculated current of aluminum dissolution on pH ((initial Cu^{2+} concentration 3.56 g/dm^3, initial amount of Al powder 1.5 g) (Reproduced from Ref. [3] with permission from The Electrochemical Society)

at pH below 10, practically does not take place in the investigated solution within the reasonable time frame (1 h).

Taking into consideration that on the surface of aluminum metal a thin protective film of Al_2O_3 is present, the deposition of Cu is unlikely to proceed at pH below 10. Under this condition, the Al_2O_3 film acts as a passivating layer. Consequently, the electrolyte containing the Cu(II) ions is not in contact with the aluminum metal, and the deposition does not occur. Even, if some deposition does occur, this can be attributed to the porosity of the Al_2O_3 film. An increase in pH above 10 leads to the dissolution of the passive Al_2O_3 film and the deposition of copper takes place. In the pH range from 10 to 12, the rate of copper deposition is relatively slow, while above the pH 12 it rapidly increases as shown by the results in Fig. 9.3.

In the acidic solutions, during the galvanic displacement deposition, a quite interesting behavior is observed. Annamalai et al. investigated the structure of galvanically deposited copper onto aluminum [4]. Copper deposition was performed using an acidic Cu(II) solution. The effects of Cu (II) and Cl(I) on ion concentrations, temperature, pH, and agitation were systematically investigated. Chloride ions are known to have a beneficial effect in the removal of the aluminum oxide film and to promote the deposition process to proceed further. Generally, with an increase in the chloride ion concentration from 5 to 20 mg/L, an increase in the rate of copper deposition is observed. A further increase in the chloride ion concentration does not have an effect on the rate of copper deposition.

Influence of the kinetics on surface morphology of copper deposited via galvanic displacement was investigated by Annamalai and Murr [5]. Deposition of copper was performed from the acidic Cu(II) solution on iron substrate. An increase in the Cu(II) concentration from 0.5 to 5 g/L led to a decrease in the rate of copper deposition.

The dependence of the rate of deposition on pH within the range between 1 and 4 passes through a maximum at pH = 2 [4]. A further increase in pH leads to a decrease in the rate of copper deposition onto aluminum. Although in the region of pH up to 1 and perhaps 2 the behavior can be quite unpredictable, depending on the system, it can be generally postulated that *during the galvanic displacement deposition in the acidic solutions, an increase in pH leads to a decrease in the rate of deposition*. In the region of pH below 2, the behavior cannot be generally predicted. It will depend on any particular system. In some systems, passivation may occur and in others a pronounced dissolution.

In the acidic solutions, generally, a decrease in pH leads to an increase in the rate of dissolution of the less noble metal as stated above. Consequently, in the acidic solutions, a decrease in pH should lead to an increase in the rate of deposition of the more noble metal. This is seen on the examples of copper cementation on iron, aluminum, or zinc, and a schematic presentation is given in Fig. 9.4.

In the acidic conditions, at lower pH, the oxide film on the substrate is easier dissolved and removed. In this way, more active sites of the less noble metal which act as a reducing agent of the more noble metal ions are produced. As a result at lower pH in the acidic solutions, a higher rate of the deposition is usually observed. With an increase in pH, dissolved metal ions from the less noble substrate (e.g., Fe, Al, Zn, or similar) are susceptible to hydrolysis. This can simply be described with the following generic reactions:

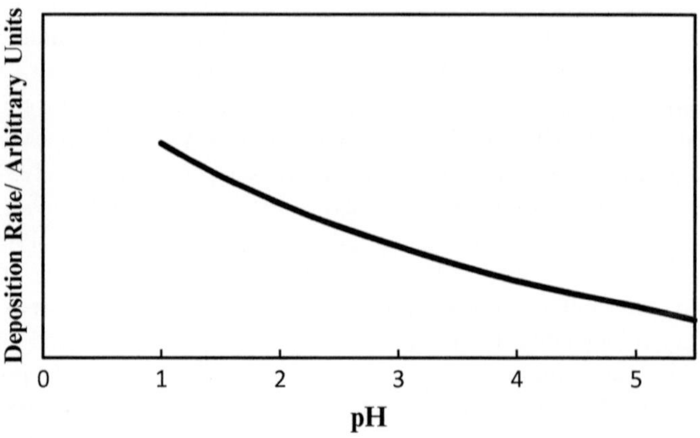

Fig. 9.4 A schematic presentation of the dependence of the rate of deposition as a function of pH in the acidic solutions

1. Dissolution of the less noble metal:

$$\mathrm{Me} \rightarrow \mathrm{Me}^{z+} + \mathrm{ze}^- \tag{9.15}$$

2. Hydrolysis of the less noble Me^{z+} ions:

$$\mathrm{Me}^{z+} + \mathrm{x\,H_2O} \rightarrow \mathrm{Me(OH)}_x^{(z-x)+} + \mathrm{x\,H^+} \tag{9.16}$$

The hydrolysis products, e.g., $Me(OH)_x^{(z-x)+}$,can be adsorbed at the less noble metal surface. Consequently, less active sites are available for the further deposition of a more noble metal. Under these conditions, the rate of deposition is significantly reduced. As a consequence, the surface morphology of the deposited more noble metal can significantly be influenced. This aspect is discussed in later sections.

As in other chemical processes, temperature significantly influences the rate of deposition during the galvanic displacement. The rate of metal deposition generally increases with an increase in temperature, which is a consequence of the Arrhenius' equation:

$$\mathrm{Rate} = A\exp\left(\frac{-E_\mathrm{a}}{RT}\right) \tag{9.17}$$

where A is the pre-exponential factor, E_a is the activation energy, R is the universal gas constant, and T is the temperature at which the reaction takes place. The logarithmic expression of Eq. (9.17) should be linear:

$$\ln(\mathrm{Rate}) = \ln A - \frac{E_\mathrm{a}}{R}\frac{1}{T} \tag{9.18}$$

which is schematically presented in Fig. 9.5.

Most of the experimental results indeed show the dependence illustrated in Fig. 9.5. However, in some cases, such as galvanic deposition of copper onto iron from the acidic solutions [5, 6], the experimental results show a different behavior. As schematically shown in Fig. 9.6, linear dependences of ln (Rate) on $1/T$ with two different slopes were experimentally obtained for the same system, i.e., galvanic deposition of copper onto iron.

Obviously, in the range of lower temperatures (up to 40 °C), a higher activation energy is required for the galvanic displacement deposition. This behavior is attributed to the possibility that the galvanic deposition may be surface reaction controlled, i.e., chemical reaction or pore diffusion [6]. In the higher temperature range, a lower activation energy is obtained. It was postulated that under these conditions the deposition reaction is boundary layer diffusion controlled.

Based on the results from Karavasteva [7], it seems that the kinetics and consequently the surface morphology of the deposited copper via the galvanic displacement reaction onto zinc, iron, and aluminum are strongly influenced by

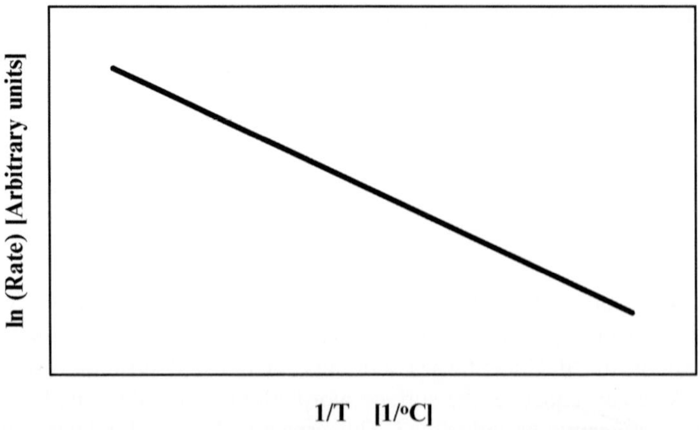

Fig. 9.5 Schematic presentation of the dependence of ln (Rate) on $1/T$, according to the Arrhenius' equation

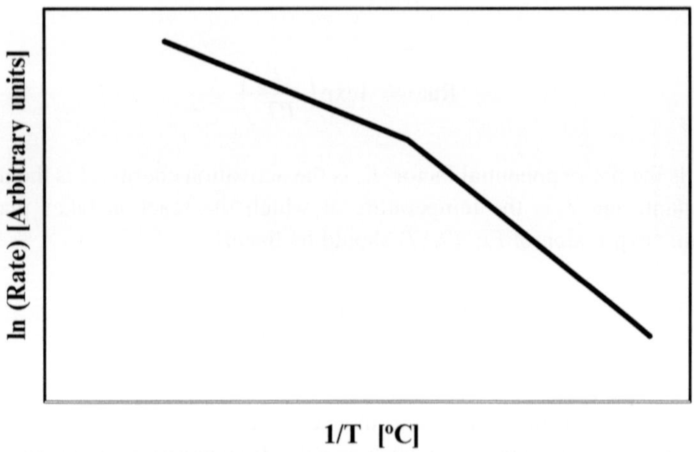

Fig. 9.6 A schematic presentation of the dependence of ln (Rate) on $1/T$ for the galvanic deposition of copper onto iron according to the results of Ref. [4–6]

the nature of the substrate. The deposition of copper on these substrates was performed from an acidic Cu(II)-containing solution at pH = 1 and 60 °C. Rates of deposition of copper onto investigated substrates, according to the results of Karavasteva, are generically presented in Fig. 9.7.

According to Fig. 9.7, the rate of deposition of copper onto different substrates decreases in the following order:

$$\text{Al} < \text{Fe} < \text{Zn} \tag{9.19}$$

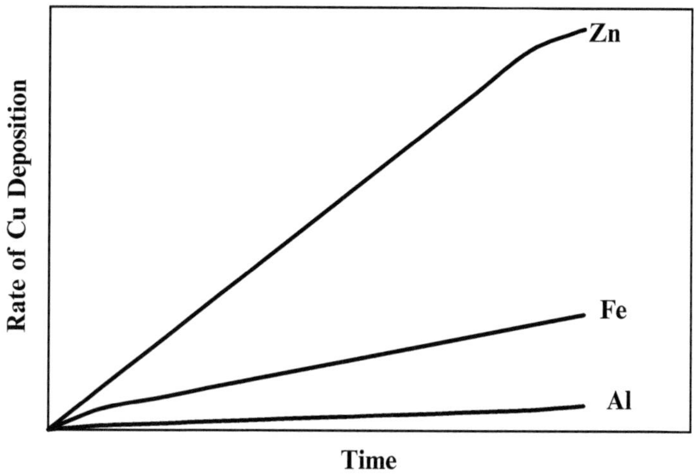

Fig. 9.7 A generic presentation of the dependence of the rate of copper deposition on time on Al, Fe, and Cu substrates according to the results of Karavasteva [7]

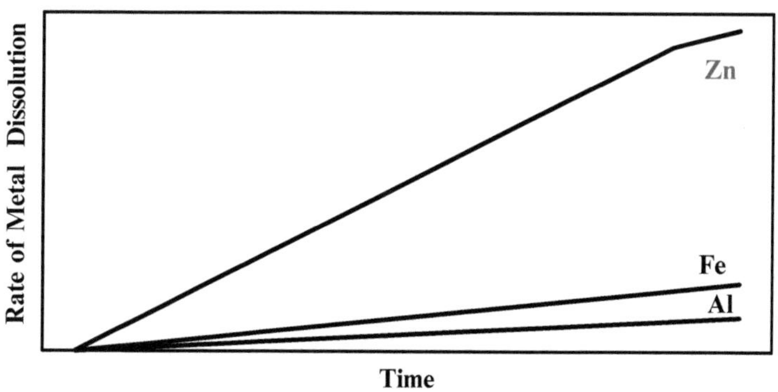

Fig. 9.8 A generic presentation of the dependence of the rates of dissolution of Al, Fe, and Cu substrates on time according to the results of Karavasteva [7]

Obviously, the rate of deposition of copper is the slowest on the aluminum surface followed by iron and zinc, respectively. The rate of copper deposition is a consequence of the rate of the dissolution of metallic substrates on which the reaction takes place. Similarly, as above, dependencies of the rate of the dissolution of aluminum, iron, and zinc on time are schematically given in Fig. 9.8.

The rate of dissolution of zinc in the investigated solution is the fastest following iron and aluminum, respectively. This is in a very good agreement with the results presented in Fig. 9.7.

9.2.1.2 Surface Morphology of Metals Produced via Galvanic Displacement Deposition

Theoretical Considerations

As explained in the previous section, the driving force for the deposition of metals via the galvanic displacement reaction is the difference in the standard electrode potentials. In these reactions, the less noble metal acts as a reducing agent of the ions of a more noble metal. For a continuation of the galvanic displacement reactions, a necessary requirement is that metal substrate (less noble metal or reducing agent) and metal ions of the more noble metal are in contact at all times. From the thermodynamics point of view, when the surface of the less noble metal, e.g., aluminum, is completely covered with the more noble metal, e.g., copper, then the galvanic displacement reaction and consequently deposition should stop. This is an ideal case, where an assumption is made that the surface of the less noble metal (e.g., aluminum) is completely covered with the more noble metal (e.g., copper). Under this condition, there is no contact between the substrate and metal ions, and a schematic presentation for this case is shown in Fig. 9.9.

The situation presented in Fig. 9.9 can only be true when the atoms of the deposited more noble metal are ideally packed on the surface of the less noble metal. In this case, the less noble metal is not anymore exposed to the electrolyte containing the more noble metal ions. As stated above, under these conditions, the further deposition does not proceed. With an ideal packing of the atoms, probably only a few monolayers of the more noble metal can be deposited.

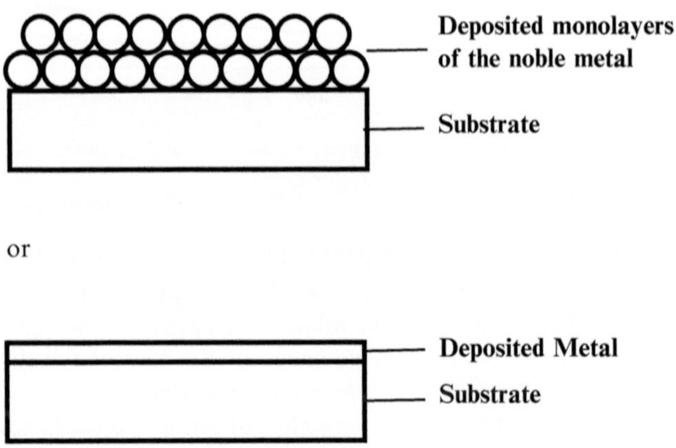

Fig. 9.9 An ideal schematic presentation when the surface of the substrate (less noble metal) is completely covered with the more noble metal

Experimental Observations

The results from real experimental settings, however, clearly show that during the galvanic displacement, reaction coatings of more noble metal significantly thicker than a *few atomic monolayers* with quite various surface morphologies are frequently obtained. The surface of the deposited more noble metal during this process can include roughening due to formation of various features, i.e., bulbous, botryoidal, dendritic, cauliflower, etc., particles. Experimental observations show that the coatings produced by the galvanic deposition are very porous. Due to porosity, the underlying, less noble metal then can easily be exposed to the electrolyte containing the ions of the more noble metal. Of course, as frequently observed in the practice, these coatings are very limited in thickness due to decreased surface area of the less noble metal exposed to the electrolyte containing the ions of the more noble metal. This is schematically presented in Fig. 9.10.

The pores are randomly distributed within the deposited metal. The existence of pores allows the electrolyte to get in contact with the substrate (less noble metal). While in contact with the electrolyte, the less noble metal dissolves producing electrons, as generically presented by the reaction (9.1). The electrons produced according to the reaction (9.1) are transferred through metal and further used for the reduction of ions of the more positive metal as described by the reaction (9.2). In this way, growth of the film of a more positive metal will further proceed. The deposition of the more noble metal and crystal growth will appear faster on the spots where pores exist compared to the pore-free sites, as schematically shown in Fig. 9.10.

In order to remove the surface oxide films and to allow the galvanic displacement deposition, an addition of the specific agents is required. In this way, the useful effect of chloride ions was reported in the literature [4]. In terms of the surface morphology, when there were no chloride ions present, galvanically deposited copper onto aluminum from acidic solutions was bulbous or botryoidal, mostly consisting of single segments [4]. This was attributed to the absence of chloride ions. Consequently, under this condition, only a few active sites are available on the

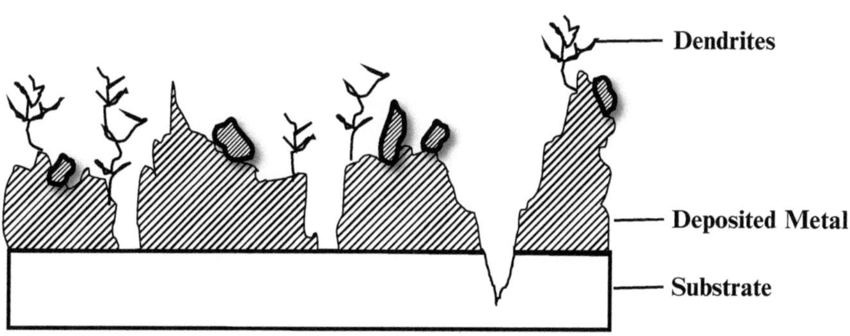

Fig. 9.10 A schematic presentation of the surface roughening during the prolonged immersion of the less noble metal substrate and deposition of the more noble metal via the galvanic displacement

aluminum surface, allowing further deposition of copper. An increase in the chloride ion concentration leads to a better removal of the aluminum oxide film and production of more uniform deposits of copper. The results of this work showed a clear transition in the surface morphology of the deposited copper bulbous or botryoidal to a crude form of dendrites and leaflike secondary arms. An increase in the agitation unavoidably leads to an increase in the rate of deposition, assuming that all other parameters, e.g., Cu(II) concentration, surface area, chloride ion concentration, and temperature, are constant. Based on the results from Annaamalai et al. [4], it seems that under these conditions various forms of copper dendrites are obtainable at different rates of agitations.

An increase in temperature leads to an increase in the rate of copper deposition from the acidic solutions onto aluminum. At temperatures up to about 40 °C, usually botryoidal or bulbous deposits of copper are obtained. A further increase in temperature leads to the deposition of the dendritic deposits. The dendrites formed above 40 °C are not compact, and they are fine and small according to the findings of Annamalai et al. [4]. The slower rate of reduction of Cu^{2+} onto iron [5] was observed at lower temperatures, and it contributed to the formation of deposits with no significant roughness. These deposits were relatively smooth. At higher rates of reduction (higher temperatures), the surface roughness of the deposited copper significantly increased, leading in production of large bulbous crystals.

The effect of copper ion concentration on the surface morphology of deposited metal was also investigated [4]. Generally, the rate of copper deposition on the concentration of Cu(II) ions passes through a maximum at about 300 mg/L [4]. The SEM images showed formation of dendritic deposits in all concentrations of Cu (II) used in this work. The authors argued that different dendrites are obtainable at various concentrations of Cu(II) ions. Presumably if the experiments of this work were properly carried out, there are no significant effects of the Cu(II) ion concentration on the surface morphology of the deposited copper: all deposits of copper onto aluminum were dendritic. During the galvanic deposition from acidic solution on iron substrate, long dendritic structures of copper were seen at higher deposition rates (lower Cu(II) concentrations) [5]. At the intermediate rates (with Cu (II) concentrations within the range 1–2 g/L), bulbous or botryoidal copper deposits were obtained. A lowest rate of copper deposition was observed with Cu (II) concentration of 5 g/L. In this case, the copper deposits exhibited a relatively flat and smooth surface.

An increase in pH in the acidic solutions, during the galvanic deposition of copper onto aluminum, leads to the formation of different deposits according to Annamalai et al. [4]. While at lower pHs, dendritic deposits were obtained, at pH close to 4, a honeycomb structure of the deposited copper is observed, evidencing that the amount of the aluminum surface is noticeably limited and, consequently, the rate of the dissolution of aluminum is significantly reduced.

It seems that the surface morphology of the deposited copper onto Al, Fe, or Zn is strongly influenced by the rate of dissolution of these metals and by the nature of these substrates [7], as noted in the previous section. Globular crystallites 1–4 μm in size are obtained on aluminum, while a coherent layer of fine crystallites less than

1 μm is deposited on iron and dendritic deposits with a high porosity are produced on zinc. The results suggest that a high rate of dissolution leads to a deposition of metals with a dendritic surface morphology. On the other hand, it seems that not only the rate of dissolution of the substrate but its structure would influence the surface morphology of the deposited metal.

A prolonged immersion of the less noble metal into the solution containing ions of the more noble metal due to deposition via the galvanic displacement reaction obviously leads to the significant surface roughening. Depending on the substrate and on the deposition product and conditions of the process (pH, temperature, composition of the electrolyte, stirring, etc.), various types of deposits, e.g., bulbous, botryoidally shapes, regular geometric shapes, cauliflower, dendritic, etc., may occur during this process. For the clarification of these issues, the reader is directed to the publications [1, 3, 8, 9] and references therein.

An experimental evidence of the surface roughening is shown in Fig. 9.11. Based on the SEM results shown in Fig. 9.11, a prolonged immersion of an

Fig. 9.11 SEM images of copper deposited onto aluminum substrates upon immersion in the alkaline Cu(II) solutions via the galvanic displacement reaction (Reproduced from Ref. [3] with permission from The Electrochemical Society)

aluminum substrate into alkaline Cu(II)-containing solutions leads to the significant roughening of the deposited copper. It seems that the surface roughening appears randomly. Probably, this is dependent on the transfer of the electrons produced due to dissolution of the less noble metal (aluminum) and on the surface concentration of the Cu(II) ions. This further suggests that the diffusion, temperature, mixing, local pH, etc., may significantly influence the surface morphology of galvanically deposited metal. Perhaps, further investigations will give a clearer picture of these issues.

Production of Metal Powders via Galvanic Displacement Deposition

Production of powders via the galvanic displacement reaction is described in literature [8]. Due to the surface roughening and unpredictable crystal growth, a prolonged immersion of a less noble metal during the galvanic displacement deposition produces particles of various sizes and shapes. A poor adhesion of the deposited metals makes it easy to separate them from the substrate. Several examples of galvanically produced powders are presented in the following figures. In Fig. 9.12 are shown dendritic powders of bismuth produced via the galvanic displacement deposition on flat aluminum substrates from a complexed Bi(III) citrate–EDTA solution at pH 2 [10].

Similarly, dendritic bismuth powders can successfully be obtained on the aluminum substrate from an acidic $K[BiI_4]$ solution at pH 1, as shown by the SEM images in Fig. 9.13.

Silver powders can be grown onto aluminum surface using the galvanic displacement reaction and different electrolytes [9]. Two examples are shown in Fig. 9.14. In Fig. 9.14 are shown the SEM images of silver powder produced onto aluminum surface from alkaline (a) and acidic (b) solutions. Based on these SEM

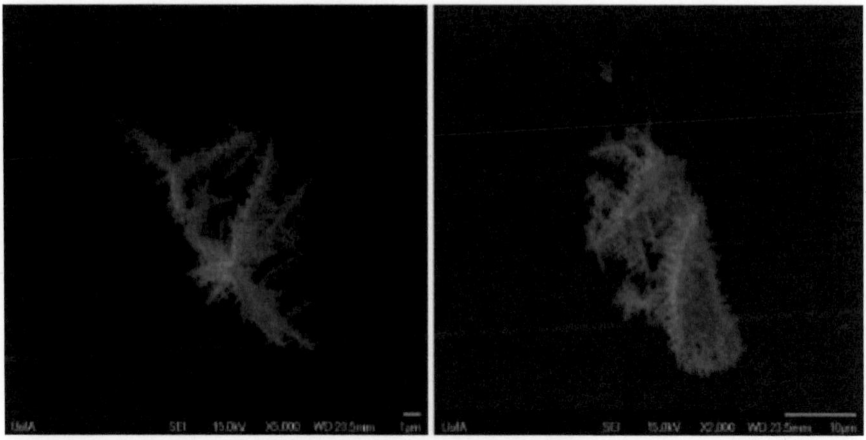

Fig. 9.12 SEM images of Bi powders produced from a complexed Bi(III)–citrate–EDTA solution at pH 2 and room temperature on aluminum metal substrate (Reproduced from Ref. [10] with permission from The Electrochemical Society)

Fig. 9.13 SEM images of bismuth powders produced on aluminum surface from an acidic $K[BiI_4]$ solution at pH 1 and room temperature via the galvanic displacement (Reproduced from Ref. [10] with permission from The Electrochemical Society)

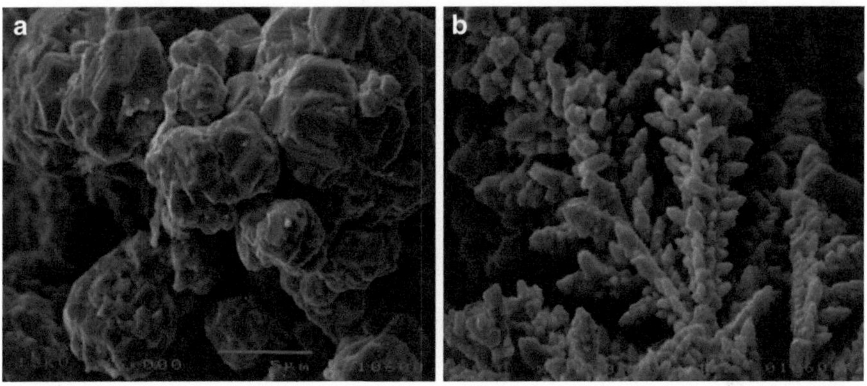

Fig. 9.14 Silver powders produced via the galvanic deposition from (**a**) 0.01 M $AgNO_3$ dissolved in diluted NH_4OH (pH 10, 22 °C) and (**b**) 0.01 M $AgNO_3$ dissolved in 0.5 M citric acid solution (pH 10, 22 °C) on an aluminum substrate (Reproduced from Ref. [9] with permission from The Electrochemical Society)

images, it seems that pH can significantly influence the shape of the particles produced. In the alkaline solutions, bulbous particles of silver were produced (Fig. 9.14a). However, as shown in Fig. 9.14b, an immersion of the aluminum foil into acidic Ag(I) solutions produces the dendritic silver powders.

In Fig. 9.15 are shown SEM images of silver produced with an immersion of niobium substrates into highly alkaline Ag(I) solutions above 90 °C. Agglomerated silver particles are clearly visible from these images. Deposition of silver onto niobium surface via galvanic displacement shows that unavoidably present oxide film at the surface of the substrate can be successfully removed. A production of powdery deposits suggests that oxide film is unevenly removed from the niobium surface.

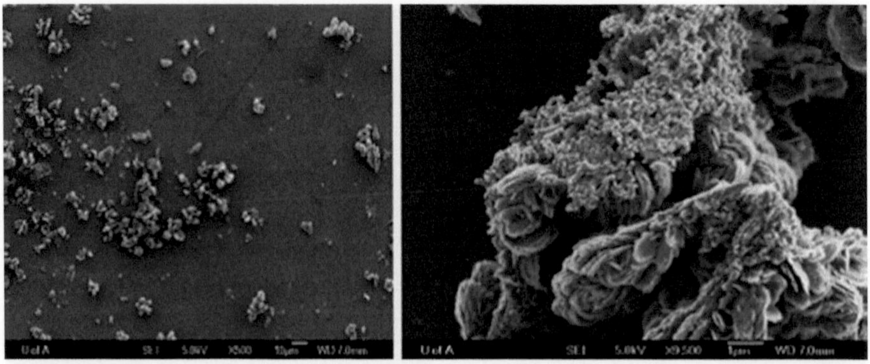

Fig. 9.15 SEM images of silver surface obtained by the immersion of Nb into alkaline Ag (I) solution for 30 s (pH 14, temperature 90–95 °C) (Reproduced from Ref. [11] with permission from The Electrochemical Society)

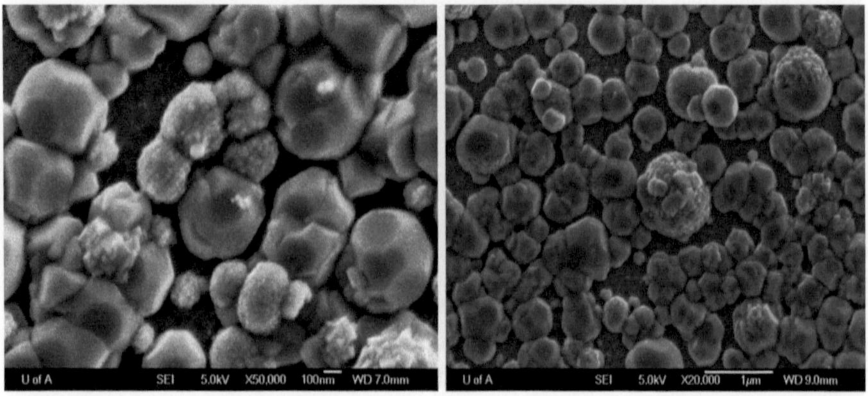

Fig. 9.16 SEM images of copper deposited onto niobium from alkaline Cu (II) solutions (pH 14, temperature 90–95 °C, deposition time 1 min)

The SEM images in Fig. 9.16 show a better surface coverage with the metal deposited (copper) in comparison with the results presented in Fig. 9.15 (silver). Bulbous copper particles (Fig. 9.16) are more densely distributed at the niobium surface than silver particles shown in Fig. 9.15. This result suggests that the niobium oxide film is more efficiently removed in the alkaline Cu(II) solutions.

Porous coatings of copper onto aluminum are shown by the SEM images in Fig. 9.17. These samples were obtained by an immersion of aluminum powders into alkaline Cu(II) solutions [3]. Due to porosity of the deposited copper, the electrolyte penetrates through pores and reacts with aluminum. In this way, simultaneous reactions of aluminum dissolution, hydrogen evolution, and copper deposition occur. A complete dissolution of aluminum and a production of hollow copper particles are shown in Fig. 9.17.

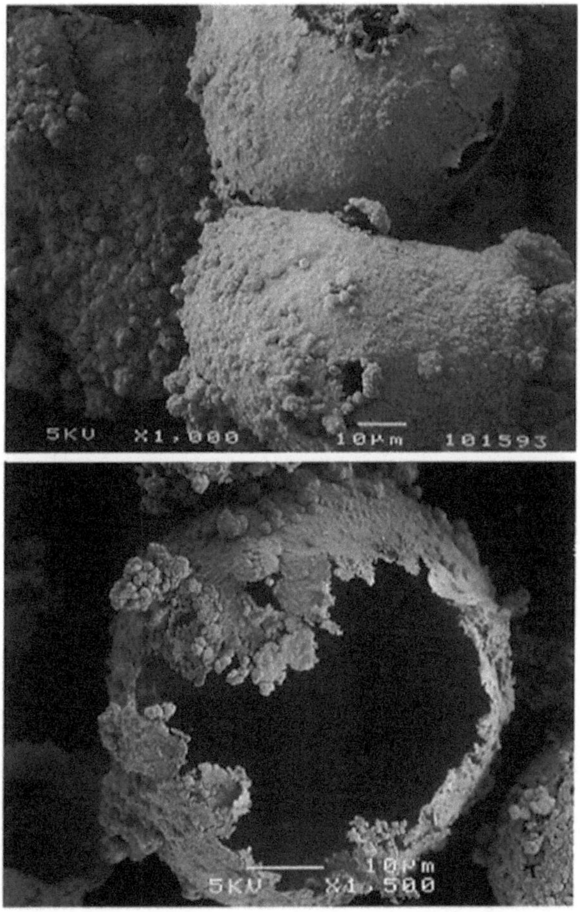

Fig. 9.17 SEM images of the hollow copper particles produced by the galvanic deposition of copper onto aluminum powder in Cu (II) alkaline solutions (pH 14, room temperature) (Reproduced from Ref. [3] with permission from The Electrochemical Society)

A sufficiently long exposure of the less noble substrate into a solution containing the ions of the more noble metal may lead to the significant surface roughening due to porosity of the deposited metal. Presence of oxide films at the surface of the less noble metal may not only delay the initial stages of the deposition but as well lead to the production of very rough deposits. A breakdown of an oxide film is usually achieved by an addition of specific additives. Due to an uneven removal of the oxide film, the deposition will not take place uniformly onto the whole surface immersed into the electrolyte. In addition, due to hydrolysis, one should take into consideration that in some electrolytes oxide films may form at the surface of the less noble metal, which further complicates deposition and formation of smooth deposits.

Changes in the operating conditions may influence the surface roughening of the deposited metal. It is obvious that a decrease in pH in the acidic solutions, for most of the metals serving as substrates, leads to an increase in their dissolution rate. In the case of the amphoteric metals, serving as substrates, an increase in pH in the alkaline solutions also leads to an increase in their dissolution rates. As well, it is generally observed that an increase in temperature leads to an increase in the dissolution rate for most of the metals. The dissolution of the less noble metal is the required condition providing the electrons for the reduction of the ions of the more noble metal and its deposition. Consequently, under the conditions where an increased rate of the dissolution of the less noble metal is observed, significant surface roughening takes place. The deposited more noble metals as can be seen from the scanning electron microscopy analysis exhibit formation of various shapes including powders, as shown in Figs. 9.11, 9.12, 9.13, 9.14, 9.15, 9.16, and 9.17.

9.2.1.3 Galvanic Displacement Deposition onto Semiconductor Substrates

The analysis presented here can further be applied to the metallization of semiconductor substrates via galvanic displacement. Metallization of semiconductors is very important in the electronic industry and as such was investigated for a relatively long time. A few examples are presented as follows.

Deposition of copper onto both p- and n- germanium substrates can be successful when the process is carried out from both acid and alkaline Cu(II)-complexed solutions [12]. The process, which proceeds via galvanic displacement, is described with the following reaction:

$$Cu^{2+} + Ge + 2H_2O \rightarrow GeO_2 + Cu + 2H^+ + H_2 \qquad (9.20)$$

with $\Delta G^o = -87.944$ kJ/mol < 0.

An XRD analysis of such produced samples showed the presence of both Cu and GeO_2 phases. The SEM images of copper-coated germanium are shown in Fig. 9.18.

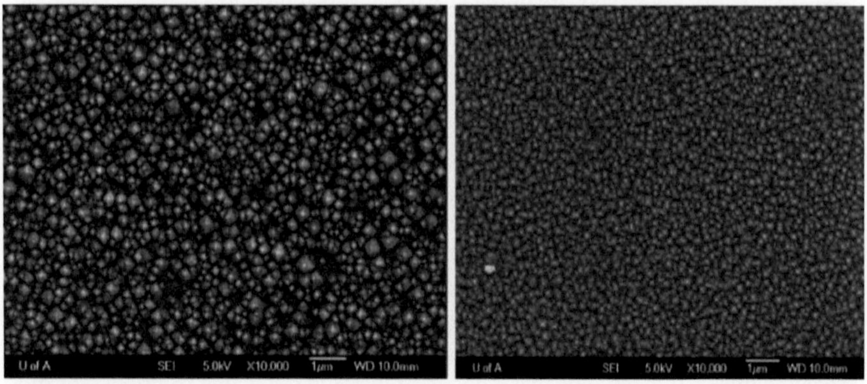

Fig. 9.18 SEM images of copper deposited from an acidic Cu(II) citrate solution (pH 2) onto n-Ge (*left*) and Ge substrate (*right*) (10 min immersion time) (Reproduced from Ref. [12] with permission from Elsevier)

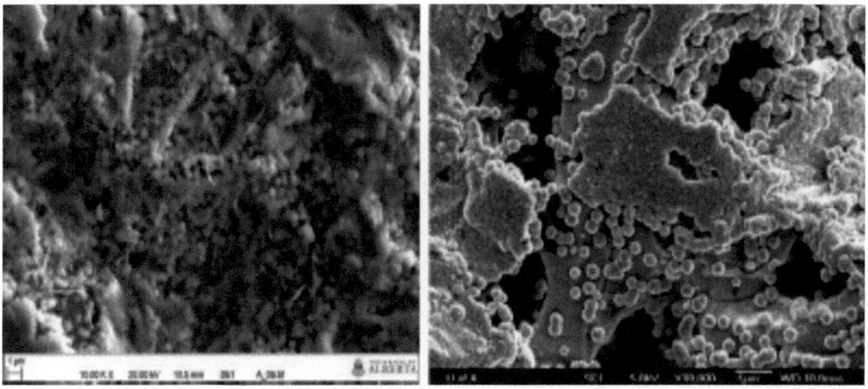

Fig. 9.19 SEM images of copper deposited from an alkaline Cu(II) solution (pH 14) on n- (*left*) and p- (*right*) Ge substrates (10 min immersion time) (Reproduced from Ref. [12] with permission from Elsevier)

As shown in Fig. 9.18, there is a quite uniform deposition of copper on both n- and p- germanium substrates. However, these images clearly show that larger particles of Cu are produced on n- germanium substrates in comparison to that of the p-Ge, leading in this way to a different surface morphology. The differences in the surface morphology of these two samples were attributed to the presence of different dopants (impurities) in the n- and p- Ge substrates [12].

The surface morphology of deposited copper onto germanium substrates is as well very much influenced by pH [12]. As shown by the SEM images in Figs. 9.18 and 9.19, the surface morphology of the deposited copper from alkaline solutions (pH 14) is quite different than that obtained from the acidic Cu (II) solutions (pH 2).

By a comparison of the SEM images from Figs. 9.18 and 9.19, it is obvious that the roughness of the deposited copper is more pronounced for the samples produced in the alkaline (pH 14, Fig. 9.19) than for those produced in the acidic solutions (pH 2, Fig. 9.18). As found by the XRD analysis [12] and as well as described by the reaction (9.20) above, GeO_2 is produced during the galvanic displacement deposition. The GeO_2 can be dissolved in the alkaline solutions which further leads to an increased rate of the Ge dissolution (oxidation). An increase in the rate of germanium dissolution, due to production of the electrons, further leads to an increase in the rate of copper deposition. Consequently under the alkaline conditions, surface roughening of the deposited copper should be pronounced as shown in Fig. 9.19.

Silver and gold can, as well, be deposited on germanium substrates [12]. It seems that the presence of GeO_2 and simultaneous hydrogen evolution (reaction (9.20)) can further influence the surface morphology of the deposited metals. Production of powdery silver deposits was demonstrated by a prolonged immersion (24 h) of the germanium substrate into acidic Ag(I) solutions. This is illustrated in Fig. 9.20, where dendritic silver powders deposited onto germanium substrates are shown.

Metallization of silicon semiconductor substrates is usually performed using the fluoride-based solutions [13]. In this case, unavoidably present SiO_2 film at the

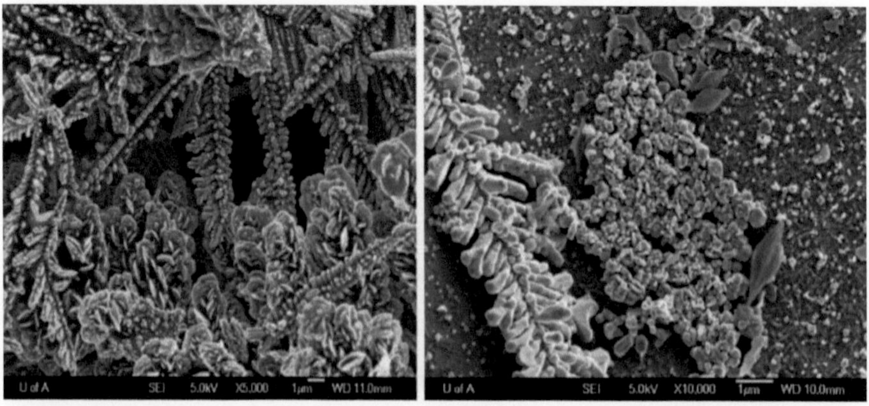

Fig. 9.20 Dendrites of silver produced on germanium substrates from an acidic Ag(I) solution (pH 2) (Immersion time 24 h) (Reproduced from Ref. [12] with permission from Elsevier)

surface of the silicon substrate is successfully removed by the fluorides present in the electrolytes. Deposition of silver onto both p-and n- silicon substrates was observed visually upon immersion into alkaline and *fluoride-free* Ag(I) solution at 85 °C [14]. Deposition of silver onto silicon proceeds via the galvanic displacement reaction which can be described by the following reaction:

$$Si + 4OH^- + 2Ag(NH_3)_2^+ \rightarrow SiO_3^{2-} + 2Ag + H_2O + 4NH_3 + H_2 \quad (9.21)$$

with $\Delta G^o = -523.958$ kJ/mol < 0.

Or, considering the presence of the SiO_2 film at the surface of silicon wafers, the overall reaction describing the deposition of silver can be written as:

$$SiO_2 + Si + 6OH^- + 2Ag(NH)_3)_2^+ \rightarrow 2SiO_3^{2-} + 2Ag + 2H_2O + 4NH_3 + H_2 \quad (9.22)$$

with $\Delta G^o = -507.176$ kJ/mol < 0.

SEM images of silver-coated silicone at 85 °C are presented in Fig. 9.21. As these results show, silver is deposited on both n- and p-silicon substrates. No significant differences in surface coverage with silver among n- and p-Si substrates are noticeable from these SEM images.

Practical applications of the galvanic displacement reaction, when smooth or continuous coatings are required, are, obviously, quite limited. It seems that only very thin coatings limited to several tens of nm with a smooth surface morphology can be achieved by this method. This aspect should be very carefully kept in mind for the modern electronics or biomedical applications. Contrarily, when hydrometallurgical applications are in question, i.e., solution purification, the galvanic displacement deposition is frequently a method of choice. This method is used

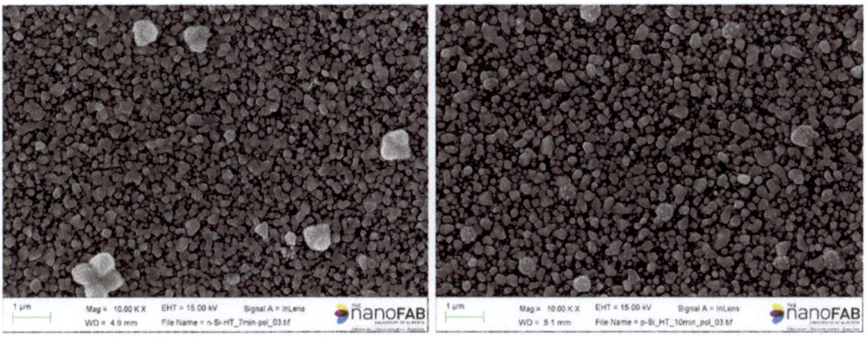

Fig. 9.21 SEM images of silver deposited at 85 °C on n-Si substrate (*Left*: 7 min immersion time) and on p-Si substrates (*Right* 10 min of immersion time) (Reproduced from Ref. [14] with permission from The Electrochemical Society)

for the removal of heavy or noble metals in the environmental applications. In both of these cases, i.e., hydrometallurgical or environmental applications, where the surface morphology of the deposited metal is not crucial, the galvanic displacement reaction can be a quite successful approach.

9.2.1.4 General Conclusions of the Galvanic Displacement Deposition in Terms of Surface Morphology

The surface morphology of deposited metals during the galvanic displacement depends on the rates of dissolution of the less noble metal and crystallization of the more noble metal. In terms of the surface morphology, following general conclusions can be made:

(a) Smooth and uniform coatings can be obtained only when the oxide film is successfully removed from the surface of the substrate.
(b) Smooth coatings can be obtained only for short deposition times and at lower deposition rates.
(c) When dealing with amorphous metals as substrates, e.g., Al, Zn, or similar in the alkaline solutions, deposition of smooth coatings is possible in solutions at lower pH (below 12).
(d) In the acidic solutions, deposition of smooth coatings takes place at higher pH (above 2).
(e) In terms of temperature, smooth coatings can be obtained at lower temperatures.
(f) A prolonged immersion of the less noble substrate into solutions containing ions of the more positive metal, due to poor adhesion and porosity of the deposited metal, unavoidably leads to the surface roughening.

(g) The surface roughening and production of powders of various sizes and shapes is strongly influenced with changes in temperature, pH, diffusion, surface concentration of the more noble metal ions, agitation, etc.

To conclude, the surface roughening during the deposition via the galvanic displacement reaction occurs either by a prolonged immersion or when an increased rate of the dissolution of the less noble metal is observed.

9.2.2 Autocatalytic Deposition

9.2.2.1 Basic Considerations

Autocatalytic deposition is the most widely used type of plating of metals from aqueous solutions, without an application of the external electrical current or potential. This process is frequently called *electroless* or *chemical deposition*, although, these terms do not precisely describe the *autocatalytic* deposition.

The autocatalytic deposition proceeds only on the catalytically active surfaces. Initiation of the autocatalytic deposition is achieved by a proper choice of catalysts, reducing agents and stable solutions containing metal ions aimed to be reduced. In many cases, such as deposition on ceramics, polymers, and other nonmetals, prior to metallization, the surfaces must be properly sensitized and/or activated in order to achieve a successful process [1, 15–17]. Similar procedures are used as well for some metallic surfaces on which the initiation of the autocatalytic deposition of a specific metal does not take place. The character of the reactions during the autocatalytic deposition and morphology of the deposit is significantly influenced by the catalytic activity of the metal particles formed in the reaction between metal ions and appropriate reducing agent.

For the autocatalytic deposition, both ions of the metal aimed to be deposited and a reducing agent are present in the homogeneous solutions in the aqueous phase. The reducing agents provide electrons for the reduction of metal ions and a consequent metal deposition. In general terms, reducing agents must be soluble and reasonably stable in the plating solution. Many reducing agents of metal ions are available in the practice. A choice of the reducing agent depends on the nature of the metal being deposited and on the properties of the final product. For example, compounds such as sodium hypophosphite ($Na_2H_2PO_2$), sodium borohydride ($NaBH_4$), and dimethyl amine borane (DMAB – $(CH_3)_2NH{\cdot}BH_3$) have quite successfully been used for the deposition of metals such as Ni, Co, Pd, Pt, etc. Hydrazine (N_2H_4) was reported as a successful reducing agent of Co(II), Ag(I), Ni(II), Pd(II), etc., ions. Formaldehyde (H_2CO) was used for a long time for the reduction of Cu(II) or Ag(I) ions. Ascorbic acid ($C_6H_8O_6$) was used for the reduction of Ag(I) or Au(I) ions. For a comprehensive review of the reducing agents used in the autocatalytic deposition, the reader is referred to the published monographs and references therein [1, 16].

9.2.2.2 Reducing Agents and Catalysts in the Autocatalytic Deposition

The reducing agent and a proper catalyst for the initiation of the autocatalytic deposition play very important roles in order to achieve the desired properties of the final product. In general terms, observations for commonly used reducing agents and respective catalysts, based on the published literature [1, 16, 17], are summarized as follows:

(a) The initiation of the deposition with borohydride (BH_4^-) as a reducing agent is realized using Pd, Pt, Ag, and Cu sites as catalysts.
(b) The initiation of the deposition with hypophosphite ($H_2PO_2^-$) as a reducing agent is realized using Ni, Co, Pd, and Pt catalysts. Silver or copper cannot be used for this purpose.
(c) The initiation of the deposition with formaldehyde (H_2CO) as a reducing agent is realized using Ag, Cu, Pd, and Pt as catalysts.
(d) Other reducing agents, e.g., hydrazine, dimethyl amine borane (DMB), ascorbic acid, etc., show in general terms quite similar behavior as that described for borohydride and hypophosphite; however, some differences are observed, depending on the metallic ion being reduced.

Many of the commonly used catalysts, e.g., Pd, Pt, Ni, etc., are considered as hydrogenation–dehydrogenation sites [16]. If the hypophosphite is used as a reducing agent, then on silver or copper surfaces dehydrogenation does not occur and consequently there is no catalytic deposition.

After the initial stage of the metal ion reduction on the catalytic sites, a further metal plating and film growth proceed via the so-called autocatalytic mechanism. Simply, the reduction of metal ions is autocatalyzed by the metal being deposited.

9.2.2.3 Kinetics and Mechanisms of Autocatalytic Deposition

The autocatalytic deposition proceeds by a controlled chemical reaction that is catalyzed by the metal or alloy being deposited [1]. This process can be presented with the following reaction:

$$M^{z+} + R^{n-} \rightarrow M + R^{z-n} \tag{9.23}$$

where M^{z+} is the ion of the metal M and R^{n-} is the reducing agent participating in the reaction. Consequently, according to the reaction (9.23), metallic ions M^{z+} are reduced to metal M, while the reducing agent R^{n-} is oxidized to R^{z-n}. Many examples of this type of deposition are described in the literature [1, 15–17].

The empirical rate law for the autocatalytic deposition according to the reaction (9.23) can be presented by the following equation:

$$-\frac{dC_{M^{z+}}^{\Delta}}{dt} = KC_{M^{z+}}^{\alpha}\ C_{R^{n-}}^{\beta}\ \exp\left(-\frac{E_a}{RT}\right) \tag{9.24}$$

where $C_M{}^{z+}$ is the concentration of the metallic ions used in the deposition, $C_R{}^{n-}$ is the concentration of the reducing agent, α and β are the reaction orders of the respective species, K is the frequency factor from the Arrhenius law, E_a is the activation energy, T is the absolute temperature, and R is the universal gas constant. The individual time derivatives $dC_M{}^{z+}/dt$ and $dC_R{}^{n-}/dt$ are referred to as the rate of reaction with respect to the species M^{z+} and R^{n-}, while $\alpha+\beta$ is the overall reaction order.

The explicit rate law must be derived for each individual system. In general terms, these equations, besides mentioned concentrations of the metal ions and reducing agents, should contain as well parameters related to the concentrations of the complexing agents, additives, inhibitors, etc. With an assumption that the concentrations of the complexing agent of M^{z+} ions, inhibitors, additives, and other constituents of an autocatalytic plating solution are ideal, the most important factors determining the rate of the deposition are the concentrations of M^{z+} and R^{n-} ions and temperature. An increase in the $C_M{}^{z+}$, $C_R{}^{n-}$, and T generally leads to an increase in the rate of deposition according to Eq. (9.24). The additives such as stabilizers, inhibitors, and similar affect the rate of deposition. The stabilizers and/or inhibitors are used to reduce the rate of deposition in order to obtain a deposit with desired physicochemical or morphological characteristics. The effects of these compounds in more details can be found in the literature [1, 16].

From the present discussion, it is obvious that at lower concentrations of both metal ion and reducing agent, or at lower temperatures, the deposition will proceed smoothly. When the concentration of reducing agent is increased and at significantly higher temperatures, the rate of deposition or, more precisely, the rate of reduction of metal ions rapidly increases and provokes the *bath instability*. In order to produce a continuous metal film/coating, the deposition in the bulk solution must be avoided.

Bath Instability

The *bath instability* is a term that is frequently used in the published literature to describe the conditions where the deposition of the desired coating is significantly diminished. Under these conditions, significant roughening of the metal deposit, and more frequently, deposition of powders, not only at the surface but, as well, in the bulk solution, is observed. Schematically, this is presented in Fig. 9.22 as a dependence of the rate of deposition on temperature. It is to be noted that dependences of the rate of deposition as function of $C_R{}^{n-}$ exhibit a similar trend.

Although, the general rate of the reduction of metal ions should increase with an increase in $C_M{}^{z+}$, $C_R{}^{n-}$, and T, even at the point of the so-called bath instability is reached, the rate of deposition of metal at the desired surfaces significantly

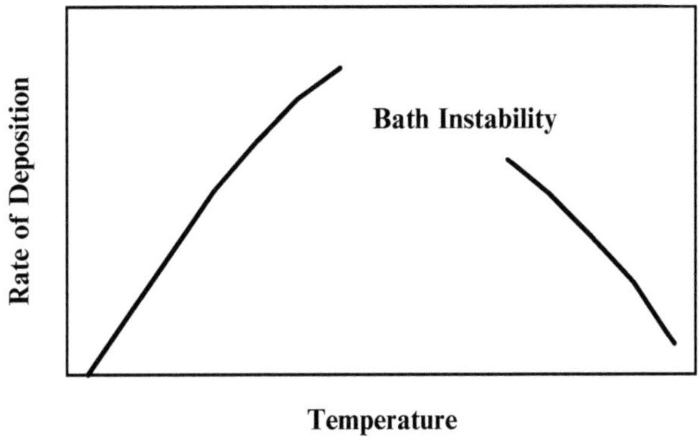

Fig. 9.22 A schematic presentation of the rate of deposition as a function of temperature

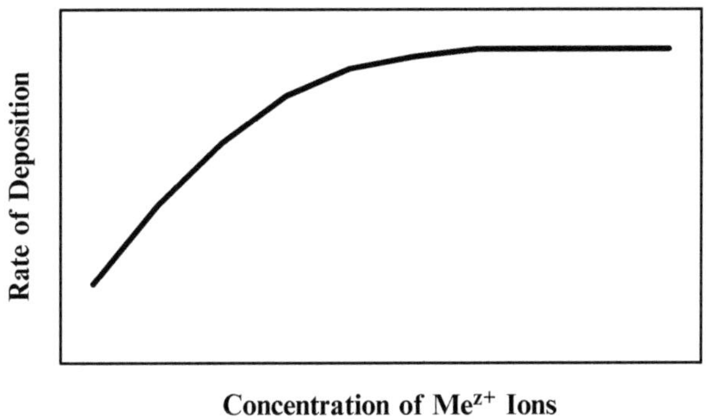

Fig. 9.23 A schematic dependence of the rate of deposition of metal Me on the concentration of its ions in solution. An assumption is made that the concentration of the reducing agent and temperature are optimal

decreases, as shown in Fig. 9.22. At the point when the bath instability is achieved, the deposition of powders in the bulk solutions takes place, rather than on the desired surface. As well, in terms of the surface morphology, significant roughening may occur under these conditions. Therefore, in considering these three variables, i.e., ${C_M}^{z+}$, ${C_R}^{n-}$, and T, on the rate of reduction of metal ions, the following can be summarized:

(a) A relatively small increase in ${C_M}^{z+}$ when ${C_R}^{n-}$ and T are kept constant may lead to a slight increase in the rate of deposition and insignificant surface roughening. A significant increase in the ${C_M}^{z+}$ may not influence the rate of reduction. This is schematically presented in Fig. 9.23.
(b) An increase in ${C_R}^{n-}$ when ${C_M}^{z+}$ and T are kept constant usually leads to the significant increase in the rate of the reduction of metal ions. As explained above, this leads to the so-called bath instability. Under these conditions, powdery deposits can frequently be obtained.
(c) When the ${C_M}^{z+}$ and ${C_R}^{n-}$ are kept constant, an increase in temperature unavoidably leads to an increase in the rate of metal ion reduction. Due to the bath instability, an increase in the surface roughening and formation of powdery deposits is easily observed.

Mechanisms of Autocatalytic Deposition

An understanding of the mechanisms of the autocatalytic deposition is crucially important in order to achieve the desired surface morphology and other physicochemical properties of the deposits produced using this approach. So far, five different mechanisms of the autocatalytic deposition have been proposed and they are listed below [1]:

- Atomic hydrogen mechanism
- Hydride ion mechanism
- Electrochemical mechanism
- Metal hydroxide mechanism
- Universal mechanism

For the review of all these mechanisms, the reader is referred to the Ref. [1]. In summary of the proposed listed mechanisms, it is, based on the experimental observations and the present knowledge, quite unlikely that the *atomic hydrogen* and *hydride ion* mechanisms are applicable in the description of the autocatalytic deposition. The *universal* mechanism is not applicable, since every single autocatalytic deposition (e.g., Ni, Co, Pd, Pt, Ag, Au, Cu, Bi, etc.) must have a specific mechanism, and a generalization is quite difficult to achieve. It seems that the *metal hydroxide* and up to some extent the *electrochemical* mechanisms are the only mechanisms that can explain most of the characteristics of the autocatalytic deposition of metals and alloys. The discussions of the mechanistic aspects of autocatalytic deposition in details is out of the scope of the present book; however, the metal hydroxide mechanism [1, 16–18] seems as the most acceptable way to explain the properties including the surface morphology of the deposits produced via the autocatalytic deposition. The metal hydroxide mechanism is based on the fact that under the conditions of autocatalytic deposition there is an unavoidable pH rise at the surface where the reaction in question takes place. Due to hydrolysis, hydrolyzed species can form and further be absorbed and/or reduced at the surface. This mechanism explains quite well the *bath instability* and formation of powders within the bulk electrolyte. When other parameters are constant, in general terms, an increase in pH of the solution leads to an increase in the rate of deposition, as schematically presented in Fig. 9.24.

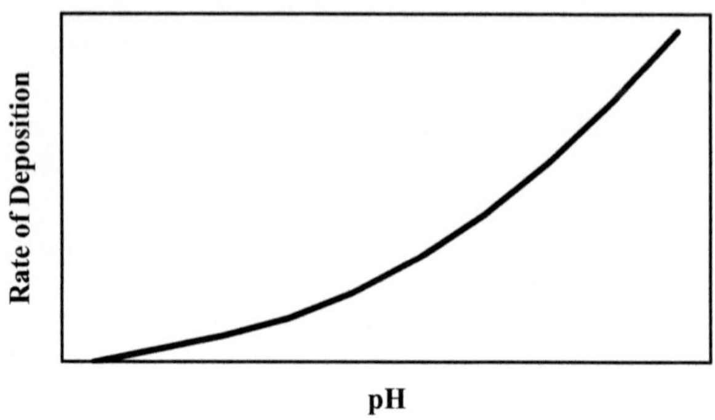

Fig. 9.24 A schematic dependence of the rate of autocatalytic deposition on pH

An increased concentration of the reducing agent, pH, and an elevated temperature significantly contribute to the hydrolysis of the metal ions causing the formation of hydroxyl complexes, hydroxides, and even oxides [19]. The autocatalytic deposition of a metal M from a homogenous liquid solution, using a reducing agent R^{n-}, is described with the reaction (9.23). In the autocatalytic deposition, the hydrolysis phenomena, as has already been published before [1, 16, 18, 19], play a very significant role in the reduction of metallic ions from aqueous solutions. Due to an increase in pH, the following reaction may occur:

$$M^{z+} + \mathrm{xOH}^- \rightarrow M(\mathrm{OH})_{\mathrm{x}}^{(\mathrm{z-x})+} \tag{9.25}$$

The hydrolyzed metallic species $\left(M\left(\mathrm{OH}\right)_{\mathrm{x}}{}^{(\mathrm{z-x})+}\right)$ are further reduced in the presence of an appropriate reducing agent (R^{n-}) to metal according to the reaction:

$$M_{\mathrm{x}}^{(\mathrm{z-x})+} + R^{\mathrm{n-}} \rightarrow M + R(\mathrm{OH})_{\mathrm{x}}^{(\mathrm{z-x-n})+} \tag{9.26}$$

These hydroxides and/or oxides, formed into the bulk solution, serve as sites, which are further reduced to the metallic state with an appropriate reducing agent. Because various sizes of particles can be produced, it is obvious that the further growth of the metallic powder is autocatalytic. Furthermore, because the formation of hydrolyzed metallic species is significantly determined by pH, it seems that the degree of hydrolysis can significantly influence the shape of powder particles.

By all means, further studies are required to clarify the mechanisms of the autocatalytic deposition. The observations suggest that every single reaction (autocatalytic deposition) must have its own mechanism. In this way, all the generalizations must be avoided, no matter how many of the autocatalytic processes are analogous or similar.

9.2.2.4 Surface Morphology of the Coatings Produced by the Autocatalytic Deposition

Most widely studied autocatalytic deposition systems include nickel, copper, silver, gold, cobalt, and palladium [1, 16]. In order to produce uniform coatings with a relatively smooth surface morphology, it is obvious that the surface coverage with the active catalysts, e.g., Pd, is uniform and sufficiently dense. A nonuniform distribution of the catalysts at the substrate will unavoidably lead to a formation of noncontinuous deposits and, frequently, produce coatings with a noticeable roughness and other surface irregularities.

With an assumption that the surface of the substrate is uniformly covered with the catalyst and that the initial stage of the metal ion reduction proceeds uninterruptedly, further film growth and its surface morphology is influenced exclusively by the parameters of autocatalytic deposition. The rate of metal ion reduction with the appropriate reducing agent at the initial stage of the deposition progressively increases with time until the continuous film is formed on the surface of the substrate. Ideally, at this point, the rate becomes constant. However, in the real experimental settings, the rate of deposition may continuously change with time and depending on the microstructure of the growing film, diffusion, and adsorption of different ions or molecules.

An increase in the deposition time leads to the increase in the grains of the deposited metal and alloy. This is frequently seen in the autocatalytic deposition of nickel, cobalt, copper, silver, etc. The effects of plating variables including the composition of the solution on autocatalytic deposition of cobalt using hydrazine as a reducing agent were systematically investigated [18]. For this purpose, cobalt was deposited on palladium-activated copper. Cobalt film produced at lower deposition rates replicated the surface morphology of the polycrystalline Cu substrate. At lower plating rates, the deposits were smooth and shiny. XRD analysis of these deposits found that Co had a hexagonal close-packed (hcp) structure. An increase in the pH, temperature, or N_2H_4 concentration or a decrease in Co^{2+} concentration, however, leads to an increase in surface roughness as shown in Fig. 9.25.

At relatively high pH values (>12.8), higher N_2H_4 concentrations, and higher temperatures (>70 °C), the cobalt reduction rate increased rapidly. Deposits obtained under these conditions exhibit significant roughness and the formation of dendrites occurs. Furthermore, due to the bath instability, the reduction of cobalt is observed in the bulk solution, even without the presence of any metallic substrate. In such a case, pure Co powder is obtained. The SEM micrographs show that the Co powder produced under these conditions was dendritic in terms of surface morphology (Fig. 9.26). XRD analysis found that the Co powder contained 70 % Co with the hcp and 30 % Co with the fcc structure.

Similarly, formation of powders of different metals with various surface morphologies during the autocatalytic deposition was reported in the literature [8, 9]. Generally, during the autocatalytic deposition, powders are produced due to the bath instability (higher pH, increased temperature, and higher concentrations of the reducing agent).

The substrate and the nature of the reducing agent may significantly influence the surface morphology of the deposited film. Autocatalytic deposition of gold onto Ni-B and Ni-P surfaces was comparatively investigated using hydrazine, hydrazine sulfate, sodium borohydride, and ascorbic acid as the reducing agents of Au(I) ions [20]. In terms of the substrate, autocatalytic deposition of gold using these reducing agents was preferably observed on Ni-B surfaces. On Ni-P substrates, only ascorbic acid and somewhat sodium borohydride were successful. Deposition of gold using hydrazine or hydrazine sulfate onto Ni-P surfaces was not observed.

While the deposition of gold using ascorbic acid as a reducing agent occurred within the temperature range from 20 to about 50 °C, with other reducing agents, the deposition was observed above 65 °C. The rate of gold deposition generally increased with temperature for all reducing agents used in this work. Using hydrazine, hydrazine sulfate, or sodium borohydride, rates of gold deposition were significantly higher in comparison with the rate of deposition when ascorbic acid was used as a reducing agent. In the case of the ascorbic acid, the bath instability as

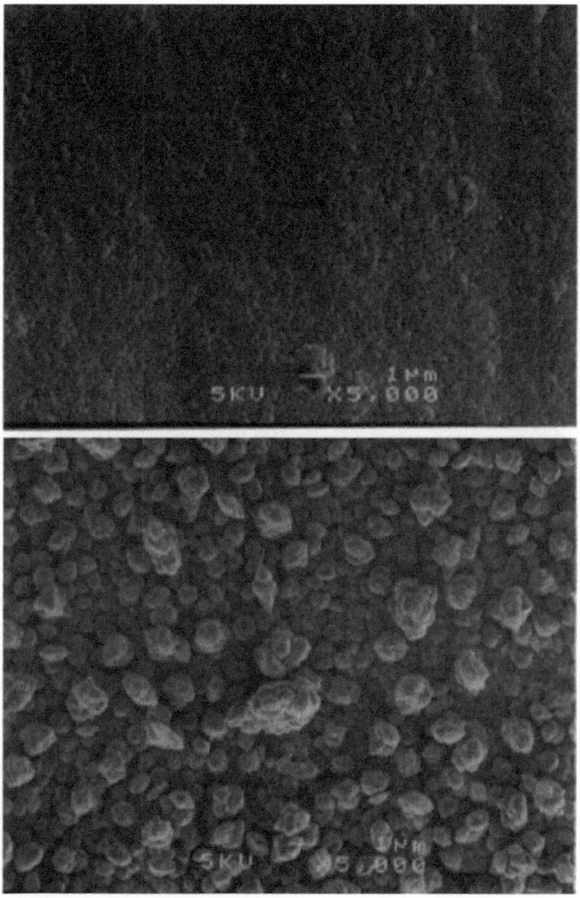

Fig. 9.25 SEM images of cobalt autocatalytically deposited onto copper activated with palladium *Top*: pH 12.3, $[N_2H_4] = 0.032$ mol/dm^3, $[Co(II)] = 0.08$ mol/dm^3, 62.5 °C, *Bottom*: pH 12.73, $[N_2H_4] = 0.32$ mol/dm^3, $[Co(II)] = 0.08$ mol/dm^3 72 °C (Reproduced from Ref. [18] with permission from The Electrochemical Society)

a function of temperature appeared at about 50 °C. For the deposition of gold with hydrazine, hydrazine sulfate, or sodium borohydride, the bath instability appeared at about 85–90 °C.

As the SEM image in Fig. 9.27 (right) shows, the Ni-B substrate was not uniformly covered with gold using sodium borohydride as a reducing agent. Rather, relatively large gold grains ranging from about 1 μm to more than 10 μm are clearly visible. The underlying Ni-B substrate with the cauliflower surface morphology and produced via autocatalytic deposition is, as well, clearly seen. At a larger magnification (Fig. 9.27 left), a significant roughness of the gold surface is visible.

A better coverage of the Ni-B substrates was achieved when hydrazine sulfate (Fig. 9.28), hydrazine (Fig. 9.29), and ascorbic acid (Fig. 9.30) were used as the reducing agents. As these figures show, there is a significant influence of the reducing agent on the surface morphology of the deposited gold. The surface roughness of gold deposited onto Ni-B substrate as a function of the reducing agent decreases in the order:

Fig. 9.26 SEM images of cobalt powder produced in bulk solution at 75 °C, pH 12.9, [Co(II)] = 0.08 mol/L, $[N_2H_4] = 0.32$ mol/L (Reproduced from Ref. [18] with permission from The Electrochemical Society)

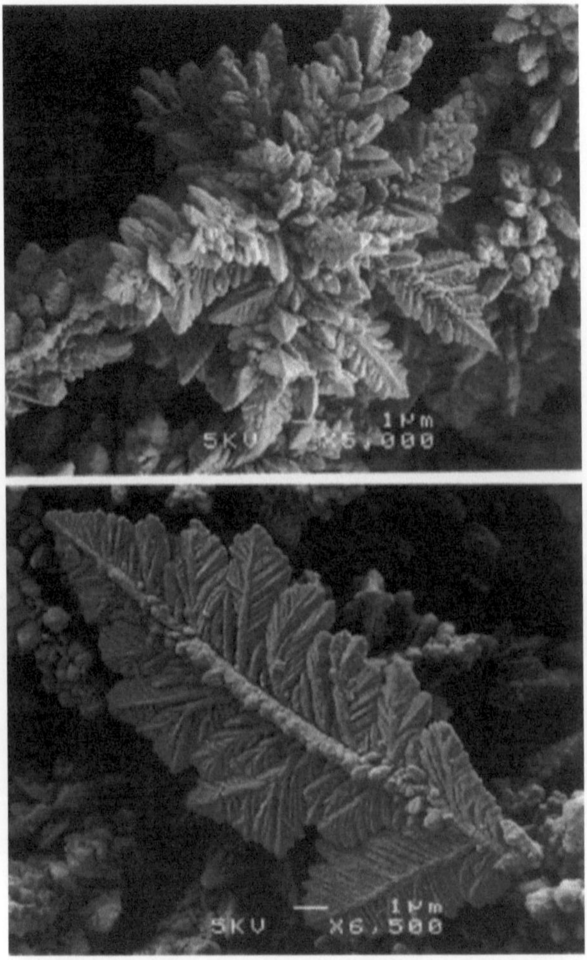

$$\text{Ascorbic Acid} < \text{Hydrazine Sulfate} < \text{Hydrazine} < \text{Sodium Borohydride} \tag{9.27}$$

It is important to note that the crystallite size, as calculated according to the XRD analysis [19], as a function of the reducing agent, follows the same order as described by the relation (9.27).

Of course, it is widely known that the additives may significantly influence the bath stability and consequently the surface morphology of autocatalytically deposited metals and alloys. According to the published literature, there are many different formulations for the autocatalytic deposition of metals and/or alloys which are used in the applications. As well, each of these formulations uses quite different additives, and there is not yet a meaningful scientific explanation and/or

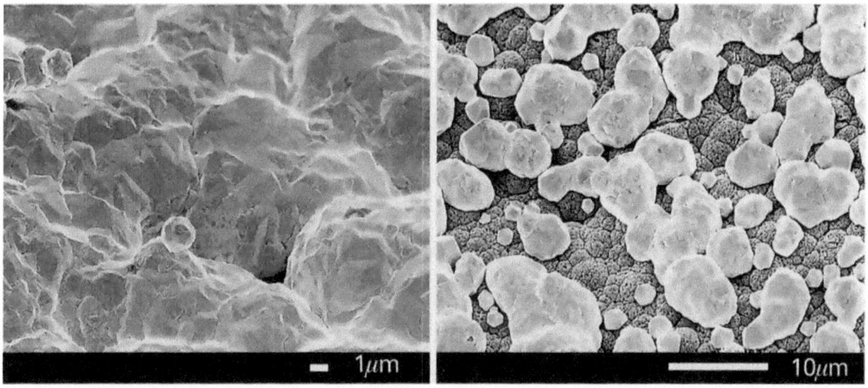

Fig. 9.27 SEM images of gold deposited onto Ni-B substrate using sodium borohydride as a reducing agent at 75 °C (Reproduced from Ref. [20] with permission from The Electrochemical Society)

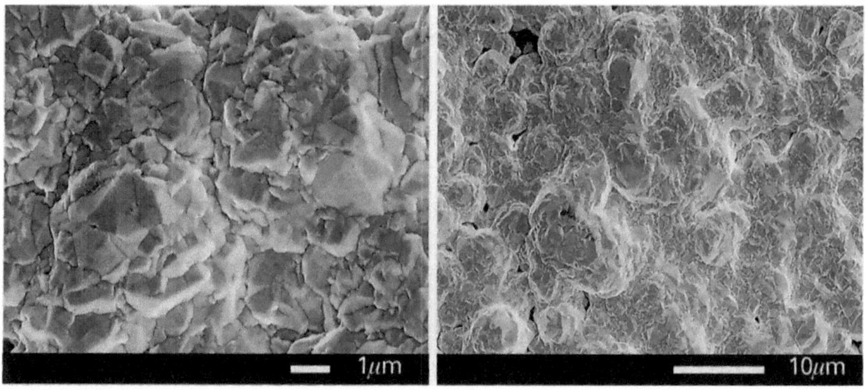

Fig. 9.28 SEM images of gold deposited onto Ni-B substrate using hydrazine as a reducing agent at 75 °C [20] (Reproduced by the permission of The Electrochemical Society Inc)

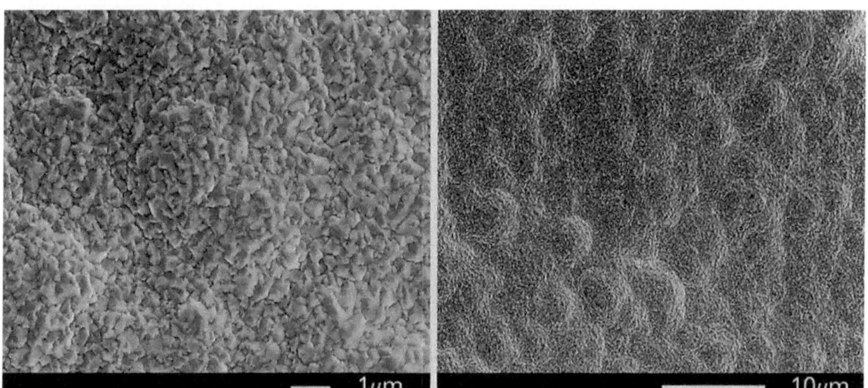

Fig. 9.29 SEM images of gold deposited onto Ni-B substrate using hydrazine sulfate as a reducing agent at 75 °C [20] (Reproduced by the permission of The Electrochemical Society Inc)

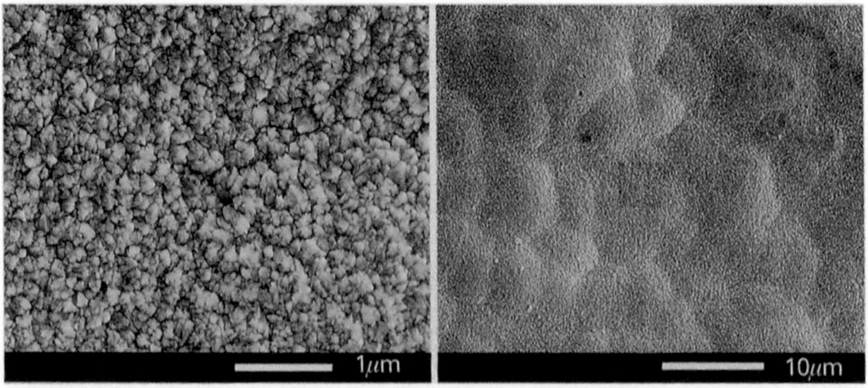

Fig. 9.30 SEM images of gold deposited onto Ni-B substrate using ascorbic acid as a reducing agent at 75 °C [20] (Reproduced by the permission of The Electrochemical Society Inc)

trend of their effects on the autocatalytic deposition. It seems that every single additive generally has a specific effect on a particular plating system. When applied to another plating system, the same additive shows a quite different behavior.

9.2.2.5 General Conclusions of the Autocatalytic Deposition in Terms of Surface Morphology

What causes the surface roughening and formation of powdery deposits? According to the observations from the practice or from the experiments, the appearance and surface morphology of metals and alloys produced during autocatalytic deposition is significantly influenced by the rate of the reduction of metal ions. These observations are briefly summarized as follows:

(a) Formation of smooth coatings can be obtained at lower concentrations of ions of the metal being deposited and/or reducing agents and at lower temperatures.
(b) An increase in the concentrations of either metal ion or reducing agent would lead to an increase in the rate of deposition and most likely to the appearance of the surface roughness at the microlevel.
(c) Significant increase in the concentration of the reducing agent if the concentration of metal ions and temperature are kept constant may lead to a noticeable surface roughening of the deposit.
(d) If both concentrations of the reducing agent and temperature are kept constant and temperature is significantly increased, an extensive surface roughening of the deposited coating and even formation of powdery deposits in the bulk solution may readily occur.
(e) If the concentration of the reducing agent and temperature are kept constant within the reasonable time frame, an increase in the concentration of the metal ions increases the rate of the reduction, but not significantly. The surface roughening may occur at the microlevel and coatings produced under these conditions are usually dull.

The influence of the deposition parameters on the surface morphology of chemically deposited metals and alloys is significantly more complex than that described for its electrodeposition counterpart. The surface morphology of chemically deposited metals and alloys depends on many parameters such as concentrations of the metal ions and reducing agent, pH, temperature, and mixing. Further studies are definitely required for a more systematic and precise description in order to achieve the desired physicochemical properties and a required surface morphology.

References

1. Djokić SS (2002) Chapter 2. Electroless deposition of metals and alloys. In: Conway BE, White RE, Bockris JO'M (eds) Modern aspects of electrochemistry, vol 35. Kluwer Academic/Plenum Publishers, New York, p 51
2. Nadkarni RM, Jelden CF, Bowles KC, Fladers HE, Wadsworth ME (1967) A kinetic study of copper precipitation on iron. Part I. Trans TMS-AIME 239:581
3. Djokić SS (1996) Cementation of copper on aluminum in alkaline solutions. J Electrochem Soc 143:1300
4. Annamalai V, Hiskey JB, Murr LE (1978) The effects of kinetic variables on the structure of copper deposits cemented on pure aluminum discs: a scanning electron microscopic study. Hydrometallurgy 3:163
5. Annamalai V, Murr LE (1979) Influence of deposit morphology on the kinetics of copper cementation on pure iron. Hydrometallurgy 4:57
6. Miller JD, Beckstead LW (1973) Surface deposit effects in the kinetics of copper cementation by iron. Metall Trans 4:1967
7. Karavasteva M (2005) Kinetics and deposit morphology of copper cementation onto zinc, iron and aluminium. Hydrometallurgy 76:149
8. Djokić SS (2012) Production of metallic powders from aqueous solution without an external current source. In: Djokić SS (ed) Modern aspects of electrochemistry, chap. 6, no. 54. Springer, New York, p 369
9. Djokić SS, Djokić NS (2011) Electroless deposition of metallic powders. J Electrochem Soc 158(4):D204
10. Djokić SS, Djokić NS, Thundat T (2012) Galvanic and chemical deposition of bismuth powders from aqueous solutions. J Electrochem Soc 159(10):D587
11. Djokić SS, Nolan L, Cadien K, Thundat T (2013) Electroless deposition of copper and silver on niobium surfaces. ECS Electrochem Lett 2(3):D16–D18
12. Djokić SS, Djokić NS, Guthy C, Thundat T (2013) Deposition of copper, silver and gold from aqueous solutions onto germanium substrates via galvanic displacement. Electrochim Acta 109:475
13. Magagnin L, Mabudian R, Carraro M (2001) Selective deposition of thin copper films onto silicon with improved adhesion. Electrochem Solid-State Lett 4:C5
14. Djokić SS, Cadien K (2015) Galvanic deposition of silver on silicon surfaces from fluoride free aqueous solutions. ECS Electrochem Lett 4(6):D11
15. Djokić SS, Magagnin L (2014) Metallization of semiconductors and nonconductive surfaces from aqueous solutions. In: Djokić SS (ed) Modern aspects of electrochemistry, chap. 7, no. 57. Springer, New York, p 341
16. Mallory O, Hajdu JB (1990) Electroless plating: fundamentals and applications. AESFS, Orlando

17. Djokić SS, Cavallotti PL (2010) Electroless deposition: theory and applications. In: Djokić SS (ed) Modern aspects of electrochemistry, chap. 6, no. 48. Springer, New York, p 251
18. Djokić SS (1997) Electroless deposition of cobalt using hydrazine as a reducing agent. J Electrochem Soc 144(7):2358–2363
19. Baes CF, Messmer RE (1976) The hydrolysis of cations. John Wiley & Sons, New York
20. Djokić SS (2000) Electroless deposition of gold using different reducing agents. In: Maltlosz M, Landolt D, Aogaki R, Sato Y, Talbot JB (eds) Fundamental aspects of electrochemical deposition and dissolution, electrochemical society proceedings, vol 99-33. Electrochemical Society, Pennington, p 381

Index

K.I. Popov et al., *Morphology of Electrochemically and Chemically Deposited Metals*, DOI 10.1007/978-3-319-26073-0

MIX
Papier aus verantwortungsvollen Quellen
Paper from responsible sources
FSC® C105338

If you have any concerns about our products,
you can contact us on
ProductSafety@springernature.com

In case Publisher is established outside the EU,
the EU authorized representative is:
Springer Nature Customer Service Center GmbH
Europaplatz 3, 69115 Heidelberg, Germany

Printed by Libri Plureos GmbH
in Hamburg, Germany